"十三五"国家重点出版物出版规划项目

现代机械工程系列精品教材

"十三五"江苏省高等学校重点教材（编号：2018-1-085）

机械工程专业英语
——交流与沟通

第2版

Specialized English for Mechanical Engineers

康　兰　编著

机 械 工 业 出 版 社

全书共分五部分。第一部分在基于如何表达产品设计信息的基础上，以工程图样内容为核心，以案例分析为主对机械设计的全过程进行了介绍；第二部分介绍了构成一台复杂机器的常用机械零件及机构，重点是描述组成一台复杂机械的基本构件及其表达方法；第三部分介绍了在机械专业领域撰写书面报告、展示口头报告、制作演示文稿等的方法与技巧；第四部分以家用轿车结构及工作原理分析为例，在较高的层次上进一步提高学习者交流与沟通的能力；第五部分详细介绍了如何撰写科技论文。各章附有相关词汇注解。

本书根据教学目标及学时的不同，可满足不同层次的教学需求，不仅适合高等院校机械类专业本科生和研究生使用，也可供高等职业院校相关专业的学生使用，还可以满足各种层次的、希望提高在机械专业领域内跨国界交流与沟通能力的人士的需求。

本书配有电子课件，向授课教师免费提供，需要者可登录机械工业出版社教育服务网（www.cmpedu.com）下载。

图书在版编目（CIP）数据

机械工程专业英语：交流与沟通/康兰编著. —2版. —北京：机械工业出版社，2018.12（2025.1重印）
“十三五”国家重点出版物出版规划项目　现代机械工程系列精品教材
“十三五”江苏省高等学校重点教材

ISBN 978-7-111-61278-0

Ⅰ.①机…　Ⅱ.①康…　Ⅲ.①机械工程-英语-高等学校-教材
Ⅳ.①TH

中国版本图书馆CIP数据核字（2018）第247204号

机械工业出版社（北京市百万庄大街22号　邮政编码100037）
策划编辑：蔡开颖　责任编辑：蔡开颖　段晓雅　林　松
责任校对：李云霞　责任印制：单爱军
北京联兴盛业印刷股份有限公司印刷
2025年1月第2版第10次印刷
184mm×260mm · 16.5印张 · 399千字
标准书号：ISBN 978-7-111-61278-0
定价：44.80元

电话服务　网络服务
客服电话：010-88361066　机　工　官　网：www.cmpbook.com
010-88379833　机　工　官　博：weibo.com/cmp1952
010-68326294　金　书　网：www.golden-book.com
封底无防伪标均为盗版　机工教育服务网：www.cmpedu.com

第 2 版前言

本书第 1 版自 2012 年 12 月出版以来，承蒙广大高校教师与学生的厚爱，已经连续印刷多次。基于近年来编者在教学中的应用实践及其他院校师生对教材的反馈意见，本次修订在尽量保持原版特色、组织结构和内容体系不变的前提下，本着“面广、实用、管用、够用”的原则，并以党的二十大提出的必须坚持人民至上、必须坚持自信自立、必须坚持守正创新、必须坚持问题导向、必须坚持系统观念、必须坚持胸怀天下为修订方向，对教材进行修订、更新和完善，修订的主要内容有：

1. 在 Part 1 中增加了与平面连杆机构及凸轮机构相关的内容，并在后面的作业中增加了与新增部分内容相关的若干练习。

2. 在 Part 3 中的 3.7 案例分析部分增加了 20 多个小型案例分析。这些案例涵盖机械工程中与材料、设计、制造、安全及专利、工作活动交流、翻译相关的许多方面，每一案例的设计以任务驱动的方式布置给学生，先由学生去完成，如果完成有困难，后面有相关的提示、背景资料或实例可以一步步帮助学生去完成相应的任务。任务可由一人完成，也可多人协同完成。这些案例的实用性和可操作性强，主要目的是提高学生在机械工程专业领域内用英文有效地进行交流与沟通的能力。

3. 将第 1 版 Part 4 科技论文写作部分改为 Part 5，新增“Part 4　How does a car work?”。Part 4 中以家用轿车这一常用而复杂的机器为综合实例，对其结构、工作原理及未来的发展方向进行分析，使学生在更高的层次上提高交流与沟通的技能。Part 4 中内容的讲解与传统教材不同，每一部分的内容讲解之后都有与此部分内容相关的任务需要学生去完成，总体设计了 22 个任务，通过任务驱动的方式让学生主动参与到学习的过程中，而不只是被动地看书学习。

4. 删除了第 1 版中“Part 5　先进制造技术”，因为学生可随时在网上搜到这些资料并进行阅读。

另外，本书编者在实际教学中与以英语为母语的专业人士合作，制作了一些全英文教学视频供学生课后学习，学生可随时扫描书中二维码观看这些教学视频。视频以问答的方式来模拟老师与学生间的交流与互动，以提高学生的听说能力，并帮助学生复习相关的专业知识。在观看视频的过程中，当老师提出问题时，学生可暂停视频，自己试着用英文去回答，然后再看视频中另一位老师是如何表达的。编者未来计划制作更多不同风格的机械工程专业领域的全英文教学视频（不一定局限于教材中的内容），逐步提供给学生使用，分享学习的乐趣。

建议教师在上课过程中通过 Part 1 和 Part 2 的讲解，先让学生掌握基本的专业词汇及表达方法，然后以 Part 3 中的案例分析为主，根据课时及专业特点，选择典型案例与学生进行交互式教学，激励学生多开口讲，这样“所学即所用”，学生更容易掌握。然后以 Part 4 中的轿车结构及工作原理分析为例，进行深度剖析和讲解，在这一过程中注重交互式教学，课堂多提问题，多鼓励学生完全用英文回答。通过 Part 4，学生可掌握用英文描述一台复杂机械结构及原理的思维方式、技能和方法，并复习和强化已学习过的专业知识。

最后，热忱欢迎读者对本书提出批评和建议，以便我们改进。

编　者

第 1 版前言

众所周知，语言与我们息息相关。尽管全世界的语言有几千种，但无论何种语言，其主要功能都是交流与沟通。英语作为世界范围内通用的语言，为全球范围内的交流与沟通架起了一座桥梁。

目前制造业的全球化使得企业在与国外同行进行合作与谈判时，迫切需要具有良好交流与沟通能力的机械工程师。这一急切的社会需求对高等教育中机械工程专业英语的教学提出了挑战。本书旨在提供一本机械工程专业的英语教材，以培养学生跨国界跨文化进行交流与沟通的能力。

本书的主要特点如下：

1）每一部分包括学习目标、学习主题、任务驱动三方面，课后相关作业或任务的完成注重小组内同学间的协同学习、交互和反思，培养团队合作能力。

2）采用案例教学，书中部分教学案例选自国外一流大学的最新教学素材，部分案例来自学生的设计作品。

3）本书根据教学目标及学时的不同，可满足不同层次的教学需求，不仅适合本科教学，也可作为研究生的学习教材，同时可满足各种层次的希望在机械专业领域内提高跨国界交流与沟通能力的人士的需求。

在学习本书时，我们建议在课程开始之时由 4~5 名同学组成一个协同学习小组，老师给每一小组布置一项课后任务，让每组选择一个与机械工程有关的主题，开展相关的研究工作。在课程结束前留出一定的课堂时间，以小组为单位将口头汇报与演示文稿相结合进行课堂展示，每位同学对自己在团队合作中的工作作一个汇报。这种研讨式的学习会给学生以激励，与别人灌输知识相比，学生记住的往往是通过自己的努力学到的知识。此法值得一试。

本书 Part 1~Part 4 由康兰副教授编写，Part 5 由许焕敏博士和周玉刚编写。在本书编写的过程中，来自澳大利亚的 Richard Porter 先生从英语为母语的读者的角度出发，严谨认真、一丝不苟地详细审阅了本书 Part 1~Part 4 的英文部分，在此表示衷心的感谢！

河海大学机电工程学院机械系主任廖华丽教授为本书在整体内容的编排方面提供了非常好的建议，机械系的王义斌老师和康兰老师指导的学生设计团队为本书的 Part 1 提供了很好的教学案例，研究生李雅编辑了部分图例。本书获河海大学机械工程及自动化专业“教育

部专业综合改革试点”项目资助，在此一并表示衷心的感谢！

感谢机械工业出版社的编审团队所给予的支持。曾在美国留学并任教多年的河海大学费峻涛教授审阅了全部的书稿，在此表示由衷的感谢！

最后，热忱欢迎读者对本书提出批评和建议，以便我们以后进行改进。

编　者

Communication Matters

目 录

Part1

Introduction to Mechanical Engineering

Objective

After completing this part, you should be able to:

- understand the essence of mechanical engineering;
- explore the engineering method for solving a problem;
- understand your responsibility as an engineer;
- clearly define and describe parts or machines by using engineering drawings;
- observe the overall engineering design process and organize your own design process in the future;
- meet the challenges of modern machine design.

1.1 Introduction

1.1.1 What Is Engineering?

Engineering is the practical and creative application of science and mathematics to solve problems, and it is found in the world all around us. Engineering technologies improve the ways that we safely travel, work, communicate and even stay healthy. One who practices engineering is called an engineer. Engineers are the innovators, planners, and problem-solvers of our society. They are always seeking quicker, better, and less expensive ways to benefit mankind. In that sense, the work of an engineer differs from that of a scientist, who would normally emphasize the fundamental discovery of physical laws rather than their application to product development. Engineering serves as the bridge between scientific discovery, commercial application, and business marketing.

1.1.2 Main Branches of Engineering

The broad discipline of engineering encompasses a range of more specialized subdisciplines, each with a more specific emphasis on certain fields of application and particular areas of technology. These disciplines concern themselves with differing areas of engineering work. Although initially an engineer will usually be trained in a specific discipline, throughout an engineer's career the engineer may become multi-disciplined, and has worked in several of the outlined areas. Engineering is often characterized as having five main branches.

Chemical engineering: The application of physics, chemistry, biology, and engineering principles in order to carry out chemical processes on a commercial scale.

Civil engineering: The design and construction of public and private works, such as infrastructure (airports, roads, railways, water supply and treatment, etc.), bridges, dams, and buildings.

Electrical engineering: The design and study of various electrical and electronic systems, such as electrical circuits, generators, motors, electromagnetic/electromechanical devices, electronic devices, electronic circuits, optical fibers, optoelectronic devices, computer systems, telecommunications, instrumentation, controls, and electronics.

Material engineering: The study of the properties of solid materials and how those properties are determined by the material's composition and structure, both macroscopic and microscopic. With a basic understanding of the origins of properties, materials can be selected or designed for an enormous variety of applications, from structural steels to computer microchips. Materials science is therefore important to many engineering fields, including electronics, aerospace, telecommunications, information processing, nuclear power, and energy conversion.

Mechanical engineering: The design of physical or mechanical systems, such as power and energy systems, aerospace/aircraft products, weapon systems, transportation products' engines, compressors, powertrains, kinematic chains, vacuum technology, and vibration isolation equipment, etc.

1.2 Mechanical Engineering

Mechanical engineering is a discipline of engineering that applies the principles of physics and materials science for analysis, design, manufacturing, and maintenance of mechanical systems. Mechanical engineering emerged as a field during the Industrial Revolution in Europe in the 18th century; however, its development can be traced back several thousand years ago around the world. Mechanical engineering has continually evolved to incorporate advancements in technology, and mechanical engineers today are pursuing developments in such fields as composites, mechatronics, and nanotechnology. Mechanical engineering overlaps with aerospace engineering, building services engineering, civil engineering, electrical engineering, petroleum engineering, and chemical engineering to varying amounts. It is one of the oldest and broadest engineering disciplines. Figure 1.1 depicts employment statistics and the distribution of engineers in the five traditional disciplines as well as several others in USA.

Mechanical engineering field requires an understanding of core concepts including mechanics,

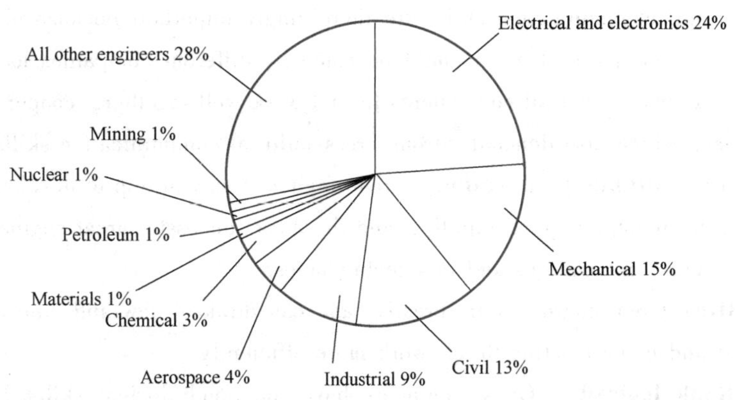

Figure 1. 1 Percentages of engineers working in various engineering

kinematics, thermodynamics, materials science, and structural analysis. Mechanical engineers use these core principles along with tools like computer-aided engineering and product lifecycle management to design and analyze manufacturing plants, industrial equipment and machinery, heating and cooling systems, transport systems, aircrafts, watercrafts, robotics, medical devices and more.

In a word, mechanical engineering is all about making useful machines.

1.3 Common Traits of Good Engineers

Engineer is a type of professions which makes society function. Engineers are responsible for some of the greatest inventions and technologies the world depends on. Everything from bridges to air conditioning systems to space shuttles requires the work of an engineer—you could say there's an engineer behind everything in your life. To be successful in the field of engineering, one must have certain qualities. Some of the top qualities include:

Good problem solving skills: Great engineers have sharp problem solving skills. Engineers are frequently called upon solely to address problems, and they must be able to figure out where the problem stems from and quickly develop a solution. Being effective problem solvers, great engineers have a firm grasp of the fundamental principles of engineering, which they can use to solve many different problems.

A strong analytical aptitude: Great engineers have excellent analytical skills and are continually examining things and thinking of ways to help things work better. They are naturally inquisitive.

Attention to detail: Great engineers pay meticulous attention to detail. The slightest error may cause an entire structure to fail, so every detail must be reviewed thoroughly during the course of completing a project.

Good interpersonal and communication skills: Great engineers have good "people skills" and communication skills that allow them to interact and communicate effectively with various people in and out of their organization. They can translate complex technical lingo into plain English and also communicate verbally with clients and other engineers working together on a project. Nowadays,

good interpersonal and communication skills are increasingly important because of the global market. For example, various parts of a car could be made by different companies located in different countries. In order to ensure that all components fit and work well together, cooperation and coordination are essential, which also demands strong cross-cultural communication skills.

Taking part in continuing education: Great engineers stay on top of developments in the industry. Changes in technology happen rapidly, and the most successful great engineers keep abreast of new research, ideas, innovations and new technologies.

Being creative: Great engineers are creative and can think of new and innovative ways to develop new systems and make existing things work more efficiently.

Ability to think logically: Great engineers have top-notch logical skills. They are able to make sense of complex systems and understand how things work and how problems arise.

Excellent math skills: Great engineers have excellent math skills. Engineering is an intricate science that involves complex calculations of varying difficulties.

Good time-management skills: Great engineers have time-management skills that enable them to work productively and efficiently.

Being a team player: Great engineers understand that they are part of a large team working together to make one project a success, therefore, they must work well as part of that team.

11. Excellent technical knowledge. Great engineers have a vast amount of technical knowledge. They understand a variety of computer programs and other systems that are commonly used during an engineering project.

1.4 What Is a Machine

A machine is a tool consisting of two or more parts that is constructed to achieve a particular goal. Machines are powered devices, usually mechanically, chemically, thermally or electrically powered, and are frequently motorized. Historically, a device required moving parts to classify as a machine; however, the advent of electronics technology has led to the development of devices without moving parts that are considered machines.

All machines are made up of elements or parts and units. Each element is a separate part of the machine and it may have to be designed separately and in assembly. Each element in turn can be a complete part or made up of several small pieces which are joined together by riveting, welding, etc. Several machine parts are assembled together to form what we call as a complete machine.

The definition of the term machine will be most useful and frequently referred in subsequent discussions in this book. Each of us is familiar with what he or she considers to be machines, and the above descriptions are our general impression about machines. There are many researchers writing about machines and giving definition of a machine. According to Franz Reuleaux's description, "A machine is a combination of resistant bodies, so interconnected that by applying force or motion to one or more of those bodies, some of those bodies are caused to perform desired work accompanied by desired motions."

Machines exist everywhere and have a very close relationship with our daily life. Figure 1.2 shows an example of a simple machine—nail clippers. The two movable parts of the nail clippers are connected to each other by a pivot in such a manner that by pressing part A, part B is caused to move relative to each other in such a manner as to do the desired cutting. Each of the two movable parts is a "resistant body" in the sense that it resists deformation sufficiently to allow it to move and work as desired when forces are applied to it.

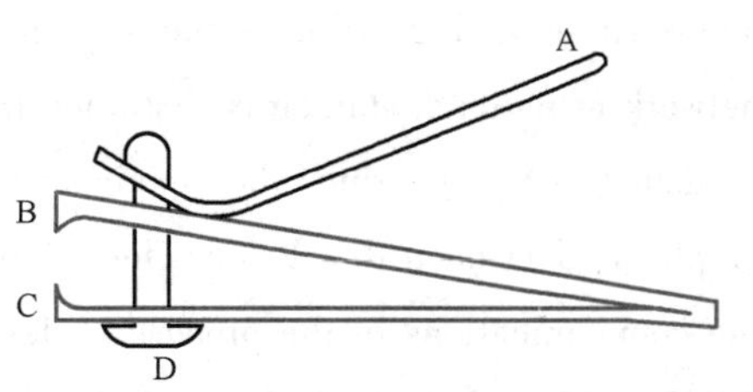

Figure 1.2 A simple machine—nail clippers

When rigid bodies connected by joints in order to accomplish a desired force and/or motion transmission, they form a simple machine or a mechanism.

Here are some examples of other complicated machines:

Lathe: It utilizes mechanical energy to cut the metals. The other types of machine tools also perform the same task.

Turbines: They produce mechanical energy.

Compressors: They use mechanical energy to compress the air.

Engines: They consume the fuel and produce mechanical energy.

Refrigerators and air-conditioners: They use mechanical energy to produce cooling effect.

Washing machines: They use mechanical energy to wash the clothes.

1.5 How to Describe a Machine?

1.5.1 Engineering Drawings

An engineering drawing, a type of technical drawing, is used to fully and clearly describe a part or a machine. The methods of description include two dimensional representation (2D) and three dimensional representation (3D).

2D engineering drawing is a two dimensional representation of three dimensional objects. In general, it provides necessary information about the shape, size, surface quality, material, manufacturing process, etc. Drawings prepared in one country may be utilized in any other country irrespective of the language spoken. Hence, the engineering drawing is called the universal language of engineers—a graphical language that communicates ideas and information from one mind to another. Any language to be communicative should follow certain rules so that it conveys the same meaning to everyone. Similarly, drawing practice must follow certain rules, if it is to serve as a means of communication.

In the United States of America, the American National Standards Institute, or ANSI, is the organization to set up the standards or the rules used in preparing the engineering documents. In the worldwide scale, the International Organization for Standardization, or ISO, is the organization to administrate and coordinate the standardization and conformity assessment system. The ISO is a network of national standards institutes from 154 countries. In order to implement the standards established for preparing the engineering documents, a scientific branch, called engineering graphics, has been developed. The subject of engineering graphics serves such a function of guiding the communications in the process of design information exchange by following the standards set by ANSI and/ or ISO. In China, National Standards (abbreviated GB) is adopted as the standard code of practice for drawings. GB was created based on ISO.

In service of the goal of unambiguous communication, engineering drawings made professionally today are expected to follow certain well-known and widely followed standards, such as GB, ANSI or a group of ISO standards that are quite similar. This standardization also contributes to internationalization, because people from different countries who speak different languages can share the common language of engineering drawings, and can communicate with each other quite well, at least as concerns the technical details of an object. Consequently, in the field of engineering design, engineering graphics is the primary medium used in developing and communicating design concepts.

For centuries, until the post-World War Ⅱ era, all engineering drawings were done manually by using pencils and pens on paper or other substrate. Since the advent of computer-aided design (CAD), engineering drawings have been done more and more in the electronic medium forms. Today most engineering drawings are done with CAD, but pencil and paper are still used.

Drawings convey the following critical information:

(1) Geometry—the shape of the object; represented as views; how the object will look when it is viewed from various angles, such as front, top, side, etc.

(2) Dimensions—the size of the object is captured in accepted units.

(3) Tolerances—the allowable variations for each dimension.

(4) Material—represents what the item is made of.

(5) Finish—specifies the surface quality of the item, functional or cosmetic.

1.5.2 Drafting Standards

Engineering drawings are prepared on standard-size drawing sheets. The correct shape and size of the object can be visualized not only from the understanding of its views but also from the various types of lines used, dimensions, notes, scale, etc. For uniformity, the drawings must be drawn as per certain standard practice. This section deals with the drawing practices as recommended by GB. These are adapted from what is followed by International Standards Organization (ISO).

1. Sheet sizes and layout

The National Standard establishes five preferred sheet sizes, as shown in Table 1.1.

The layout of a drawing sheet is shown in Figure 1.3.

Table 1. 1 Sheet sizes

<table>
<tr><th rowspan="2">Code</th><th rowspan="2">Size(B/mm)×(L/mm)</th><th colspan="3">Margin/mm</th></tr>
<tr><th>a</th><th>c</th><th>e</th></tr>
<tr><td>A0</td><td>841×1189</td><td rowspan="5">25</td><td rowspan="3">10</td><td rowspan="2">20</td></tr>
<tr><td>A1</td><td>594×841</td></tr>
<tr><td>A2</td><td>420×594</td><td rowspan="3">10</td></tr>
<tr><td>A3</td><td>297×420</td><td rowspan="2">5</td></tr>
<tr><td>A4</td><td>210×297</td></tr>
</table>

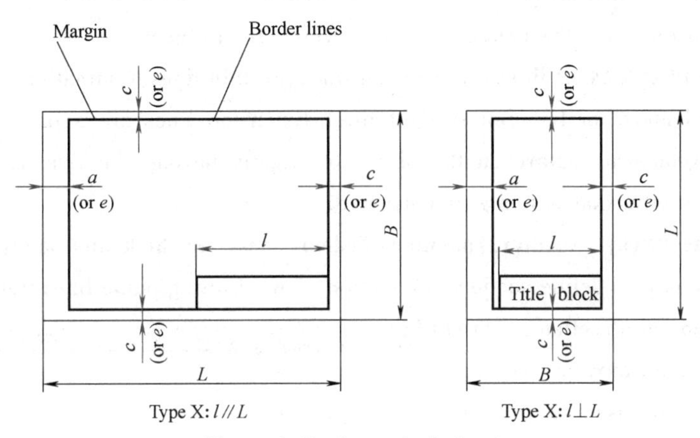

Figure 1. 3 Layout of sheet

2. Scales

If the actual linear dimensions of an object are shown in its drawing, the scale used is said to be a full-size scale. Wherever possible, it is desirable to make drawings full-size.

The scale is the ratio between the measurement on the drawing and the actual size. Listed in Table 1. 2 are the scales used in technical drawings.

Table 1. 2 Scales

Full-size	1 : 1
Reduction scales	(1 : 1.5) 1 : 2 (1 : 2.5) (1 : 3) (1 : 4) 1 : 5 (1 : 6) 1 : 10 etc.
Enlargement scales	2 : 1 (2.5 : 1) (4 : 1) 5 : 1 10 : 1 etc.

Note: It is permissible to choose the scales shown in brackets, if necessary. Whatever scale is used, the dimensions on the drawing indicate the true size of the object, not of the view.

3. Types of lines

Technical drawings use several different line types to help convey the shape and size of a physical object. Types of lines are as follows (see Figure 1. 4):

Part outlines (or simply visible lines): are thick (or heavy) solid lines used to depict edges and outlines of geometric features directly visible from a particular angle.

Section lines: are thin (or light) angled lines in a pattern (pattern determined by the mate-

rial being "cut" or "sectioned") used to indicate surfaces in section views resulting from "cutting". Section lines are commonly referred to as "cross-hatching". Notice that if a single metal part is sectioned, all the resulting section lines will be at the same angle and spacing. If several assembled metal parts are sectioned, each part would have section lines at different angles or spacing. The space between the section lines varies with the size of the object being sectioned, but a rule of thumb is to space the section lines about 1 to 3mm apart.

Hidden lines: are thin dashed lines used to represent the edges or outlines that are not directly visible from the direction of the viewer.

Center lines: are alternately long-and short-dashed lines used to locate the center of a geometric feature or to represent the central axis of a symmetrical object.

Dimension and extension lines: dimension lines are thin lines terminating in arrow heads with a dimension placed above or between straight lines. Extension lines are a pair of thin solid lines extending from a geometric feature on the object to slightly beyond the dimension line. Extension lines help clarify exactly what is being dimensioned.

Cutting plane lines: are thin, medium-dashed lines, or thick alternately long-and double short-dashed lines used to define sections for section views. Cutting plane lines indicates the location and viewing direction of a section being cut or special view such as an auxiliary view.

Break lines: are used to show that a portion of an object has been left out. They can be either thick/thin irregular freehand lines used with short breaks or thin line segments connected by a zigzag used for long breaks.

Phantom lines: (not shown) are alternately long-and double short-dashed thin lines used to represent a feature or component that is not part of the specified part or assembly, e. g. billet ends that may be used for testing, or the machined product that is the focus of a tooling drawing.

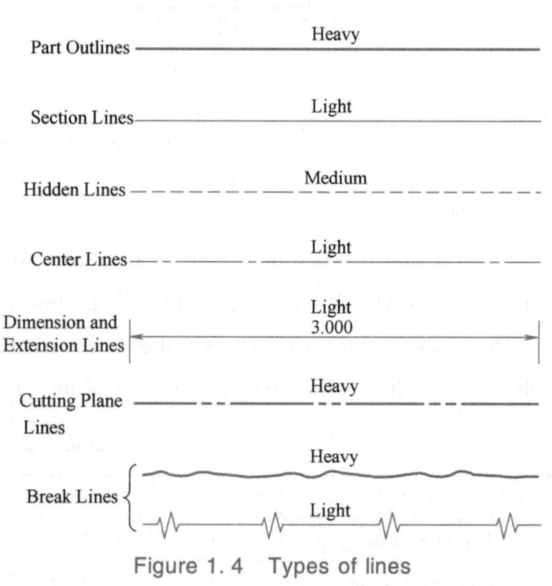

Figure 1.4 Types of lines

Lettering is defined as writing of titles, sub-titles, dimensions, etc., on a drawing. The basic rules for lettering are defined in detail in GB. These rules are not mentioned in this section.

1.5.3 Views

1. The orthographic projection

The orthographic projection shows the object as it looks from the front, right, left, top, bottom, or back, and are typically positioned relative to each other according to the rules of either first-angle or third-angle projection. The origin and vector direction of the projectors (also called projection lines) differs, as explained in Figure 1. 5.

In first-angle projection, the projectors originate as if radiated from a viewer's eyeballs and shoot through the 3D object to project a 2D image onto the plane behind it. The 3D object is projected into 2D "paper" space as if you were looking at a radiograph of the object: the top view is under the front view; the right side view is at the left side of the front view. First-angle projection is the ISO standard and is primarily used in China and Europe.

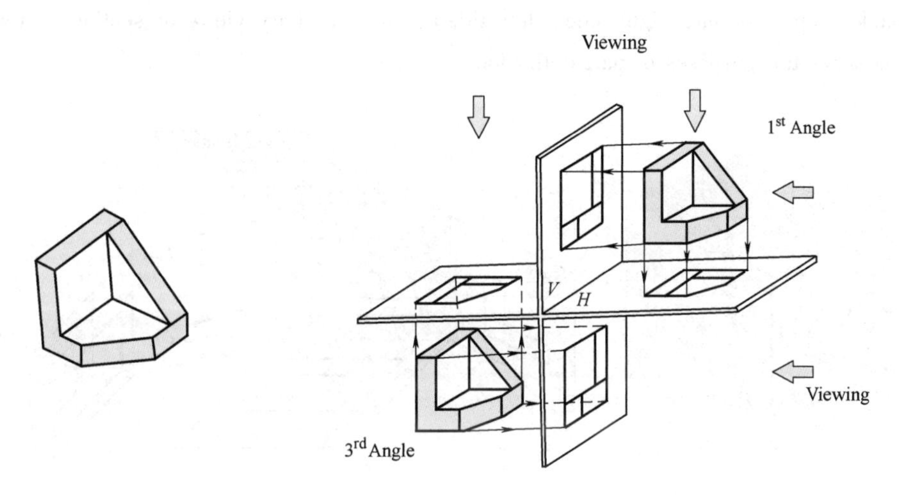

Figure 1. 5 1st and 3rd angle projections

In third-angle projection, the projectors originate as if radiated from the 3D object itself and shoot away from the 3D object to project a 2D image onto the plane in front of it. The views of the 3D object are like the panels of a box that envelopes the object, and the panels pivot as they open up flat into the plane of the drawing. Thus the left side view is placed on the left and the top view on the top; and the features closest to the front of the 3D object will appear closest to the front view in the drawing. Third-angle projection is primarily used in the United States and Canada, where it is the default projection system according to British Standard BS 8888 and ASME standard ASME Y14. 3M.

To ensure that those reading the drawing know whether it is 1st or 3rd angle projection, use one of two symbols illustrated in the title block, as shown in Figure 1. 6.

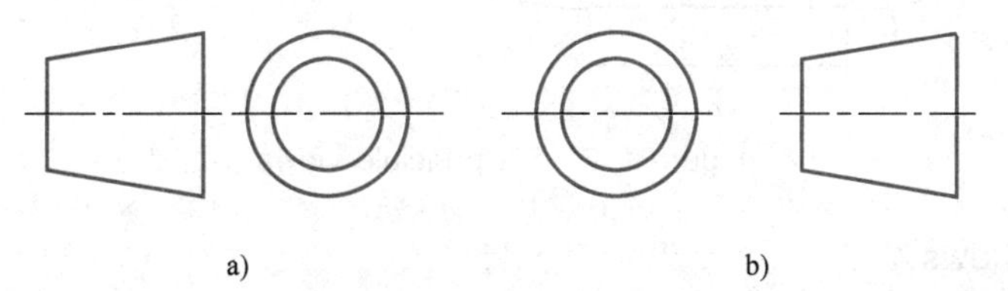

Figure 1. 6 Projection symbols

a) 1st angle projection symbol b) 3rd angle projection symbol

2. Multiple views

Six principle views: Here let's assume an object is placed inside of an imaginary transparent

box, then project the object onto six planes respectively based on the orthographic projection, thus six principle views are obtained. They are front, back, top, bottom, right side, left side views, as shown in Figure 1.7.

Not all views are necessary to illustrate an object. Generally only as many views are used as are necessary to convey all needed information clearly and economically. The front, top, and left side views are commonly considered the core group of views included by default, but any combination of views may be used depending on the needs of the particular design. In addition to 6 principle views (front, back, top, bottom, right side, left side), any auxiliary views or section views may be included as serve the purposes of part definition.

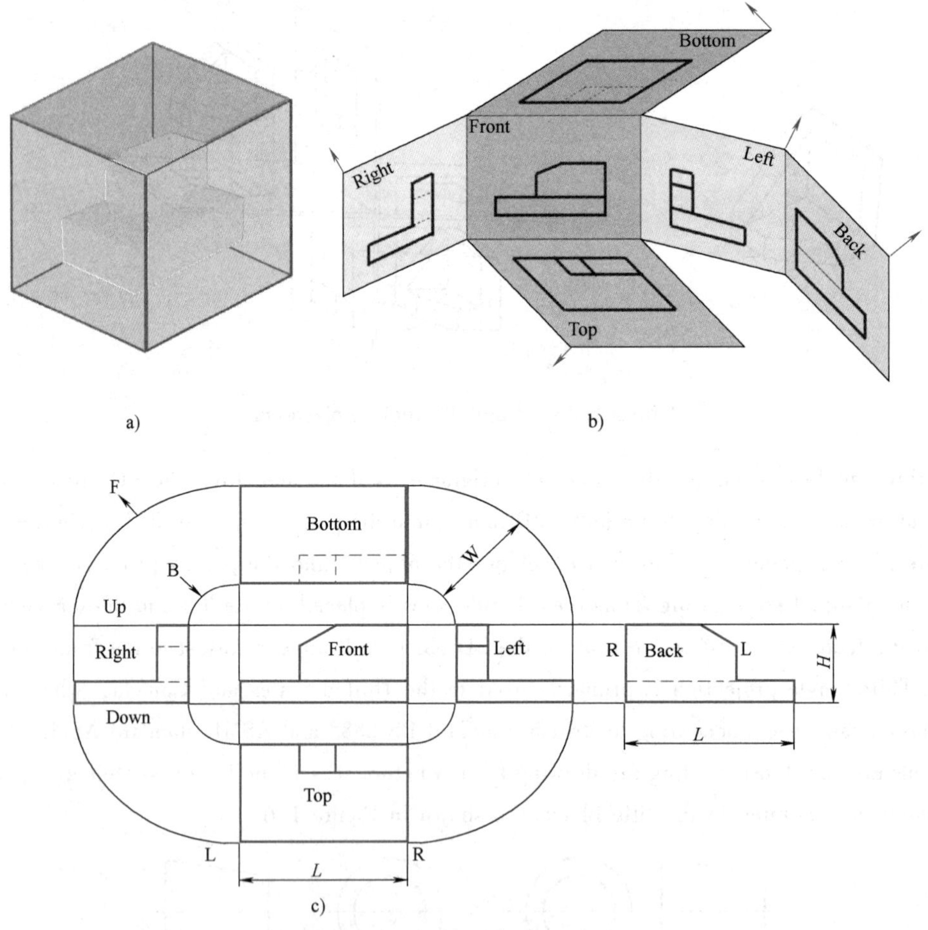

Figure 1.7　Six principle views

3. Auxiliary views

An auxiliary view is an orthographic view that is projected onto any plane other than one of the six principle views. These views are typically used when an object contains some sort of inclined plane. Using the auxiliary view allows for that inclined plane (and any other significant features) to be projected in their true size and shape. The true size and shape of any feature in an engineering drawing can only be known when the line of sight is perpendicular to the referenced plane, as shown in Figure 1. 8.

Isometric view of the object is also used as an auxiliary view in the engineering drawing. The isometric projection shows the object from angles in which the scales along each axis of the object are equal, as shown in Figure 1. 9. Isometric projection corresponds to rotation of the object by ±45° about the vertical axis, followed by rotation of approximately ±35. 264° [= arc sin(tan30°)] about the horizontal axis starting from an orthographic projection view. "Isometric"comes from the Greek for "same measure". One of the factors that makes isometric drawings so attractive is the ease with which 60° angles can be constructed with only a compass and straightedge. Isometric projection is a type of axonometric projection and it is commonly used.

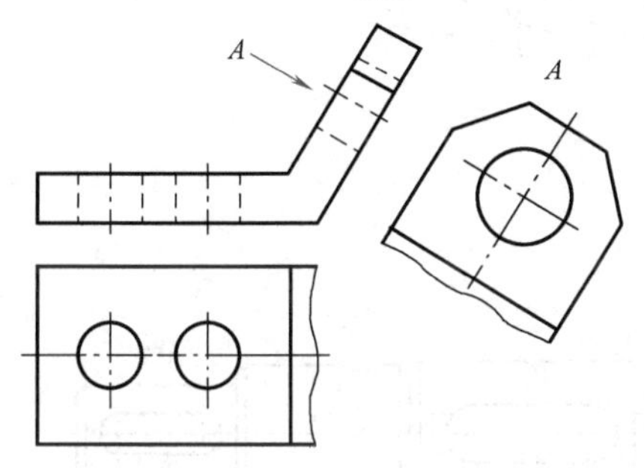

Figure 1. 8 An auxiliary view

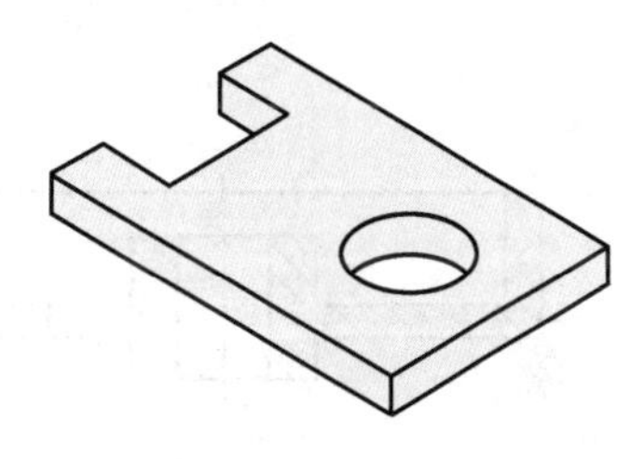

Figure 1. 9 Isometric view

4. Section Views

Section views are projected views (either auxiliary or orthographic) which show a cross section of the source object along the specified cut plane. These views are commonly used to show internal features with more clarity than those by using regular projections or hidden lines, as shown in Figure 1. 10.

In order to improve visualization of interior features, section views are used when important hidden details are in the interior of an object. These details appear as hidden lines in one of the orthographic principal views; therefore, their shapes are not very well described by pure orthographic projection. Section views show how an object would look if a cutting plane cut through the object and the material in front of the cutting plane was discarded.

Types of section views are as follows: full sections, half sections, offset sections, removed sections, revolved sections, broken-out sections, as shown in Figure1. 10.

View lines or section lines (lines with arrows marked "*A*—*A*", "*B*—*B*", etc.) define the direction and location of viewing or sectioning. Sometimes a note tells the reader in which zone (s) of the drawing to find the view or section.

In assembly drawings, solid parts (e. g. nuts, screws, washers, shafts, ribs, spokes) are typically not sectioned when a cutting plane passes through their longitudinal axis, as shown in Figure 1. 11.

1. 5. 4 Working Drawings

Working drawings, also known as production drawings, provide the complete size and shape

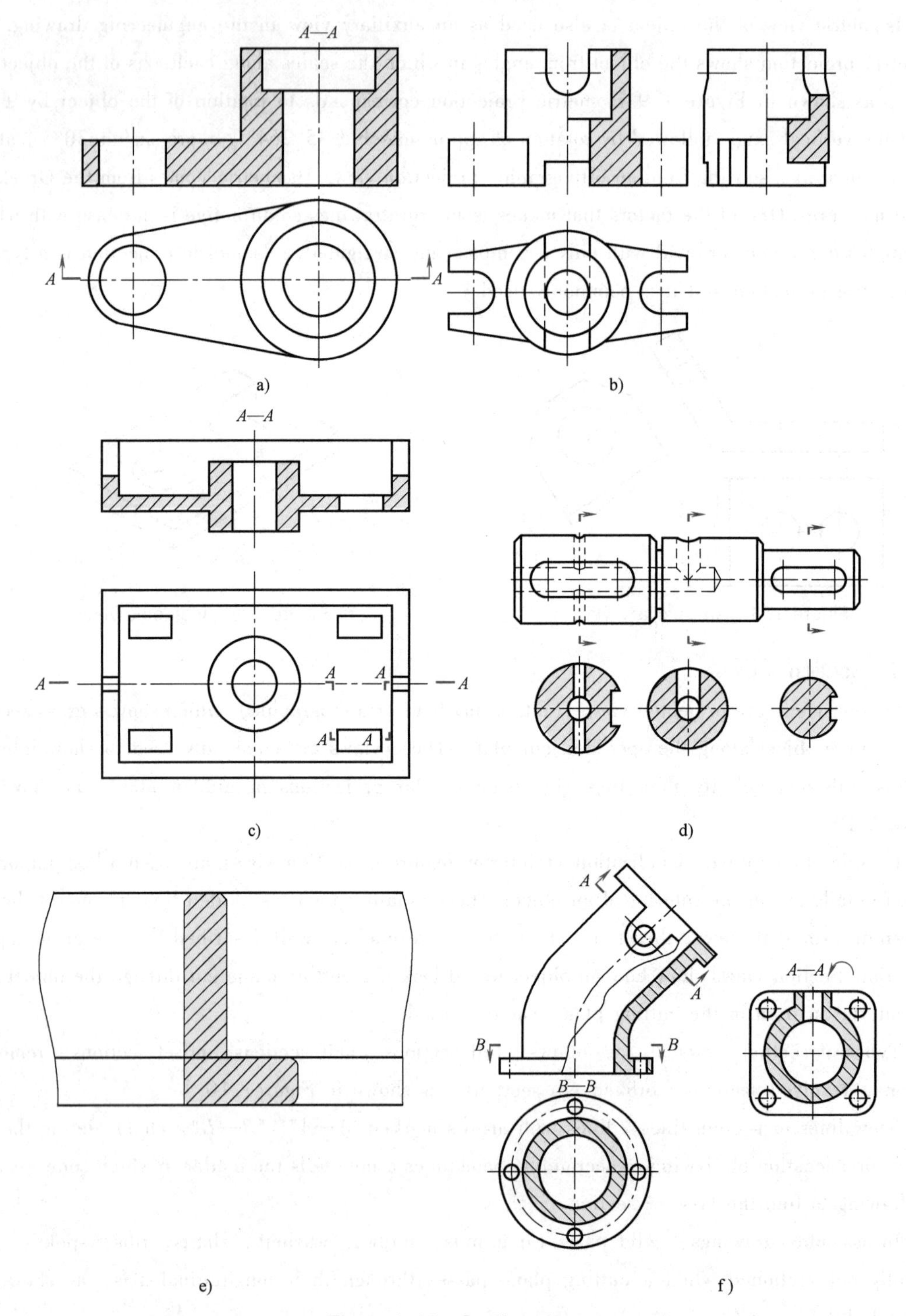

Figure 1.10 Section views

a) Full section b) Half section c) Offset section d) Removed section

e) Revolved section f) Broken-out section

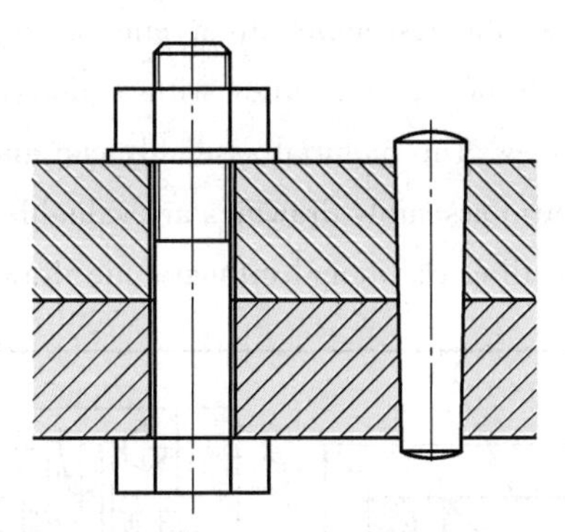

Figure 1. 11 Solid parts cut along their axis shown without section lines

description of a part or group of parts so that the part can be manufactured according to specification or assembled in the correct sequence.

1. Detail working drawings/detail drawings

The detail working drawing delineates the exact shape and size of an individual part. The part is fully described using dimensions, notes, possible finish marks and GD&T (Geometric Dimensioning and Tolerance) symbology. Figure 1. 12 illustrates the detail working drawing of a spur gear.

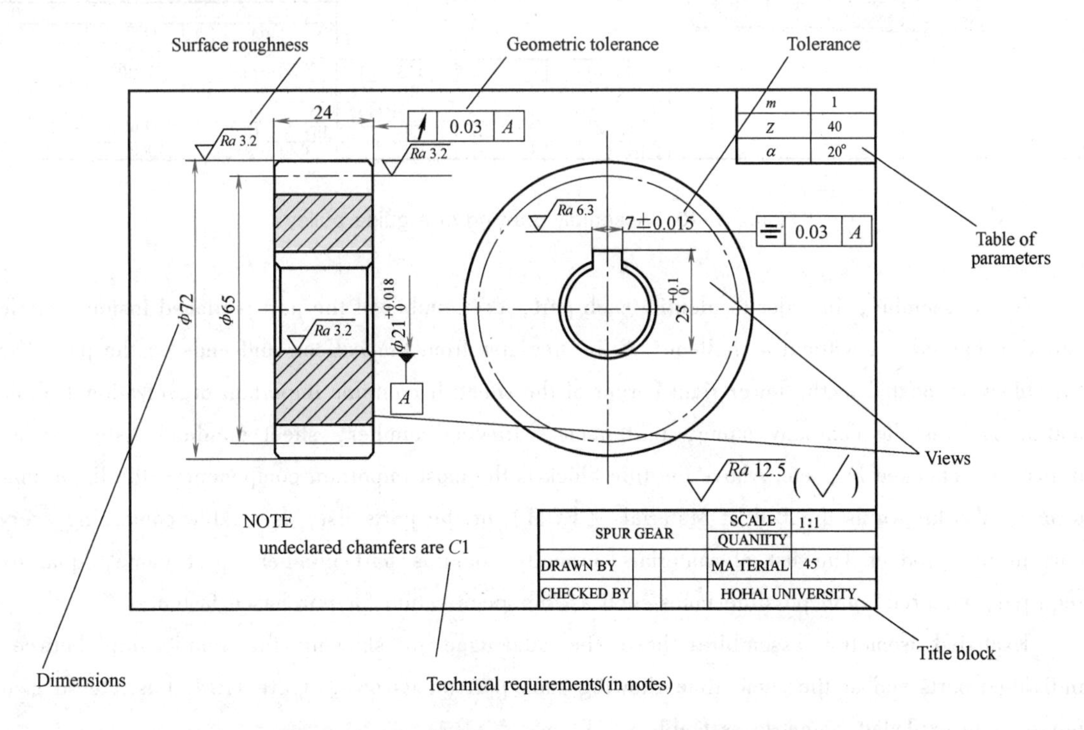

Figure 1. 12 Detail working drawing of a spur gear

2. Assembly drawings

An assembly drawing shows a completely assembled product and supplies the information about the location and relationship of the parts. An assembly drawing is intended to show how parts are related to one another. There are two basic categories of assembly drawings: orthographic and pictorial. Orthographic assemblies view the assembly from an orthographic viewpoint. Figure 1.13 illustrates an orthographic assembly of a guide pulley with a pictorial assembly. Pictorial assemblies are normally drawn as isometric view. The pictorial assembly can also be viewed as a section. Isometric assembly and sectioned isometric assembly drawings are valuable when trying to explain a product to an audience unfamiliar with reading traditional orthographic drawings.

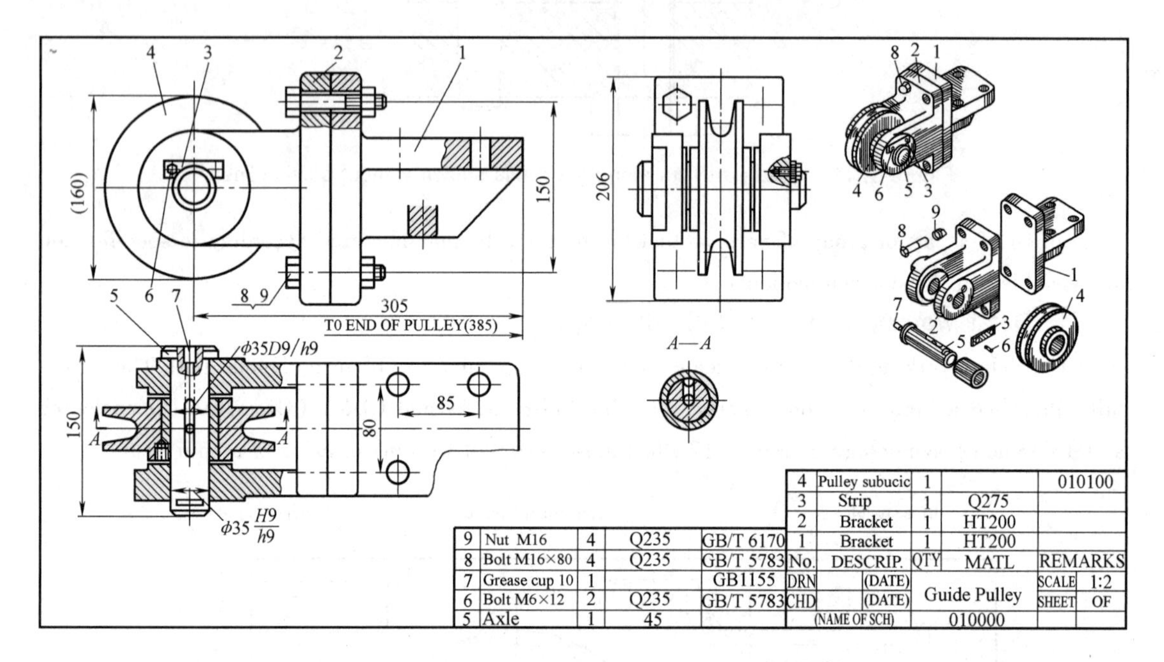

Figure 1.13 Assembly drawing of a guide pulley

In an assembly, in order to identify each part, the number of the part is placed inside a circle called a "balloon" or without a "balloon". A line radiates from the balloon and ends on the part. The title block is located in the lower right corner of the sheet. It contains important organizational information such as the company name, part name, drawing number, sheet number, date, scale, drawn by, checked by, etc. Above the title block is the most important components—the list of materials, also known as the Bill of Materials (BOM) or the parts list, is a table containing every part in the product. The list of materials normally contains part number, part name, quantity required, material, and possible notes such as the specification for purchased fasteners.

Exploded isometric assemblies have the advantage of showing the relationship between individual parts and at the same time showing each part on its own. Figure 1.14 illustrates a gear pump as an exploded isometric assembly.

1.5.5 Feature-Based 3D Models

Nowadays, the new design environment, characterized by feature-based modeling with the embedded parametric and associate capability, is widely used in mechanical design. These 3D systems are capable of capturing the design intents while producing a quality database that can be used for

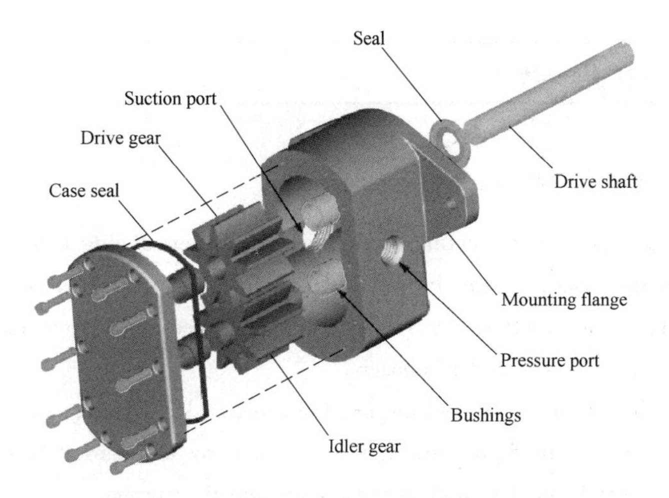

Figure 1. 14 Exploded view drawing of a gear pump

several purposes including documentation, engineering analysis, the integration of design and manufacturing and even further more product lifecycle management. Features on a model are directly related to other features. Changing the base model will automatically update features added to it. Likewise, a completed part is directly related to assemblies to which it is linked. Part changes then flow through the design from the original concept to the completed assembly as well as to the assembly drawing.

Today, CAD systems have become more intelligent and are more focused on system integration. 3D CAD models which are created by these systems are all in the form of a geometric database. Communications for information and data exchange are thus greatly facilitated. Therefore, feature-based CAD models have become a very effective and important way to describe objects in mechanical engineering. Figure 1. 15 shows 3D CAD models of plastic injection molds created by using 3D CAD system.

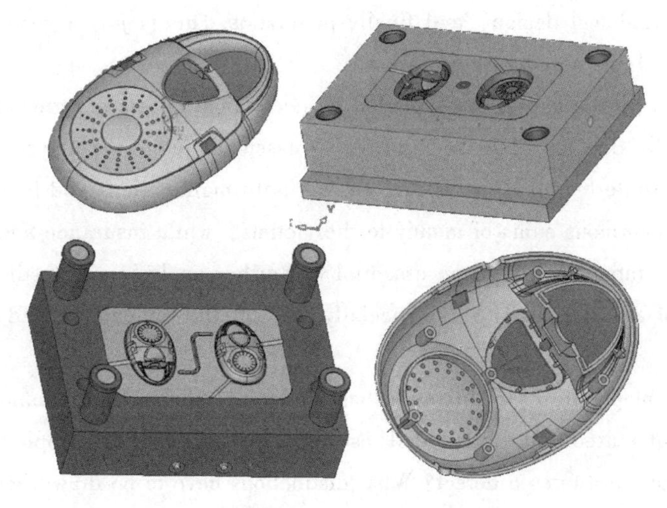

Figure 1. 15 3D CAD models of plastic injection molds

1.6 The Design Process

1.6.1 Introduction to Design Process

To engineers, design means creating something new by enhancing existing designs, altering them to perform new functions, or simply introducing new concepts. Design is not restricted to engineers, of course, but it is practiced by a large body of professionals from fashion and industrial designers to architects, sculptors and composers.

A design is usually produced to meet the need of a particular person, group or community. It is driven by the consumers, shaped by the users and priced by the market. Therefore, to design a product, one needs to establish the problem constraints and then propose a solution that will operate within those constraints. Constraints are algebraic equations that are functions of the design parameters. They are usually in the form of equality and inequality conditions that must be met by the objective function. In the process of designing something, it will become apparent in most cases that more than one solution exists, although in some cases no solution exists. Hence a question one would ask is,"What is the best or optimal design solution?" To answer this question, the engineer might need further information in terms of social and economic variables pertaining to the use of the product being designed. We often face decisions about factors involved in the design when specifics are not known. This requires experience and professional judgment.

Traditionally, the design process involves draftspersons and design engineers, who, once they have completed their jobs, usually present the blueprints (layouts) of the product to the manufacturing or production division. The latter employs machinists, welders and manufacturing engineers who will try to produce the product according to the specifications given by the design group.

The engineering design process is a multi-step process including the research, conceptualization, feasibility assessment, establishing design requirements, preliminary design, detailed design, production planning and tool design, and finally production. This is just a general summary of each step of the mechanical engineering design process.

In recent years, the litigation cases related to mechanical failure are growing. Product performance failure is usually due to a lack of analysis. Professional judgment played a major role in the early days as a substitute for such analysis. It was not until malfunctions and failures started to cause the industry to pay enormous sums of money to the victims, while insurance and liability costs were increasing at a rapid rate, so it became essential for further analysis and testing to be done before the final approval of any design. This is especially true for the automobile and commercial aircraft industries.

This ever-present possibility of litigation has profoundly affected the manner in which products are designed, manufactured and marketed. Those involved have had to grapple with such seemingly simple issues as what constitutes a defect? What distinctions have to be drawn between manufacturing flaws and design defects? Is it ever acceptable to market a defective product? What sort of warnings would make an unacceptable product acceptable? This leads ultimately to consideration being given to the Engineering Ethics—a field examines and sets the obligations by engineers to society, to their clients, and to the profession. As a scholarly discipline, it is closely related to subjects such as the philosophy of science, the philosophy of engineering, and the ethics of technology.

As the designers for a machine, engineers should hold paramount the safety, health and welfare of the public in the performance of their professional duties. During the design process, engineers should formulate and follow a rigorous design process which plays a critical role in ensuring the production of reliable, profitable and environmentally safe products that meet the needs of customers.

Therefore, the total design is of interest to us in this section. How does it begin? Does the engineer simply sit down at his or her desk with a blank sheet of paper and jot down some ideas? What factors influence or control the decisions which have to be made? Finally, how does this design process end?

The complete process, from start to finish, is often outlined in Figure 1. 16. The process begins with recognition of a need based on market research and a decision to do something about it (i. e. problem identification), finally ends with the product prototyping. In the next several sections, these steps will be examined in the design process in detail. The objective is to mark out a reference model that translates needs and ideas into technical prescriptions to transform the most suitable resources into useful material products. Although not even one model exists that includes the huge variety of possible processes for product development, and each process could be considered unique, it is possible to identify activities and elements held in common.

In general, the design process is viewed as an integration of main six stages, i. e. problem definition, conceptual design, preliminary design, detailed design, final design presentation and product prototyping, as shown in Figure 1. 16.

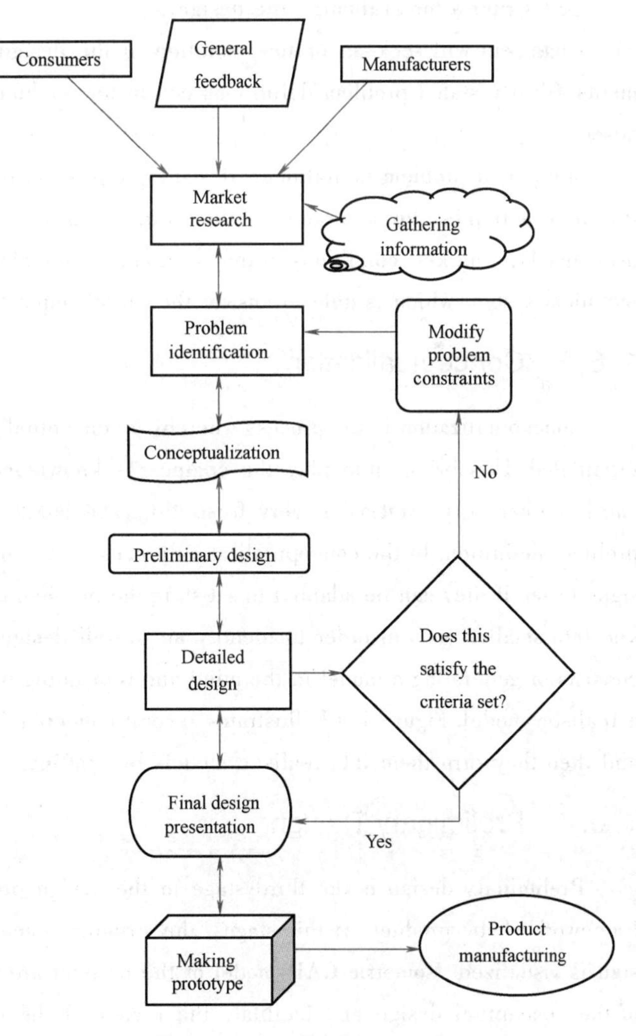

Figure 1. 16 Design process

1.6.2 Problem Definition

Problem Definition is a process
to do market research and write down or list all information to describe the problem. When the information about the problem has been collected, engineers need to define the real problem.

A well-defined problem is the key to a successful design solution. The design process involves many stages requiring careful thinking; the problem definition helps everyone focus on the objectives of the problem and the things that must be accomplished. Engineers should not overlook the importance of this crucial first step in design, nor should they act hastily in stating the problem. The problem definition must be broadened in a reasonable fashion to make sure that the ultimate solution is the desired one. The problem definition should include the following:

(1) A statement of objectives and goals to be achieved.

(2) A definition of constraints imposed on the design.

(3) Criteria for evaluating the design.

Engineers will seek an optimal solution to the design problem in keeping with these requirements. Clearly stated problem definitions can foster productivity by keeping the engineer's efforts focused.

With good problem definitions, designing a product becomes a process of transforming information from a request for a product outline with certain characteristics (requirements, constraints, user needs, market conditions, and available technology) to the complete description of a technical system which is able to answer the initial request.

1.6.3 Conceptualization

Conceptualization is the process whereby a conceptual design satisfying the problem definition is formulated. This brings into play the engineer's knowledge, ingenuity and experience. This phase can be either very exciting or very frustrating. The latter experience could be the result of a poor problem definition. In the conceptualization process, it is often desirable to look at some existing designs to see if they can be adapted to satisfing the problem definition. Complex problems must be broken into smaller ones in order to identify an overall design solution. In any case, conceptualization consists of generating a model in the mind and translating it back into forms and shapes to conform to a realistic model. Figure 1.17 illustrates several conceptual models which are in designers'mind first and then they turn them into realistic models by drafting.

1.6.4 Preliminary Design

Preliminary design is the third stage in the design process; it focuses on creating the general framework of the product. At this stage, the structure necessary in implementing the conceptual design is visualized. Isometric CAD model of the product are made to capture the major characteristics of the conceptual design and facilitate the review of the new product, and also to ensure spatial compatibility. At this stage, a successful and effective design relies a great deal on the synthesis as-

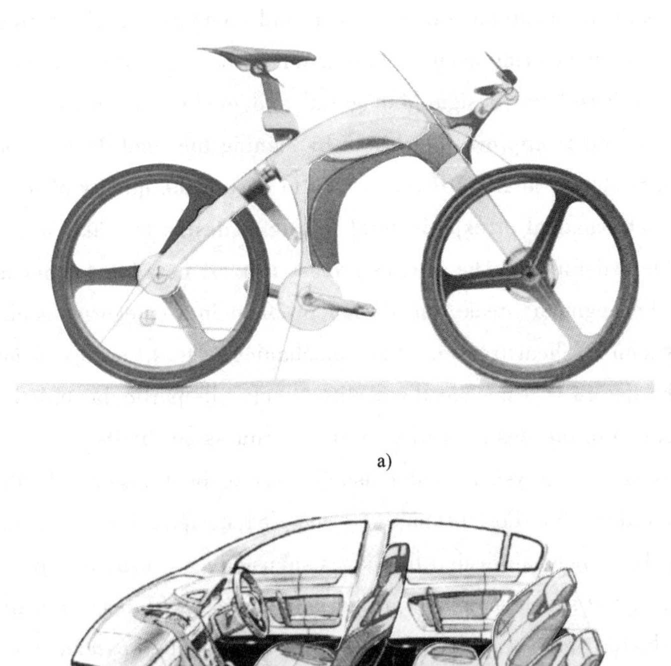

a)

b)

Figure 1. 17 Conceptual design models

a) A bike b) Inside of a car

pect of the process—the process of taking elements of the concept and arranging them in the proper order, sized and dimensioned in the proper way. This is one of the most challenging tasks which an engineer faces. In this stage, the information required for the proposed conceptualization is organized and a plan is devised for achieving that design. To achieve a viable synthesis decision, all the elements affecting the design, including product configuration, cost and labor, must be considered. Modifications to the design of key components are made until the entire design team reaches an agreement to pursue detailed design.

1. 6. 5 Detailed Design

The fourth stage in the process is "detailed design". In this stage, a complete engineering description of the product is developed. Detailed design involves the specification of dimensional and geometric tolerances, the analytical evaluation of the system and components, the confirmation of the cost and quality of the manufacturing processes, and the examination of assembly and component drawings. Completion of the design documentation is the essential element in the detailed design

stage. It should be noted that computer-aided design and computer graphics provide exceptional capabilities for design and engineering documentation. When the detailed design is complete, a conference is convened to "release" the design. Managerial and engineering personnel meet to conduct a thorough design review and to approve the design by signing the final documents.

At detailed design stage, analysis is a key step to ensure the quality of the product. Analysis is concerned with the mathematical or experimental testing of design to make sure it meets the criteria set forth in the problem definition. The engineer must test all possible factors important to the design. For instance, the engineer breaks the design problem into categories such as stress analysis, vibration, thermodynamics, heat transfer, fluid mechanics, etc. In each category, the design as a whole or a part of the design is tested for the ability to serve its particular function. A safety factor is usually added to make sure the design works within certain safety limits.

Before the analysis, the physical model usually has to be transformed into analysis model in order to reduce computing time. For instance, a complex design, like an automobile, is divided into realistic models. If we need to design the car suspension, we usually represent the body of the car by a mass M, and the suspension by springs and dampers (linear or nonlinear). A vibration analysis is then conducted for extraction of the spring and damper parameters that yield the most comfortable ride. Figures 1. 18 and Figures 1. 19 illustrate possible model used to simulate the vibration response of a car seat subjected to different road conditions.

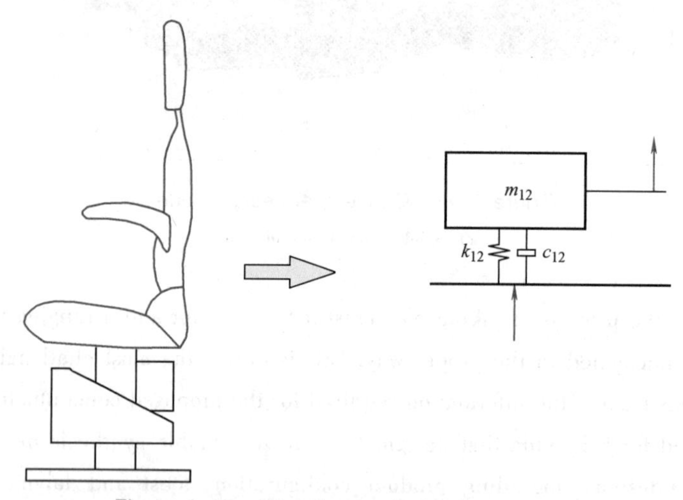

Figure 1. 18 Vibration analysis of a car seat

Developing models requires ingenuity and experience. It is important that the models are developed to be realistic, simple and mathematically testable. If a model is too complex, it is likely to take a long time to analyze and hence to cost more. But if it is too simple, it might be unrealistic; that is, predictions from its analysis might not be typical of the proposed design. Simplicity is important, but the model must exhibit behavior that is close to that of the actual design, based on good engineering judgment. Ultimately, for all analyses, the models we develop must be adequate and exhibit the salient characteristics of the design under consideration.

In addition to mathematical and experimental verification in analysis, it is customary to

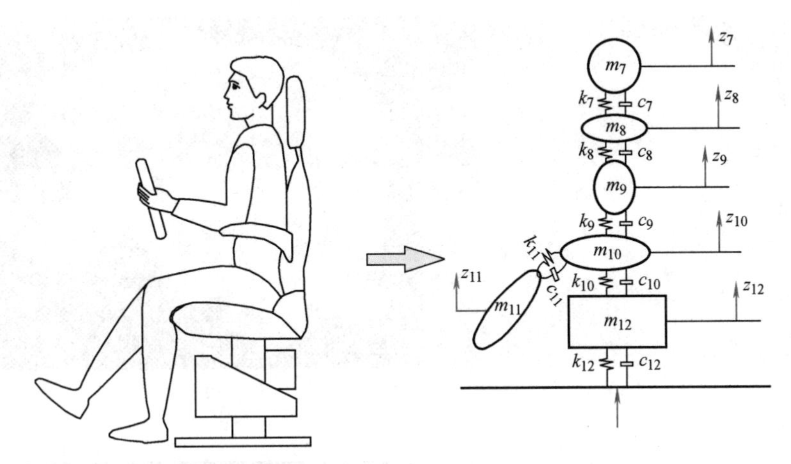

Figure 1.19 Vibration analysis of an occupied car seat

examine prototypes for comfort, cushion, rigidity and so on. Engineers use their experience to judge whether the product will meet customers' requirements.

The advancement of computer-aided engineering, or CAD program has made the detailed design phase more efficient. This is because a CAD program can provide analysis and optimization. It can also calculate stress and displacement using the finite element method to determine stresses throughout the part. In advanced CAD systems, the analysis tools are equipped with graphics tools. It is possible to conduct an analysis and predict performance in the design stage. It is common practice that many numerical analyses involved in the design are performed on computer, leaving the designers free to make decisions based on the output from these analyses and their knowledge-based judgment. Figure 1.20 illustrates an example of Computer-Aided Analysis (CAA).

The model in Figure 1.20c is rebuilt based on its point cloud model, which is acquired from the part by using non-contact measuring device, as shown in Figure 1.20 a, b. The reconstructed surface may not have passed through all data points and there would be an error which associated the surface with respect to the point cloud model. After completing the reconstruction of CAD model, how do we evaluate the errors of reconstruction and then give a reasonable report for further modification if the model cannot meet the design specifications? It's almost impossible to do such analysis without the help of computers.

Today it's an easy operation to make the analysis on computers. Designers can import the reconstructed CAD model into reverse engineering software or import point cloud model into CAD system to investigate the precision of the reconstructed CAD model. By computing the Euclidean distances between each of the data points in point cloud model and the corresponding point on the surface of CAD model, the computer gives an error report graphically in seconds, as shown in Figure 1.20d. Judging from the figure the error is within the error allowance except those noisy points, hence the error could meet the need. This error may be measured as the maximum value or average value of the errors between each of the data points and the corresponding point on the surface of the CAD model; it depends on the requirement of designers.

Figure 1.21 illustrates the filling analysis result of an air-conditioner grille on computer.

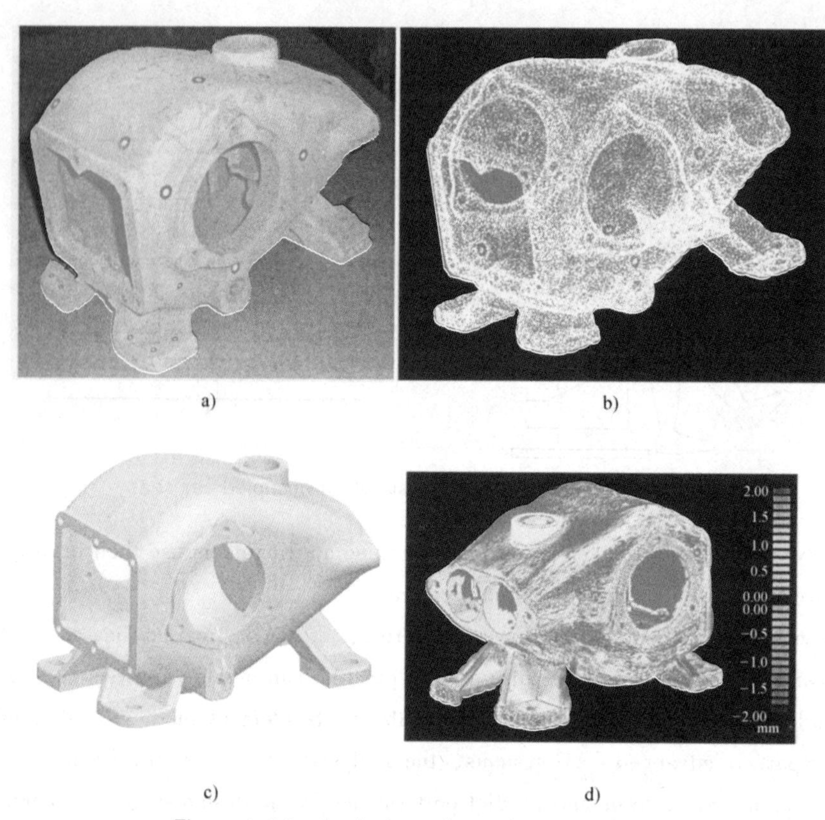

Figure 1.20 Analysis performed on computers

a) Part b) Point cloud model c) Reconstructed CAD model

d) Error analysis of reconstruction

According to its original design, the product has 18 injection gates. 76MPa injection pressure can assure that the cavity is able to be filled, as we perceive in Figure 1.21.

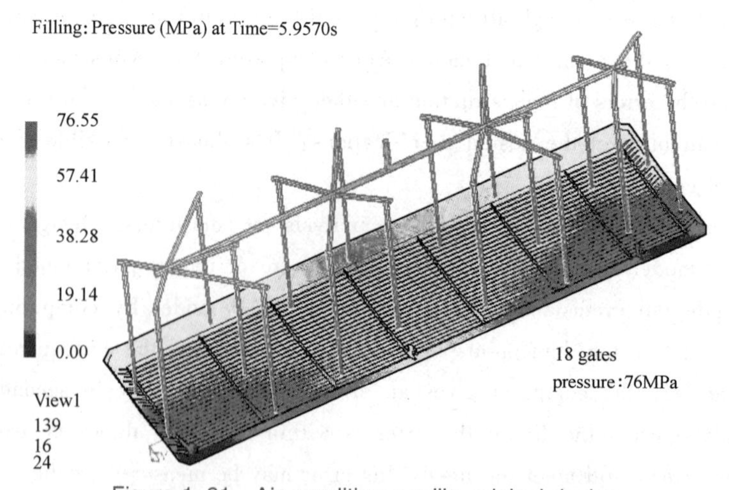

Figure 1.21 Air-conditioner grille-original design

Figure 1.22 illustrates the filling analysis result after 18 gates are reduced to 8 gates. As we can see that the analysis shows the cavity is still able to be filled under 75MPa injection pressure. Therefore, the second revised design is a better option.

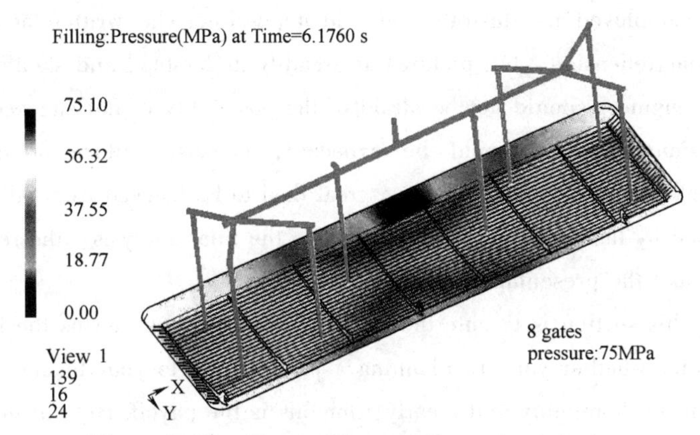

Figure 1. 22 Air-conditioner grille-revised design

1. 6. 6 Final Design Presentation

Communicating the design to others is the final, vital presentation step in the design process. Undoubtedly, many great design inventions and creative works have been lost to posterity simply because the originators were unable or unwilling to explain their accomplishments to others. Presentation is a selling job. The engineers, when presenting a new solution to administrative, managerial or supervisory persons, are attempting to sell or prove to them that this solution is a better one. Unless this can be done successfully, the time and effort you have spent on obtaining the solution have been largely wasted. When designers sell a new idea, they also sell themselves. If they are repeatedly successful in selling ideas, designs, and new solutions to management, they begin to receive salary increases and promotions; in fact, this is how anyone succeeds in his or her profession.

Basically, there are three means of communication. They are comprised of the written, the oral, and the graphical forms. A successful engineer should be technically competent and versatile in all three forms of communication; otherwise competent person who lacks ability in any one of these forms is severely handicapped. If ability in all three forms is lacking, no one will ever know how competent that person is! The three forms of communication—writing, speaking and drawing—are skills, that is, abilities which can be developed or acquired by any reasonably intelligent person. Skills are acquired only by practice—hour after monotonous hour of practice [if necessary] . Musician, athletes, surgeons, typists, writers, dancers, aerialists, and artists, for example, are skillful because of the number of hours, days, weeks, months, and years they have practiced. Nothing worthwhile in life can be achieved without work, often tedious, dull, and monotonous, and lots of it; and engineering is no exception.

The ability in writing can be acquired by writing letters, reports, memos, papers and articles. It does not matter whether or not the articles are published—the practice is the important thing. Ability in speaking can be obtained by participating in professional activities. This participation

provides abundant opportunities for practice in speaking. To acquire drawing ability, pencil sketching should be employed to illustrate every idea possible. The written or spoken word often requires study for comprehension, but pictures are readily understood and should be used freely.

The competent engineer should not be afraid of the possibility of not succeeding in a presentation. In fact, occasional failure should be expected, because failure or criticism seems to accompany every really creative idea. There is a great deal to be learned from failure, and the greatest gains are obtained by those willing to risk defeat. In the final analysis, the real failure would lie in deciding not to make the presentation at all.

The purpose of this section is to note the importance of presentation as the key step in the design process. No matter whether you are planning a presentation to your teacher or your employer, you should communicate thoroughly and clearly, for this is the payoff. Helpful information on report writing, public speaking, and sketching or drafting is available from countless sources, and you should take advantage of these sources. In Part 3 communication skills and techniques will be discussed in detail.

1.6.7 Product Prototyping

A prototype is an early sample or model built to test the design model or process or to act as a thing to be replicated, and it is designed to test and trial a new design to enhance precision by system analysts and users. Therefore, prototyping serves to provide specifications for a real, working system rather than a theoretical one. Tasks to complete in product prototyping include selecting the material, selection of the production processes, determination of the sequence of operations, and selection of tools, such as jigs, fixtures and tooling. This task also involves testing a working prototype to ensure the created part meets qualification standards.

With the completion of qualification testing and prototype testing, the engineering design process is finalized. The next process should move onto mass production.

1.7 Modern Machine Design Challenges

1.7.1 Mechatronics

Intense competition is putting pressure on machine and device designers to deliver systems with higher throughput, reduced operating cost, and increased safety. Machine designers have switched from rigid, single-purpose machines relying purely on mechanical gears and cams to flexible multipurpose machines by adopting modern control systems and servomotors. Although these improvements have made machines flexible, they have also introduced a significant amount of complexity to the machines and subsequently to the machine design process.

More machine designers today are in positions to deal with complex machine systems, as shown in Figure 1.23. Figure 1.23 shows a typical product of complex machines. This system is a multidisciplinary field of engineering.

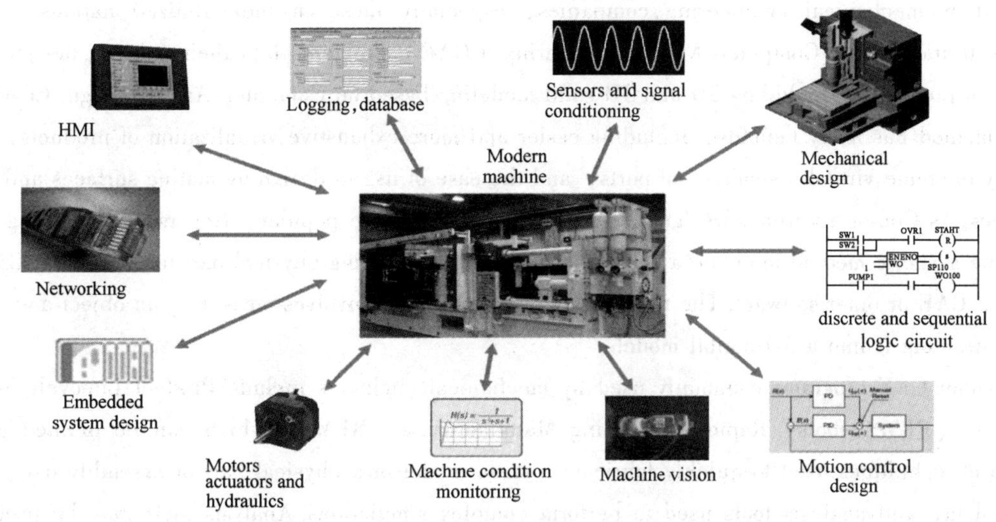

Figure 1. 23 A typical product of complex machines

Meeting these multidisciplinary engineering challenges requires improvements in all machine design areas. Mechatronics, gaining in popularity as a way to describe this evolution, represents an industry-wide effort to improve the design process by integrating the best development practices and technologies to streamline machine design, prototyping and deployment. Mechatronics is an integration of electronics, control engineering, computer engineering, software engineering and mechanical engineering. A mechatronics-based approach can lower the risks associated with machine design early in the design cycle and help machine designers meet the key challenges they face today.

In the past, the mechatronics design approach was difficult. But developments in the market, such as the integration between best-in-class software packages used for different design phases, the evolution of flexible and modular programmable automation controllers (PACs), intuitive graphical system design software, and advanced control design techniques, have helped machine designers successfully implement the mechatronics design approach.

1. 7. 2 The Impact of Digital Technology on Machine Design

Over the past 40 years, computers have evolved from specialized and limited information-processing and communication machines into ubiquitous general-purpose tools. Whereas once computers were large machines surrounded by peripheral equipment and tended by technical staff working in specially constructed and air-conditioned centers, today computing equipment can be found on the desktops and in the work areas of secretaries, factory workers, and shipping clerks, often alongside the telecommunication equipment that links home offices to suppliers and customers. In the course of this evolution, computers and networks of computers have become an integral part of the research and design operations of most enterprises. In the last two decades, moreover, microprocessors have allowed computers to escape from their boxes, embedding information processing in a growing array of artifacts as diverse as greeting cards and automobiles, thereby extending the reach of this technology into new territory.

Many mechanical engineering companies, especially those in industrialized nations, have begun to incorporate Computer-Aided Engineering (CAE) programs into their existing design and analysis processes, including 2D and 3D solid modeling based on Computer-Aided Design (CAD). This method has many benefits, including easier and more exhaustive visualization of products, the ability to create virtual assemblies of parts, and the ease of use in designing mating surfaces and tolerances. As Computer-Aided Design (CAD) has become more popular, Reverse Engineering has become a viable method to create a 3D virtual model of an existing physical part for use in 3D CAD, CAM, CAE or other software. The reverse engineering process involves measuring an object and then reconstructing it into a 3D digital model.

Other CAE programs commonly used by mechanical engineers include Product Lifecycle Management (PLM) tools, Rapid Prototyping Manufacturing (RPM), which can be defined as a group of techniques used to quickly fabricate a scale model of a physical part or assembly using 3D CAD data, and analysis tools used to perform complex simulations. Analysis tools may be used to predict product response to expected loads, including fatigue life and manufacturability. These tools include Finite Element Analysis (FEA), Computational Fluid Dynamics (CFD), Computer-Aided Testing (CAT), Computer-Aided Process Planning (CAPP) and Computer-Aided Manufacturing (CAM). Figure 1.24 shows a typical modern design process in mechanical engineering based on digital technologies.

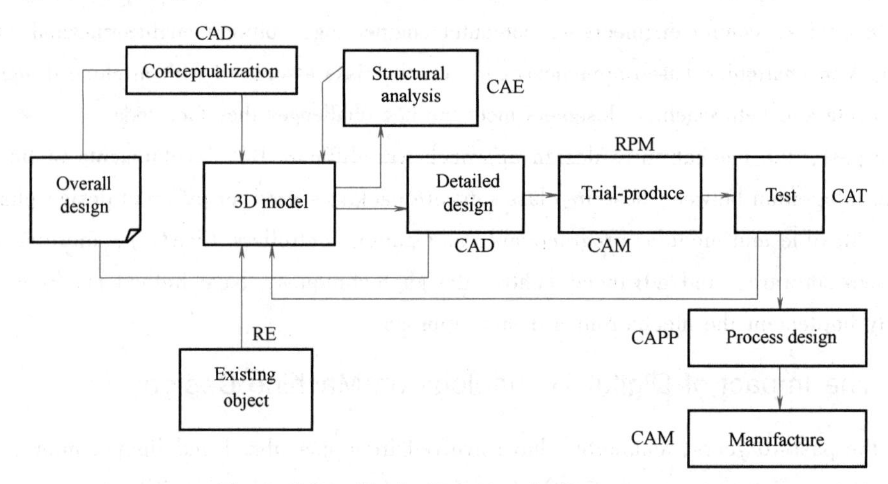

Figure 1.24 Typical modern design process

Using CAE programs, a mechanical design team can quickly and cheaply iterate the design process to develop a product that better meets cost, performance and other constraints. No physical prototype need to be created until the design nears completion, allowing hundreds or thousands of designs to be evaluated, instead of a relative few. In addition, CAE analysis programs can model complicated physical phenomena which can't be solved by hand.

As mechanical engineering begins to merge with other disciplines, as seen in mechatronics, Multidisciplinary Design Optimization (MDO) is used with other CAE programs to automate and improve the iterative design process. MDO tools often cover existing CAE processes, allowing product evaluation to continue even after the analyst goes home for the day. They also utilize sophisticated optimization algorithms to more intelligently explore possible designs, often finding better innovative solutions to difficult multidisciplinary design problems.

Nowadays, the above digital technologies as shown in Figure 1.24 have been widely accepted and used in machine design and manufacturing processes. Contemporary designers have to accept this challenge and cooperate with these new digital technologies.

1.8 Design Case Study

1.8.1 Introduction

In this section we trace the progress of a students'team as they design a table. The design steps are followed, and a practical application is used to illustrate how the design is achieved when they develop a new table. Although the table itself is not very complex, the design process includes major steps in designing a mechanical product.

During design process the students get started with studying the need of consumers, get general feedback from market, gather information related to the product, define problems to be solved, collect innovative ideas by individual brainstorming as well as group brainstorming, and form anoverall design scheme. With the help of CAD system, they build the product's model and test its function virtually, do a lot of analyses and optimization in order to determine whether the goals set forth in the problem definition step have been reached, go back and forth between the early stages of the process, and finally the project proceeds toward a final design and manufacture. The students practice the overall product development strategy during this process.

1.8.2 Design Motivation

Creative design lies at the center of the mechanical engineering profession, and an engineer's ultimate goal is to produce a new product that solves one of the technical problems in the society. The initial idea of designing a product always comes from a specific motivation. In this case, the original motivation comes from the following awkward situations; most of us may have had such kind of experiences in our daily life. Figure 1.25 illustrates two of them.

If you are an engineer, what would you think about these situations? What would you do? If you are a very devoted and innovative engineer, these situations may inspire you to seek a better solution. This could be your original motivation for creating a new product or making improvement to the existing product.

A need is identified based on the above experiences or complains received from other people. The team puts forward an idea to make a new table with adjustable size according to the actual demand. The task, to design an extensible table, serves as the "identification of a need". This is the stage of problem definition. The problem definition should include a definition of constraints imposed

Figure 1.25 Awkward situations in our daily life

on the design and criteria for evaluating the design. Several constraints and criteria for evaluating the table are roughly outlined; for instance, the product must be functional, safe, reliable, maneuverable, competitive, usable, manufacturable and marketable. In this design these words are meant to convey the following meanings.

Functional: The table must perform to meet its intended need and customers' expectation.

Safe: The table is not hazardous to the user, bystanders, or surrounding property. Hazards which can not be "designed out" are eliminated by a protective enclosure; if that is not possible, appropriate directions or warnings should be provided.

Reliable: Reliability is the conditional probability, at a given confidence level, that the table will perform its intended function satisfactorily.

Maneuverable: The table should be easy to be manipulated.

Competitive: The table is expected to have the potential of being a contender in its market.

Usable: The table is "user-friendly", accommodating to human size, strength, posture, reach, force, power and control.

Manufacturable: The table has been reduced to a "minimum" number of moving parts, suited to mass production, with dimensions, distortion and strength under control.

Marketable: Designers should take marketable factors into consideration. In this case, it's a design assignment for students, so less consideration is required before going into production.

Under the guidance and encouragement of teachers, the team begins to do this project. The first step is to get started with market research.

1.8.3 Market Research

Market research is a key factor to get advantage over competitors. Market research provides important information to identify and analyze the market need, market size and competition. Market research is for discovering what people want, need or believe. It can also involve discovering how they act. Through market information one can know the price of different tables in the market, their structures and functions as well as the supply and demand situation. Once the research is completed, it can be used to determine how to market your product.

Information about the market can be obtained from different sources, including market place, the Internet, etc. In this case students go to supermarkets and furniture markets to gather information. They search data related to tables on the Internet and also check patent states of tables through the website of state intellectual property office of P. R. C. They study many kinds of existing tables with variable size in the market and conclude that current expandable tables can be divided into the following three categories, as shown in Figure 1. 26.

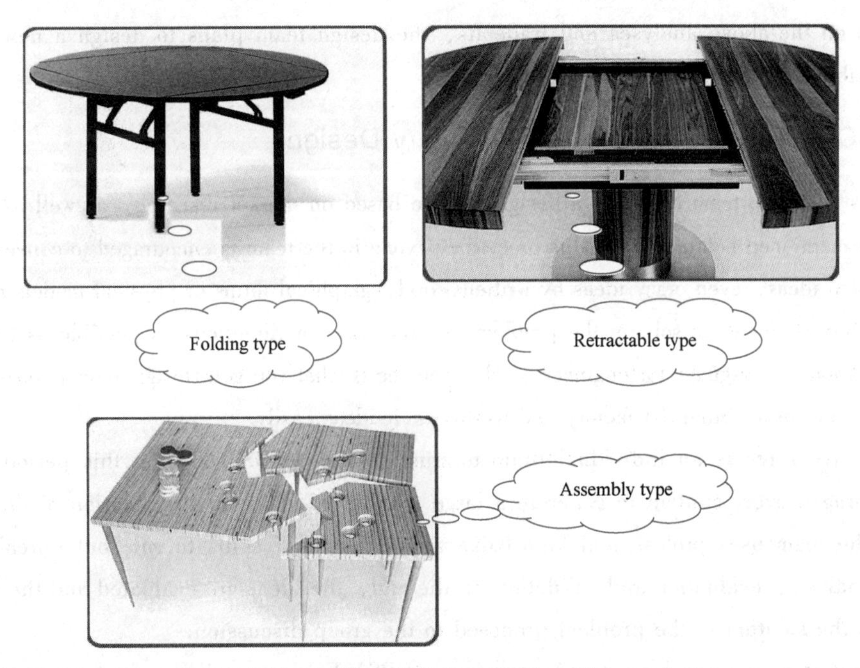

Figure 1. 26 Different kinds of table in the market

Through studying these existing tables, students observe the following disadvantages of these products: inconvenient to operate; hard to keep a rectangular or round shape before or after expanding them; some tables contain unsafe factors for kids' use.

When we compare round tables with rectangular tables, it can be seen that round tables offer the most efficient use of space and are the preferred shape for conversation since there are no awkward angles in either line of sight or seating as shown in Figure 1. 27. Some disadvantages of current round tables are that it's difficult to find a round table that can be expanded and still stay round. Rectangular tables offer a traditional dining experience and can be found in a huge variety of sizes—from small ones that will fit in a kitchen nook to oversized grand estate tables to seat 20. One disadvantage of a rectangular or square table is that leg room could be confined by the legs of table, and the sharp corners of the table have potential danger of hurting users, as shown in Figure 1. 28.

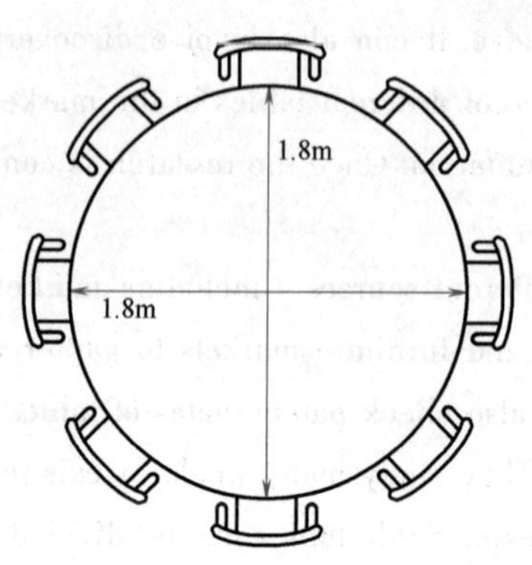

Figure 1.27 Round tables

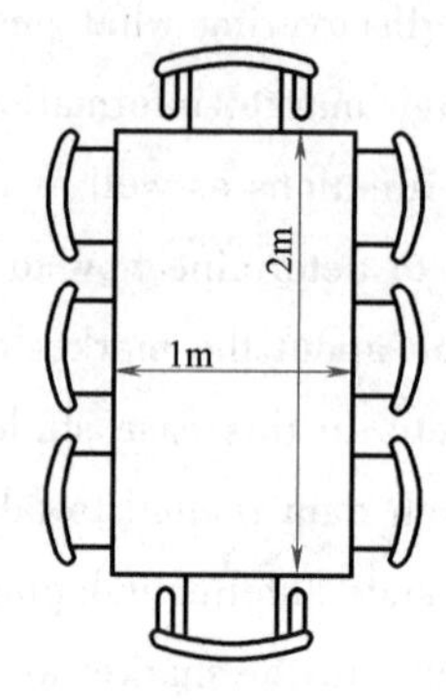

Figure 1.28 Rectangular tables

Based on the above analyses and tradeoffs, the design team plans to design a new round expandable table to improve the current situation.

1.8.4 Conceptualization and Preliminary Design

At this stage the team contrives a design scheme based on market research, as well as constraints and criteria mentioned before. During this process everyone in the team is encouraged to convey their conceptual design ideas, even crazy ideas by written, oral, graphical forms or physical models made of old materials. Students focus on solving the problem in more ways or directions as possible as they can. 3D drawing is a natural language for engineers, what you see is what you get. Designers can convey their design ideas freely in a virtual 3D factory and modify their ideas easily.

The above stage is an individual brainstorming session (IBS). During this period, teachers have encouraged every student to generate a large number of ideas for the solution of the problem, and then the team uses professional knowledge and intellectual skills to rule out unrealistic ideas based on analysis, evaluation and validation. In the end, the ideas are evaluated and the best one is selected as the solution to the problem proposed to the group discussion.

The next stage is group brainstorming sessions (GBS). Everyone writes a short summary in words, by drawing or presenting a physical model to illustrate the general approach that he or she uses, they also liste the major concepts, assumptions that they have been expected to use. Students have to express the design ideas clearly during group discussion. During the GBS stage, different ideas and design plans coming from the IBS are shared, blended and extended in a criticism free environment. Two goals are designed for GBS: one is to collect ideas, even unusual ideas; the other one is to select or generate alternative ideas or solutions for problems. In final stage of GBS, the most promising and optimal idea or design plan is selected or generated based on IBS for final trade offs sessions.

At final trade offs sessions, all designers, advisors, including workers (if it's possible) who are responsible for manufacturing process attend the meeting, analyze and evaluate the final optimal design plan at the meeting through question and answer discussion. All group members are encouraged to give their critical and constructive suggestions concerning design plans, and the plans are evaluated and reshaped through collective talk, well-grounded arguments and counter-arguments. In the end, a more optimal design plan is expected to be obtained and the team moves on to the next design process: detailed design—synthesis, analysis and optimization.

The phase of obtaining an optimal design plan is one of the most difficult, challenging and time-consuming stage in the design process. For students, after more and more brain exercises, even much more painful exercises, their ideas really begin to turn into reality. This is also an interesting, exciting and rewarding process. Even in the end their ideas might not be adopted, they enjoy the teamwork process and also creative ideas proposed by other designers stir their passion for creativity and nourish their minds.

Creativity is involved in every step of the design process, even during period of detailed design the best idea generated at the above steps would probably be discovered to be flawed, a return to any previous step would be necessary, and iteration would be required within the entire process. Such iteration makes more problems even troubles for students, challenges their intelligence again and again and also gives them a good opportunity to do more brain exercises. Students become smarter and more creative through different levels of discussion and trade offs as well as iteration.

Figures 1.29-1.32 illustrate some of their original conceptual design ideas in regard to the table.

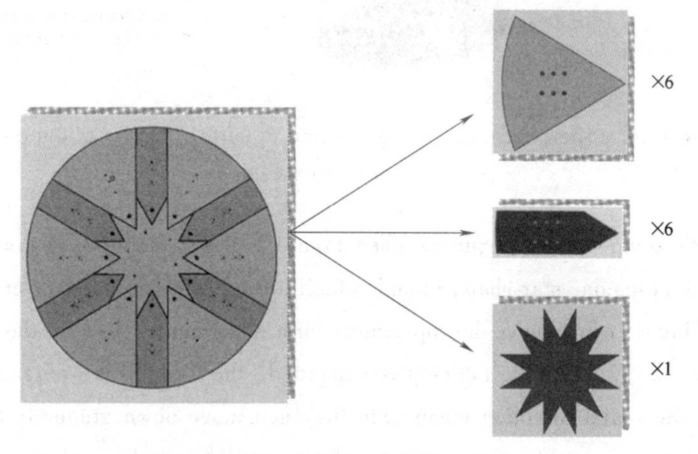

Figure 1.29 Division of the table top

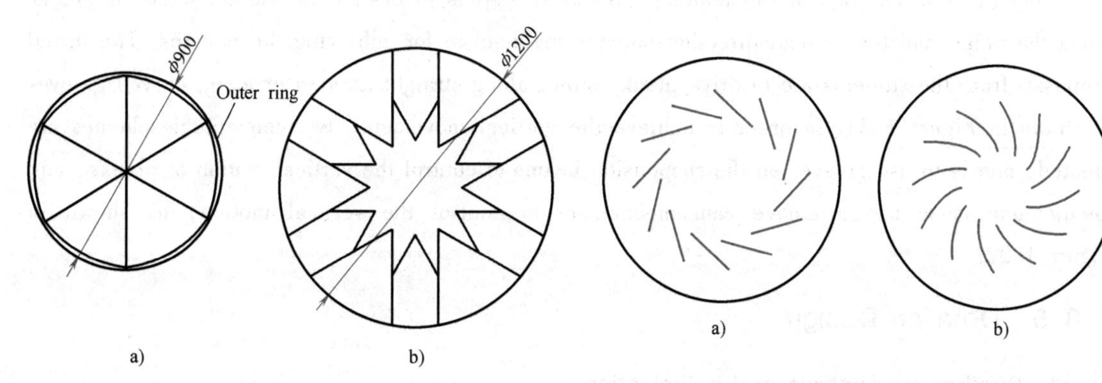

Figure 1.30 Size of the table

a) Before expansion b) After expansion

Figure 1.31 The radial motion path that table planks are expected to follow

a) Along straight grooves b) Along curved grooves

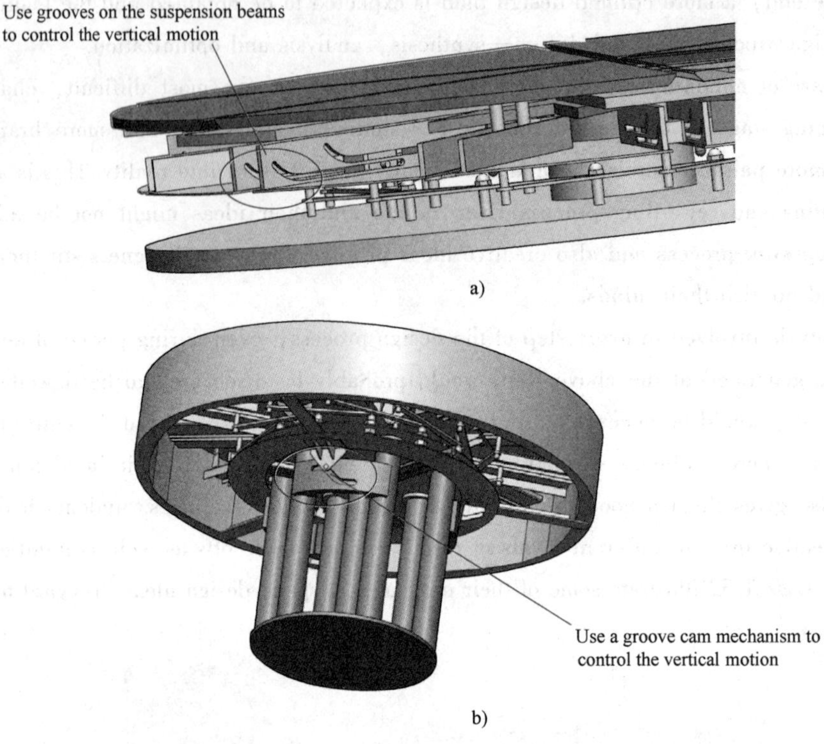

Figure 1.32 The conceptual design related to vertical motion of the desktop

a) Scheme one b) Scheme two

The tabletop is divided into 13 pieces (see Figure 1.29), including 6 fan-shaped planks, 6 arrow-shaped planks plus one star-shaped plank which are on three different layers (from top to bottom) respectively before expansion; the top planks form a perfect circle with the outer ring decoration (see Figure 1.30a). When the tabletop is expanded, the top and the next layer planks are expected to move in the radial direction meanwhile they can move down gradually at different heights until three layer planks are on the same height to form a perfect circle as shown in Figure 1.30b.

After the segmentation of the tabletop, the next step is to design the motion paths of planks along the radial and the vertical direction and the mechanism for achieving the motions. The initial proposals from the students are to drive planks either along straight grooves or along curved grooves as shown in Figure 1.31. In order to achieve the vertical movement, two conceptual schemes are created, one is to use grooves on the suspension beams to control the vertical motion of planks, and another one is to use a groove cam mechanism to control the vertical motion, as shown in Figure 1.32.

1.8.5 Detailed Design

1. Synthesis, analysis and optimization

As it can been seen that from the stage of conceptual design, an overall design scheme has been obtained and multiple solutions for detailed structures are sketched out. At detailed design stage, designers have to find the best solution from all feasible solutions based on synthesis, analysis and optimization.

Synthesis is the process of combining some initial ideas developed into a form or concept, which offers a potential solution to the design requirement. Some analysis, evaluation and optimization should be made at this stage to reduce the number of concepts requiring further work. Once a concept has been proposed, then it can be analyzed to determine whether constituent components can meet the demands placed on them in terms of performance, manufacture, cost and any other specified criteria. Inevitably there are conflicts between requirements. In the case of the table design, the team has to consider many factors, for instance: size, manoeuvrability, cost, ease of use, stability, etc. These considerations form the optimization of the product through producing the best or most acceptable compromise between the desired criteria.

Detailed design of the system is the last design activity before implementation begins. All design problems must be addressed by the detailed design or the design is not complete. The final design should be detailed enough to ensure that the digital model is a precise mapping of product instead of a rough description. In other words, each part has to have precise sizes, and precise position must be designed as to where the parts must be placed, how far and in what directions the parts must move, which parts must be connected to other parts and how they must be connected. When all parts are moving according to their design trajectories, check whether there are movement interferences or entity interferences among moving parts by a series of analysis, meanwhile make an analytical evaluation of the product and components, make a confirmation of the cost and quality of the manufacturing processes, and examine all assembly and component drawings. When the detailed design is complete, a thorough design review and check are needed to approve the final documents.

After the design release, the next stage is to produce a full-sized prototype to verify performance.

2. Final Design

During the design process, a lot of trade-offs in terms of function, material, complexity of the mechanism, cost and manufacturability are made in order to produce a prototype. For instance, the original ideas about motion paths of planks and the driving mechanism are modified and developed further. Figures 1.33-1.36 show the final design of the table. The size of the tabletop is 900mm before the expansion and 1200mm after the expansion as shown in Figure 1.33. 12 Slider-crank mechanisms are employed to drive 6 fan-shaped planks and 6 arrow-shaped planks which are connected to 12 sliders respectively as shown in Figure 1.34-1.35, among them the top layer connected to 6 fan-shaped planks can travel at 20mm vertically along the groove through a groove cam mechanism, the second layer connected to 6 arrow-shaped planks can travel at 10mm vertically along another groove so that when the table is rotated these two layers will reach to the same height of the star-shaped plank (the third layer) as shown in Figure 1.36. This is the fundamental idea behind the mechanism which drives the table planks move along radial as well as vertical directions, so that all planks will be on the same layer or on different layers when the table is expanded or folded.

The prototype made by the team is shown in Figure 1.37, and functionally the mechanism under the table works very well. We have got a patent for innovation for this design, and all original

files of 3D CAD models created by Solidworks, as well as a video shows how it works will be provided as supplementary teaching materials.

In this case, the team gets started with market research, problem identification, and then goes through conceptualization, preliminary design and detailed design phases with the help of CAD

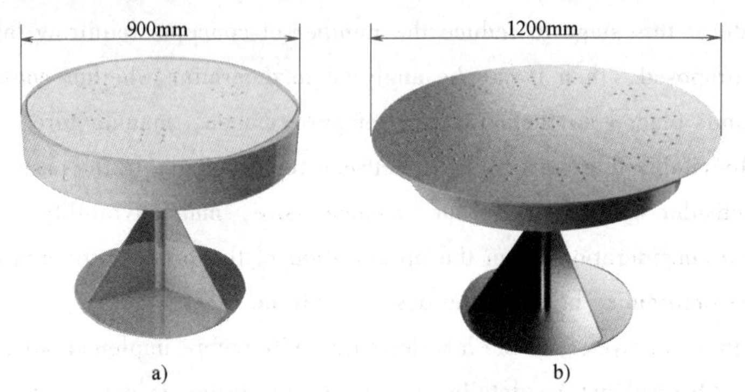

Figure 1.33 The appearance of the table

a) Before the expansion b) After the expansion

Figure 1.34 The inner structure of the table

Figure 1.35 Slider crank mechanism Controlling the radial motion of desktop

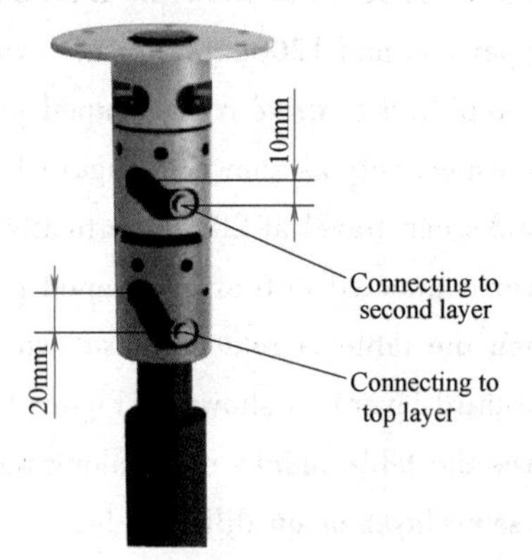

Figure 1.36 Groove cam mechanism controlling the vertical motion of the desktop

design system, finally makes a prototype by using some waste materials in order to test the function of the mechanism. However, many improvements are still needed to be done in terms of precision, material, and so on. In this case some old plastic materials are cut off and covered with papers to form the tabletop, so it looks like uneven comparing with the digital model. Similarly, the supporting base also looks different from the digital model. However, these flaws don't affect the main function of the table. In the future, more delicate modifications will be needed before the table goes to product manufacturing phase.

a) b)

Figure 1. 37 The prototype of the table made by the team

a) Assembly process b) The final prototype

Assignments

Suggestions: Encourage students to collaborate with each other, working and learning in a group. Working together to learn the material in this book will make it easier, and it's very likely that you'll remember it longer.

1. 1 Can you name one thing in the room in which you are sitting (excluding yourself, of course) that is not developed, produced or delivered by an engineer?

1. 2 Since engineering is evidently very important, can you now define the engineering method for solving a problem?

1. 3 What is design? What comes to your mind when you consider the mechanical design?

1. 4 Standards and codes have been developed over the years by various organizations to ensure product safety and reliability in services. Why do we need standards, for example, GB, in mechanical engineering drawing?

1. 5 What are the three engineering projections that are most commonly used? Prepare the three projections for an object you would like to present. You may sketch the projections by hand or

by computer software.

1.6 Describe the functions and the contents of an assembly drawing.

1.7 How do you value teamwork in mechanical engineering? Try to identify the specific characteristics of the team that made it effective. Please list these characteristics and tips for working successfully in a group.

1.8 Imagine you accept an assignment to design a bicycle for three-year-old children, how will you begin with it? Try to draw a design flow chart.

1.9 Based on your personal experience or observation, describe the importance of documentation in engineering design. Why do computers and information technology play an important role in today's engineering design community?

1.10 Write a brief report detailing the development of safety belts in cars. When was the first safety belt designed? Which was the first car manufacturer to incorporate safety belts as standard items in its cars? You can search the Internet for background information.

1.11 Have you ever used finite element analysis software or any other analysis software, such as kinematics or dynamics analysis software? If so, introduce the functions and characteristics of the software to your group members, or share your analysis work that you have done before (if it's possible).

1.12 Search the Internet with the following URLs to get the latest information about the CAD software systems.

CATLA	http://www.catia.com
I-DEAS	http://www.sdrc.com
Pro/Engineering	http://www.ptc.com
UNGRAPHICS	http://www.unigraphics.com
SOLIDWORKS	http://www.solidworks.com
AutoCAD	http://www.autodesk.com

Based on the above search or your personal learning experience, select one of the software systems and write a one-page summary report on the characteristics of it.

1.13 This is a virtual brainstorming session. The purpose is to collect ideas on the following topic: what kind of blackboard will be used in the future? The ground rules are: no criticism of ideas as they are being presented. Don't worry about looking foolish, record all your ideas, even crazy ideas. Every group records all ideas from your group members and writes them down. After brainstorming identify the promising ideas. Don't evaluate the ideas in detail yet! Discuss ways to improve a promising idea or promising ideas. Choose and list the ideas for detail evaluation. Evaluate the most promising ideas! Finally choose one as the most promising idea in the group. You can gather all promising ideas from other groups together and repeat the above process. In the end there will be one or more promising ideas obtained. Do this practice, maybe your ideas could change our way of teaching and learning. It's worth doing.

1.14 Try to explain the following working drawing (see Figure 1.38). Work in pairs, A and B. Each of you has to describe your explanation to your partner.

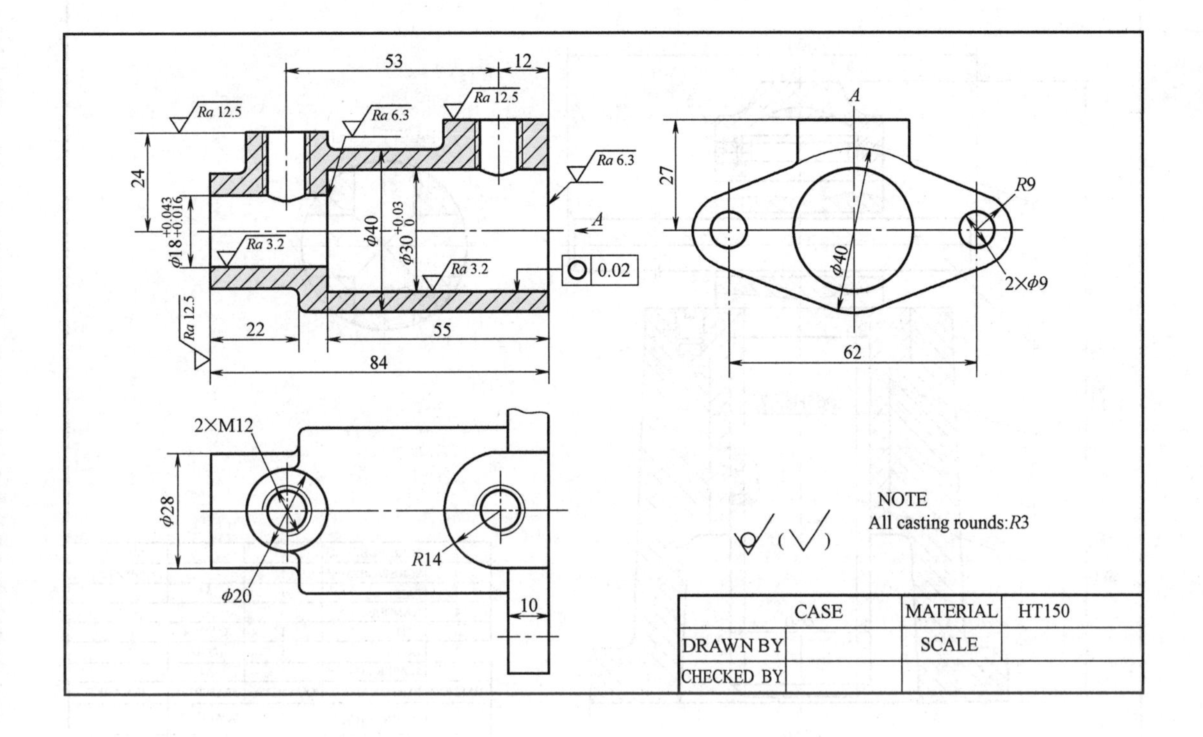

Figure 1. 38 Working drawing of a case

Hints:

(1) Explain the drawing in detail, such as the function of this drawing, the name of the part, its size, shape and material. Describe views, technical requirements, etc.

(2) The following texts will help you with the vocabulary you need.

The name of the part: case or body.

Limits and fits on drawings:

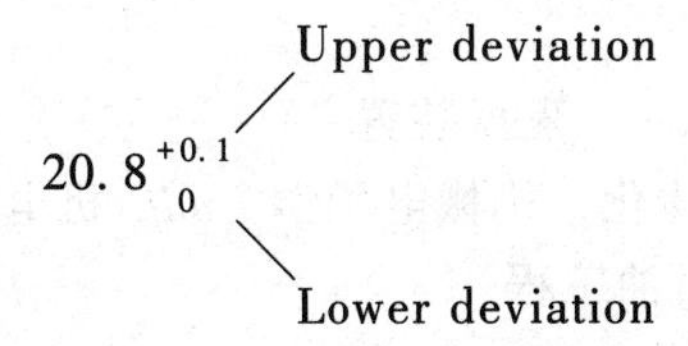

1. 15 Try to explain how the following lifting jack operates using whatever English you know (see Figure 1. 39). Write your explanation down, and then work in pairs, A and B. Each of you has to describe your explanation to your partner.

Hints:

(1) Texts contained in the drawing will help you with the vocabulary you need.

(2) A lifting jack is a device used to change angular motion into linear motion and usually to transmit power.

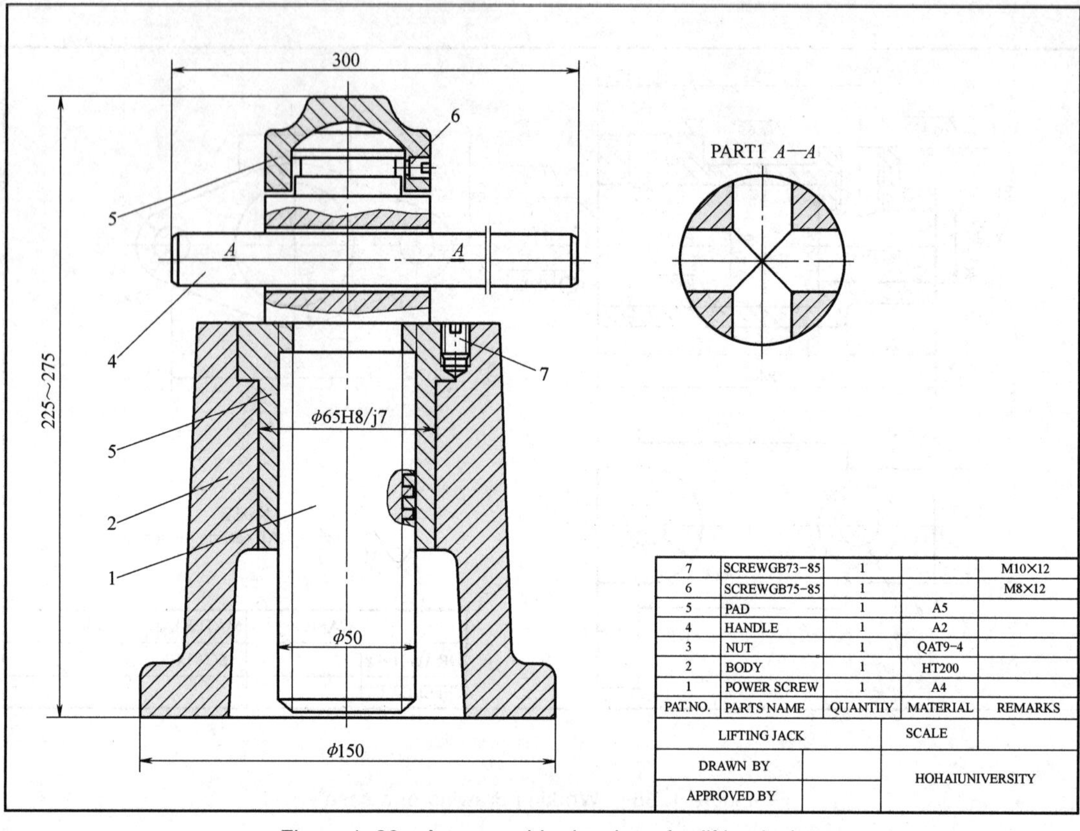

PAT.NO.	PARTS NAME	QUANTIIY	MATERIAL	REMARKS
7	SCREWGB73-85	1		M10×12
6	SCREWGB75-85	1		M8×12
5	PAD	1	A5	
4	HANDLE	1	A2	
3	NUT	1	QAT9-4	
2	BODY	1	HT200	
1	POWER SCREW	1	A4	

LIFTING JACK	SCALE	
DRAWN BY		HOHAIUNIVERSITY
APPROVED BY		

Figure 1.39　An assembly drawing of a lifting jack

相关词汇注解

1　subdiscipline　*n.* 学科的分支，子学科
2　compressor　*n.* 压缩机
3　powertrain　*n.* 动力系统
4　kinematic chain　运动链
5　vibration isolation equipment　隔振装置
6　mechatronic　*n.* 机电一体化，机械电子学　*adj.* 机电一体化的
7　petroleum engineering　石油工程
8　mechanics　*n.* 力学，结构
9　kinematics　*n.* 运动学
10　thermodynamics　*n.* 热力学
11　watercraft　*n.* 船只
12　people skills　人际交往能力
13　rigid body　刚体
14　mechanism　*n.* 装置，机制
15　engineering graphics　工程图学

16 substrate *n.* 基材
17 vellum *n.* 牛皮纸
18 mylar *n.* 聚酯薄膜
19 tolerance *n.* 公差
20 drafting standards 制图标准
21 sheet size and layout 图纸大小（规格）及布局
22 linear dimension 线性尺寸
23 full-size scale 原值比例
24 reduction scale 缩小比例
25 enlargement scale 放大比例
26 orthographic projection 正投影法
27 vector *n.* 矢量
28 1st angle projection 第一分角投影
29 2nd angle projection 第二分角投影
30 multiple view 多视图
31 six principle views 六个基本视图
32 auxiliary view 辅助视图
33 isometric view 等轴测图
34 compass *n.* 圆规
35 straightedge *n.* 直尺
36 axonometric projection 轴测投影
37 section view 剖视图
38 full section 全剖视图
39 half section 半剖视图
40 offset section 阶梯剖
41 revolved section 重合断面剖
42 removed section 移出断面剖
43 section line 剖面线
44 broken-out section 局部剖视图
45 detail working drawing 零件图
46 assembly drawing 装配图
47 surface roughness 表面粗糙度
48 technical requirements 技术要求
49 title block 标题栏
50 spur gear 直齿圆柱齿轮
51 geometric tolerance 几何公差
52 finish mark 表面粗糙度符号
53 pictorial drawing 立体图
54 bench vice 台虎钳
55 exploded view 零部件分解图

56 gear pump 齿轮泵
57 case seal 密封圈
58 suction port 吸入口
59 pressure port 泄压口
60 drive shaft 驱动轴
61 bushing *n.* 衬套
62 idler gear 惰轮，空转齿轮
63 mounting flange 法兰凸缘，安装用法兰
64 feature-based 3D Model 基于特征的三维模型
65 plastic injection mold 注塑模具
66 optimal design solution 优化设计解决方案
67 machinist *n.* 机械师
68 draftsman *n.* 绘图员
69 specifications *n.* 说明书
70 conceptual design 概念设计
71 preliminary design 初步设计
72 detailed design 详细设计
73 reverse engineering 逆向工程，反求工程
74 error analysis 误差分析
75 point cloud model 点云模型
76 injection pressure 注射压力
77 product prototyping 产品成形
78 multidisciplinary *adj.* 多学科的
79 trial-produce *v.* 试制，试生产
80 maneuverable *adj.* 容易操作的，机动的
81 manufacturable *adj.* 可制造的
82 marketable *adj.* 有销路的，有市场前景的
83 reliability *n.* 可靠性
84 brainstorming session 集体讨论会，头脑风暴讨论会
85 trade off *n.* 权衡
86 optimization *n.* 优化
87 groove *n.* 凹槽
88 suspension beam 悬支架，悬支梁
89 entity interference 实体干涉

Part2 Machine Elements and Mechanisms

Objective

After completing this part, you will be able to:

- be familiar with special properties and terminology in regard to fasteners, springs, bearings, clutches, brakes, shafts, belt drives, chain drives and gear drives, etc.;
- understand and better describe to others the anatomy and function of common machines;
- select machine elements and components that are best suited to a particular application;
- describe a machine clearly and concisely.

2.1 Introduction

Machine elements refer to an elementary component of a machine. These elements consist of three basic types: ①structural components such as bearings, axles, keys, fasteners, seals, and lubricants, ②mechanisms that control movement in various ways such as gear trains, belt or chain drives, linkages, cam and follower systems, including brakes and clutches, and ③control components such as buttons, switches, indicators, sensors, actuators and computer controllers, they are generally not considered to be a machine element, the shape, texture and color of covers are important parts of a machine that provide a styling and operational interface between the mechanical components of a machine and its users.

Machine elements are basic mechanical parts and features used as the building blocks of most machines, as shown in Figure 2.1. Most of them are standardized to be used as interchangeable parts, such as bolts, screws, nuts, keys, pins, bearings, etc. These parts are usually called standardized parts. Gears are not standardized parts, but standard gears have standardized profiles, for instance, the involute gear profile is the most commonly used system for gearing. In an involute gear, the profiles of the teeth are involutes of a circle. Machine elements may be features of a part (such as screw threads or integral plain bearings) or they may be discrete parts in and of themselves such as wheels, axles, pulleys, rolling-element bearings or gears. All of the simple machines may be described as machine element and many machine elements incorporate concepts of one or more simple machines. For example, a leadscrew incorporates a screw thread, which may be defined as an inclined plane wrapped around a cylinder. Simple machines provide a "vocabulary" for understanding more complex machines.

Many mechanical design, invention and engineering tasks involve knowledge of various machine elements and an intelligent and creative combining of these elements into a component or assembly that meets a need (serves an application). Fasteners, springs, bearings, clutches, brakes, shafts, belt drives, chain drives and gears, etc., each has special properties and terminology, they are basic elements and components in machines. Mechanical engineers need to be familiar with such "building blocks" in order to select the component that is best suited to a particular application. Meanwhile, engineers should be familiar with common vocabulary and expressions in mechanical engineering in order to make a good communication with other engineers who speak English as their first language.

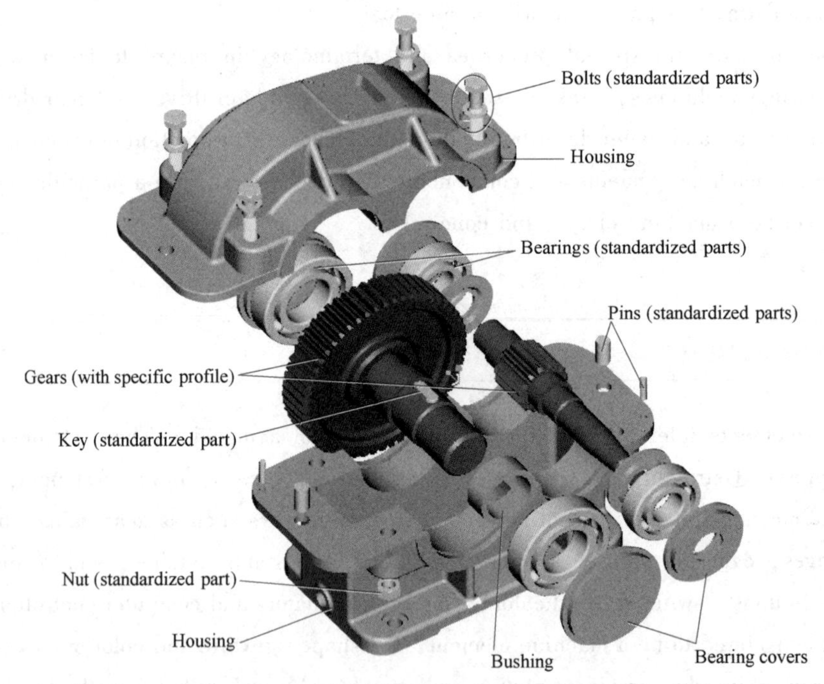

Figure 2.1 Machine elements are the building blocks of most machines

2.2 Fasteners

2.2.1 Introduction

A fastener is a device used to connect or join two or more components. Traditional forms of fastener include nuts, bolts, screws and rivets. They are frequently used and most of them have been standardized with their shapes and sizes.

It has been estimated that there are over 2 500 000 fasteners in a Boeing jumbo jet. For more modern airliners, such as the Airbus Industrie A340, this figure has been dramatically reduced, but the total is still significant. This type of example illustrates the importance of fasteners to the designer. There are thousands of types of fastener commercially available from specialist suppliers. The most common include threaded fasteners such as bolts, screws, nuts, studs and rivets. In addition, there is the range of permanent fastener techniques by means of welding, brazing, soldering and use of adhesive. Given the wide range of fastener techniques available and the costs involved in sourcing, stocking and assembly, the correct choice of fastener is very important. This section serves to introduce a variety of types of fasteners and their selection based on appropriate analysis and consideration of the intended function.

2. 2. 2 Thread Terminology

There is a large variety of fasteners available using a threaded form to produce connection of components. The common element of screw fasteners is a helical thread that causes the screw to advance into a component or nut when rotated. Screw threads can be either left-handed or right-handed, depending on the direction of rotation desired for advancing the thread as illustrated in Figure 2. 2. Generally, right-hand threads are normally used. The detailed aspects of a thread and the specialist terminology used are illustrated in Figure 2. 3 and defined in Table 2. 1.

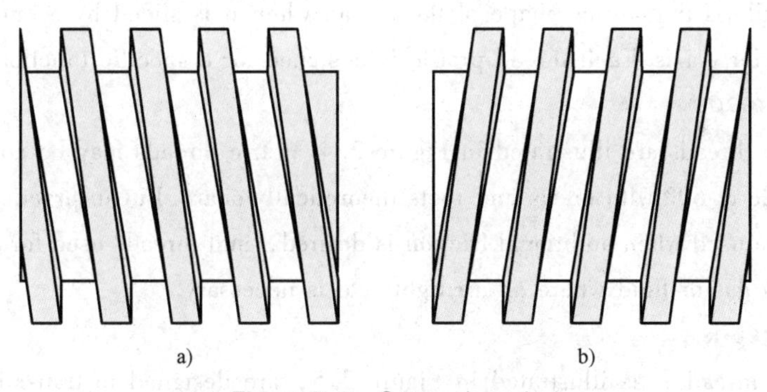

Figure 2. 2 Screw threads

a) Right-hand thread b) Left-hand thread

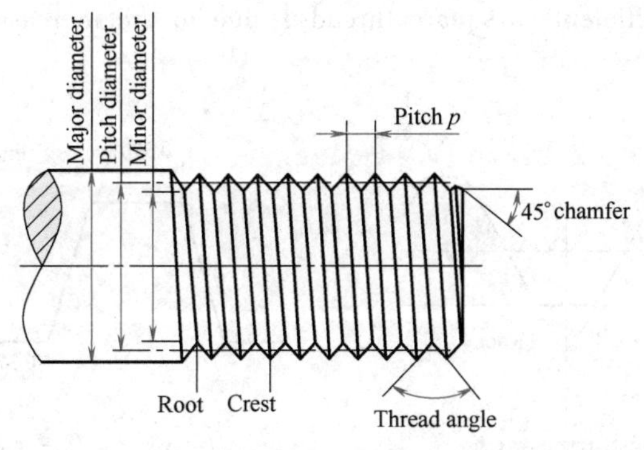

Figure 2. 3 Specialist terminology used for describing threads

Table 2. 1　Thread terminology

Term	Description
Pitch	The thread pitch is the distance between corresponding points on adjacent threads. Measurements must be made parallel to the thread axis.
Major diameter	The outside or major diameter is the diameter over the crests of the thread measured at right angles to the thread axis.
Crest	The crest is the most prominent part of thread, either external or internal.
Root	The root lies at the bottom of the groove between two adjacent threads.
Flank	The flank of a thread is the straight side of the thread between the root and the crest.
Root diameter	The root, minor or core diameter is the smallest diameter of the thread measured at right angles to the thread axis.
Effective or pitch diameter	Effective diameter is the diameter that is halfway between the major and minor diameters.
Lead	The lead of a thread is the axial movement of the screw in one revolution.

2. 2. 3　Thread Forms

A thread profile is the outline shape of the thread when it is sliced by a cutting plane passing the threaded cylinder's axis. Each thread profile is designed for a specific function.

1. Metric threads

Metric profile threads are illustrated in Figure 2. 4. Metric threads may be coarse or fine, they have a thread angle of 60° with crests and roots theoretically sharp. But in practice a very small flat is formed. They are used when additional friction is desired, and threads used for adjustment, or on piping transmitting gas or fluid where an air tight seal is necessary.

2. Acme threads

Acme profile threads, as illustrated in Figure 2. 5, are designed to transmit power along the axis of the threaded cylinder. The lead screw of a metal lathe is an application example of acme profile thread. Acme threads have a 29° thread angle, which is easier to machine than square threads. They are not as efficient as square threads, due to the increased friction induced by the thread angle.

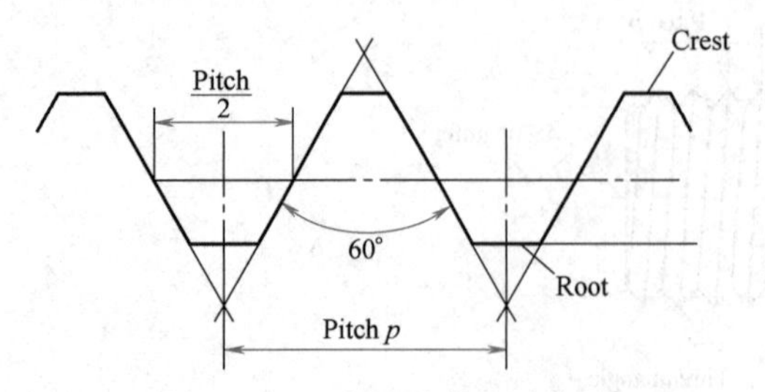

Figure 2. 4　Metric thread form

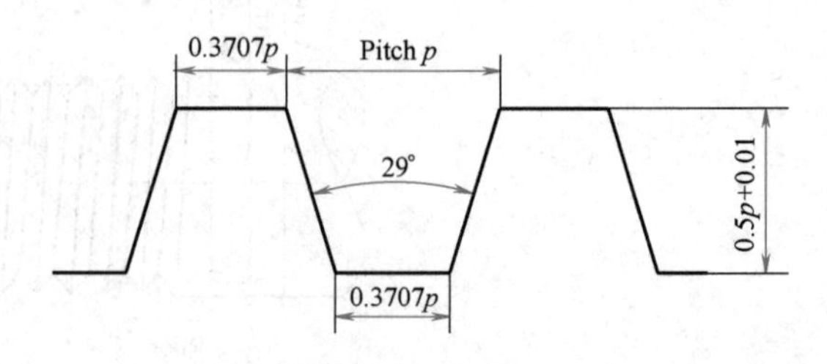

Figure 2. 5　Acme thread form

3. Square threads

Square profile threads, as illustrated in Figure 2. 6, are designed to transmit power along the axis of the threaded cylinder. Square threads are named after their square geometry. They are the most efficient and they have the least friction, so they are often used for screws that carry high power. But they are also the most difficult to machine and thus the most expensive.

4. Buttress threads

The buttress thread has a standardized profile composed of a series of non-equilateral trapeziums with an angle of 45°, as shown in Figure 2. 7. This is designed to transmit heavy forces in one direction only. Buttress threads are used where the load force on the screw is only applied in one direction. They are as efficient as square threads in these applications, but are easier to manufacture.

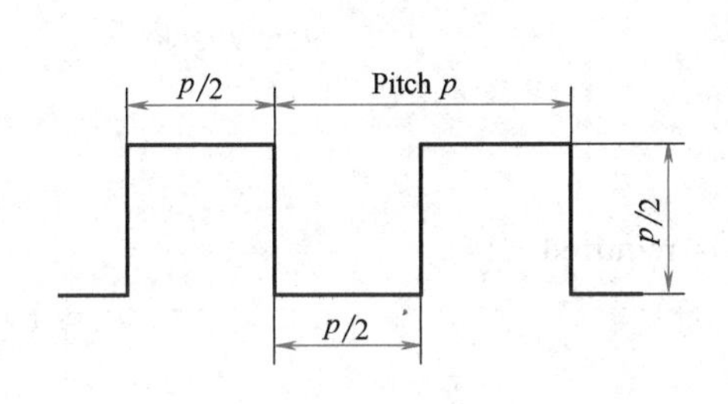

Figure 2. 6 Square thread form

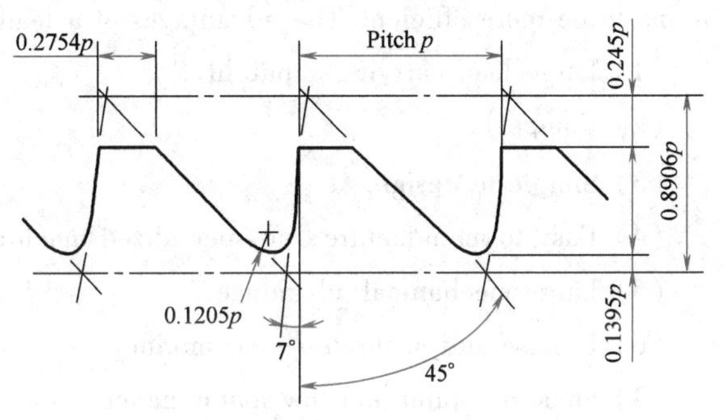

Figure 2. 7 Buttress thread form

5. Pipe thread

Threaded pipes can provide an effective seal for pipes transporting liquids, gases, steam and hydraulic fluid. These threads are now used in materials other than steel and brass, including PTFE, PVC, nylon, bronze and cast iron. 55° angle remains commonly used today. The thread profile is shown in Figure 2. 8.

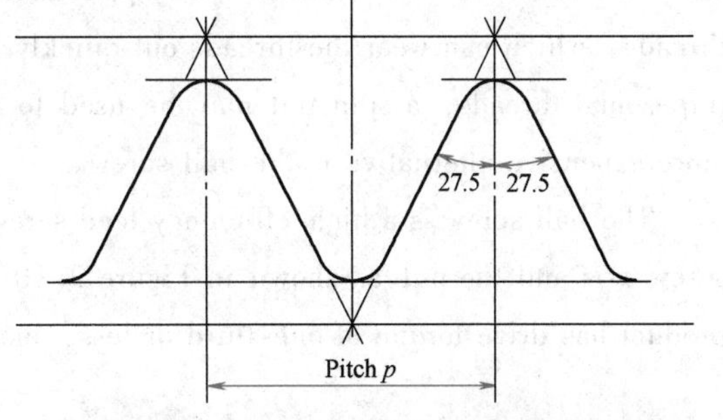

Figure 2. 8 Basic pipe thread with 55° angle

2. 2. 4 Power screws

A power screw, also known as a leadscrew (or lead screw) or translation screw, is a device used in machinery to change angular motion into linear motion and usually to transmit power. Common applications are linear actuators, machine slides (such as in machine tools), vises, presses, and jacks.

Leadscrews are manufactured in the same way as other thread forms, for instance, square thread, acme thread, buttress thread. However, V-threads or Metric threads are less suitable for leadscrews than others such as acme, because they have more friction between the threads and their threads are designed to induce this friction to keep the fastener from loosening. Leadscrews, on the other hand, are designed to minimize friction. Therefore, in most commercial and industrial use, V-threads are avoided for leadscrew use. Nevertheless, V-threads are sometimes successfully used as leadscrews, for example on microlathes and micromills.

An application of power screws to a power-driven jack is shown in Figure 2. 9. You should be able to identify the worm, the worm gear, the screw and the nut.

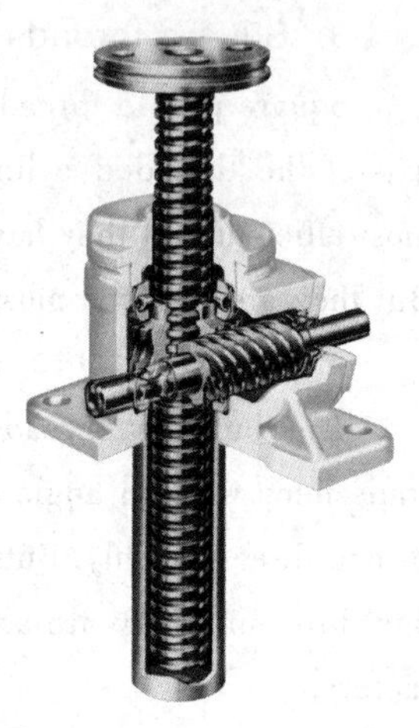

Figure 2. 9 A worm-gear screw jack

A leadscrew nut and screw mate with rubbing surfaces, and consequently they have a relatively high friction and stiction compared to mechanical parts which mate with rolling surfaces and bearings. Leadscrew efficiency is typically between 25% and 70%, with higher pitch screws tending to be more efficient. The advantages of a leadscrew are:

(1) Large load carrying capability.

(2) Compact.

(3) Simple to design.

(4) Easy to manufacture; no specialized machinery is required.

(5) Large mechanical advantage.

(6) Precise and accurate linear motion.

(7) Smooth, quiet and low maintenance.

(8) Minimal number of parts.

(9) Most are self-locking.

The disadvantages are that most are not very efficient. Due to the low efficiency they can't be used in continuous power transmission applications. They also have a high degree for friction on the threads, which can wear the threads out quickly. For square threads, the nut must be replaced; for trapezoidal threads, a split nut may be used to compensate for the wear. A higher performing but more expensive alternative is the ball screw.

The ball screw is a high-efficiency feed screw with the ball making a rolling motion between the screw axis and the nut, as shown in Figure 2. 10. Compared with a conventional sliding screw, this product has drive torque of one-third or less, making it most suitable for saving drive motor power.

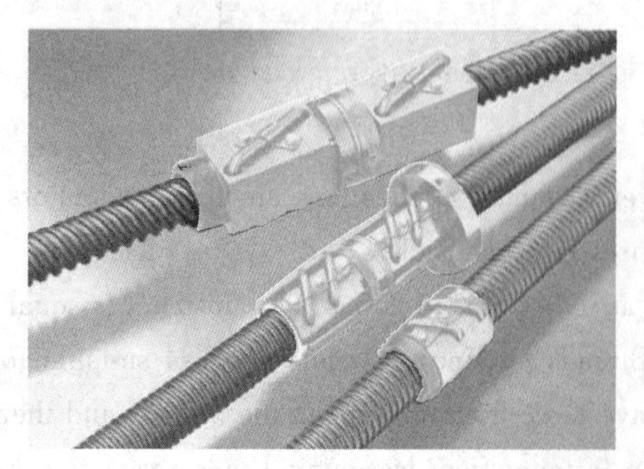

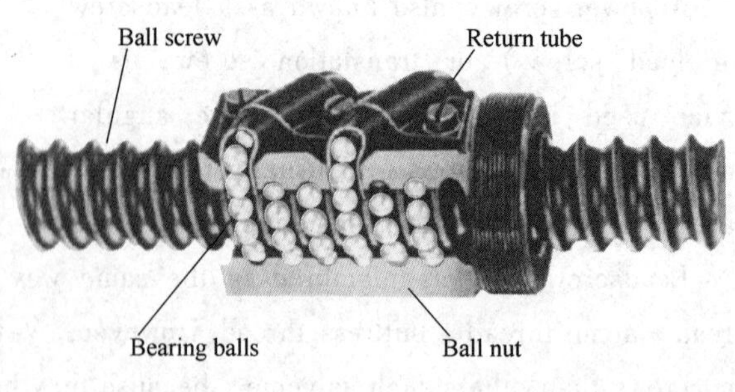

Figure 2. 10 Ball screws

2.2.5 Bolts, Studs, Nuts and Plain Washer

1. Bolts

Bolts are normally tightened by applying torque to the head or nut, which causes the bolt to stretch. The stretching results in bolt tension, known as preload, which is the force that holds a joint together.

Figure 2.11 illustrates detailed aspects of a bolt, the specialist terminology as well as drawing proportion.

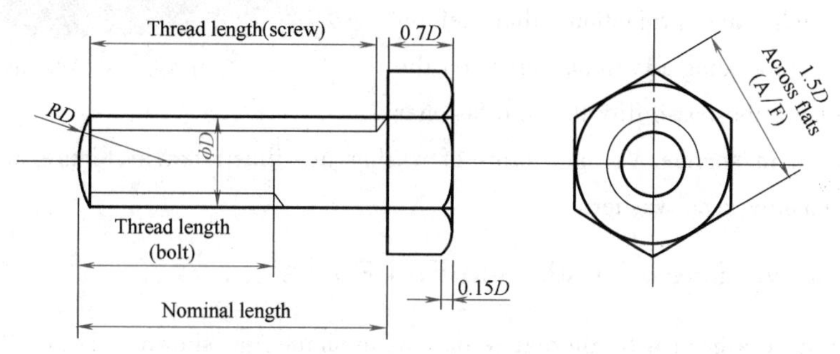

Figure 2.11 Bolts and screws

2. Studs

Studs are the threaded fasteners that hold on the two parts together. One side of the stud is semi-permanently mounted directly to one of the parts, on the other side of the stud a nut is fastened over to the stud to hold the part.

Figure 2.12 illustrates detailed aspects of a stud, the specialist terminology as well as drawing proportion.

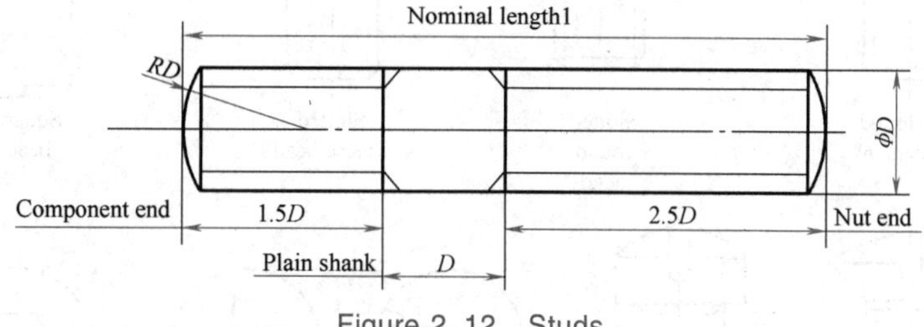

Figure 2.12 Studs

3. Nuts

Nuts are almost always used opposite a mating bolt to fasten a stack of parts together. The two partners are kept together by a combination of their threads, a slight stretch of the bolt and compression of the parts. In applications where vibration or rotation may work a nut loose, various locking mechanisms may be employed: adhesives, safety pins or lockwire, nylon inserts, or slightly oval-shaped threads. The most common shape is hexagonal, as illustrated in Figure 2.13.

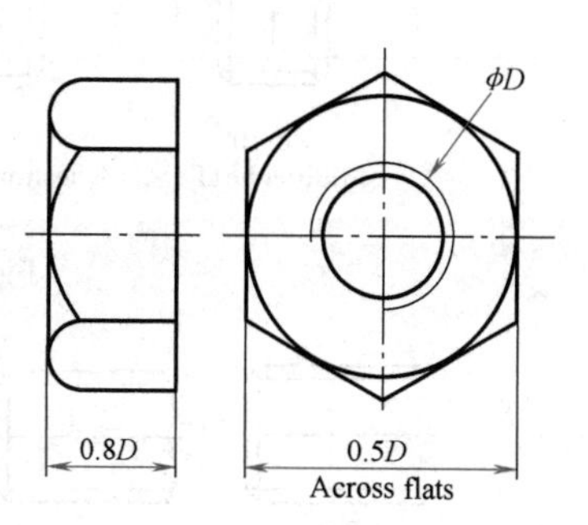

Figure 2.13 Standard nuts

4. Washers

Figure 2. 14 Washers

Washers can be used either under the bolt head or the nut or both in order to distribute the clamping load over a wider area and to provide a bearing surface for rotation of the nut. The most basic form of a washer is a simple disc with a hole through which the bolt or screw passes. There are, however, many additional types with particular attributes such as lock washers, which have projections that deform when compressed, producing additional forces on the assembly, decreasing the possibility that the fastener assembly will loosen in service. Various forms of washer are illustrated in Figure 2. 14, the plain washer is the commonly used washer.

2. 2. 6 Alternative Screw Heads and End Points

Screws often have a head with alternative options of shape, as shown in Figure 2. 15, which is a specially formed section on one end of the screw that allows it to be turned or driven. Common tools for driving screws include screwdrivers and wrenches. The head is usually larger than the body of the screw, which keeps the screw from being driven deeper than the length of the screw and to provide a bearing surface. Screws also have alternative screw points, as illustrated in Figure 2. 16.

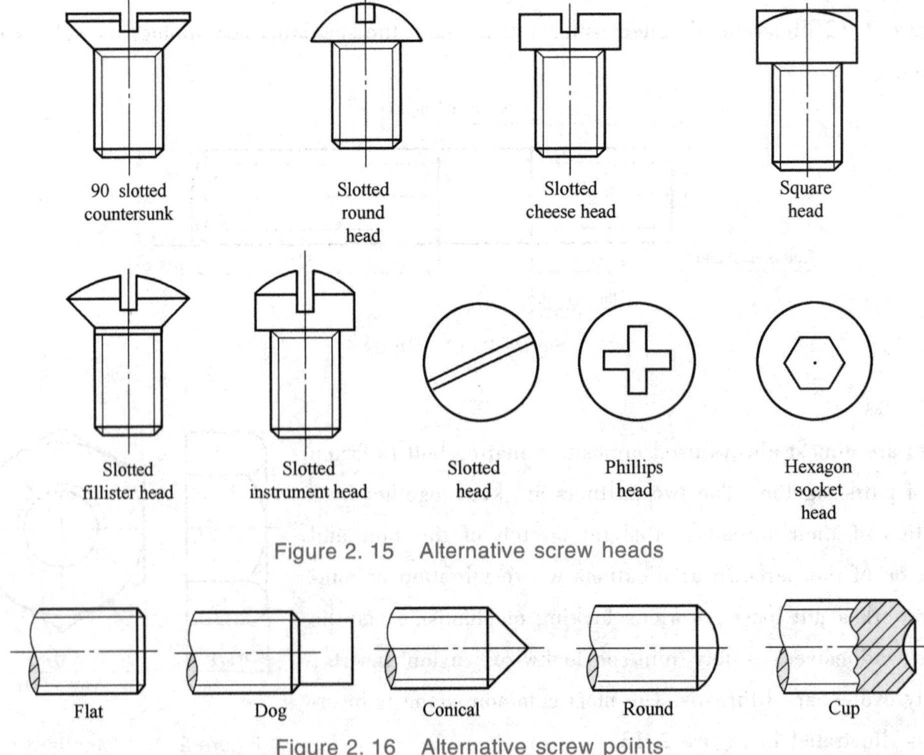

Figure 2. 15 Alternative screw heads

Figure 2. 16 Alternative screw points

2. 2. 7 Selection of Screwed Fasteners

Screws and bolts may be made from a wide range of materials, steel is perhaps the most common in many varieties. Where great resistance to weather or corrosion is required, stainless steel, titanium, brass, bronze, monel or silicon bronze may be used. Some types of plastic, such as nylon or polytetrafluoroethylene (PTFE), can be threaded and used for fasteners requiring moderate strength and great resistance to corrosion or for the purpose of electrical insulation. Often a surface coating is used to protect the fastener from corrosion. Selection criteria for the screw materials include temperature, required strength, resistance to corrosion, joint material and cost as well as assembly considerations.

2. 2. 8 Systems for Specifying the Dimensions of Screws

Screws are most commonly used in machines. They are produced by specialist manufacturers based on systems for specifying the dimensions of screws. There are many systems. As a skilled international engineer, an engineer should be familiar with these different systems when they make correct selection of screws. Currently, ISO metric screw thread preferred series has displaced other old systems. Other relatively common systems include the British Standard Whitworth, BA system (British Association) and the Unified Thread Standard.

1. ISO metric screw thread

The basic principles of the ISO metric screw thread are defined in international standard ISO 68-1 and preferred combinations of diameter and pitch are listed in ISO 261. The smaller subset of diameter and pitch combinations commonly used in screws, nuts and bolts is given in ISO 262, as illustrated in Table 2. 2-2. 3. The most commonly used pitch value for each diameter is the coarse pitch. For some diameters, one or two additional fine pitch variants are also specified, for special applications such as threads in thin-walled pipes. ISO metric screw threads are designated by the letter M followed by the major diameter of the thread in millimeters (e. g. M8). If the thread does not use the normal coarse pitch (e. g. 1. 25mm in the case of M8), then the pitch in millimeters is also appended with a multiplication sign (e. g. "M8×1" if the screw thread has an outer diameter of 8 mm and advances by 1 mm per 360° rotation). Currently, this system of thread is adopted in China.

Table 2. 2 Thread sizes (ISO 261)

Nominal diameter D/mm		Pitch P/mm	
1st choice ISO 262	2nd choice	coarse	fine
1		0. 25	
1. 2		0. 25	
	1. 4	0. 3	
1. 6		0. 35	
…			

Table 2.3 Thread sizes (ISO 262)

Nominal diameter D/mm		Pitch P/mm	
1st choice	2nd choice	coarse	fine
16		2	1.5
	18	2.5	2 or 1.5
20		2.5	2 or 1.5
	22	2.5	2 or 1.5
…			

2. British Standard Whitworth

The first person to create a standard (in about 1841) was the English engineer Sir Joseph Whitworth. Whitworth screw sizes are still used, both for repairing old machinery and where a coarser thread than the metric fastener thread is required. Whitworth became British Standard Whitworth, abbreviated to BSW (BS 84: 1956) and the British Standard Fine (BSF) thread was introduced in 1908 because the Whitworth thread was too coarse for some applications. The thread angle was 55°, and the depth and pitch varied with the diameter of the thread (i.e. the bigger the bolt, the coarser the thread), as illustrated in Table 2.4.

Table 2.4 Thread sizes

Whitworth size/in①	Core diameter/in	Threads per inch	Pitch/in	Tapping drill size
1/16	0.0411	60	0.0167	Number Drill 56 (1.2mm)
3/32	0.0672	48	0.0208	Number Drill 49 (1.85mm)
1/8	0.0930	40	0.025	Number Drill 39 (2.55mm)
5/32	0.1162	32	0.0313	Number Drill 30 (3.2mm)
…				

① 1in = 0.0254m。

3. British Association screw thread

A later standard established in the United Kingdom was the British Association (BA) screw threads, named after the British Association for Advancement of Science. Screws were described as "2BA", "4BA", etc., the odd numbers being rarely used, except in equipment made prior to the 1970s for telephone exchanges in the UK.

While not related to ISO metric screws, the sizes were actually defined in metric terms, a 0BA thread having a 6mm diameter and 1mm pitch as illustrated in Table 2.5. Although 0BA has the same diameter and pitch as ISO M6, the threads have different forms and are not compatible. BA threads are still common in some applications.

Table 2. 5 Thread sizes

BA	Outer diameter	Threads per inch	Threads per mm	Tap drill
0	0. 2362in/6mm	25. 38	1	5. 1mm/7gauge
1	0. 2087in/5. 3mm	28. 25	1. 112	4. 5mm/16gauge
2	0. 1850in/4. 7mm	31. 35	1. 234	4. 0mm/21gauge
3	0. 1614in/4. 1mm	34. 84	1. 372	3. 4mm
…				

4. Unified Thread Standard

The Unified Thread Standard (UTS) is most commonly used in the United States of America, but it is also extensively used in Canada and occasionally in other countries. It has the same 60° profile as the ISO metric screw thread used in the rest of the world, but the characteristic dimensions of each UTS thread (outer diameter and pitch) are chosen as an inch fraction rather than a millimeter value. The UTS is currently controlled by ASME/ANSI in the United States of America.

The size of a UTS screw is described using the following format: X-Y, where X is the nominal size (the hole or slot size in standard manufacturing practice through which the shaft of the screw can easily be pushed) and Y is the threads per inch (TPI). For sizes 1/4 inch and larger the size is given as a fraction; for sizes less than this an integer is used, ranging from 0 to 12, as illustrated in Table 2. 6. The integer sizes can be converted to the actual diameter by using the formula 0. 060 + 0. 013×number. For example, a #3 screw is 0. 060+0. 013×3 = 0. 0990 inches in diameter. For most size screws there are multiple TPI available, with the most common being designated a Unified Coarse Thread (UNC or UN) and Unified Fine Thread (UNF or UF).

Table 2. 6 Unified screw thread—UCN, UNF and UNEF

Major diameter /in(mm)	Threads per inch(Pitch)			Tap drill size—Preferred sizes		
	Coarse(UNC)	Fine(UNF)	Extra fine (UNEF)	Coarse	Fine	Extra fine
#0 = 0. 0600(1. 5240)	—	80			3/64 in	
#1 = 0. 0730(1. 8542)	64	72		#53	#53	
#2 = 0. 0860(2. 1844)	56	64		#50	#50	
#3 = 0. 0990(2. 5146)	48	56		#47	#45	
…						

2. 2. 9 Applications of Screwed Fasteners

Figure 2. 17 illustrates an application example of screwed fasteners.

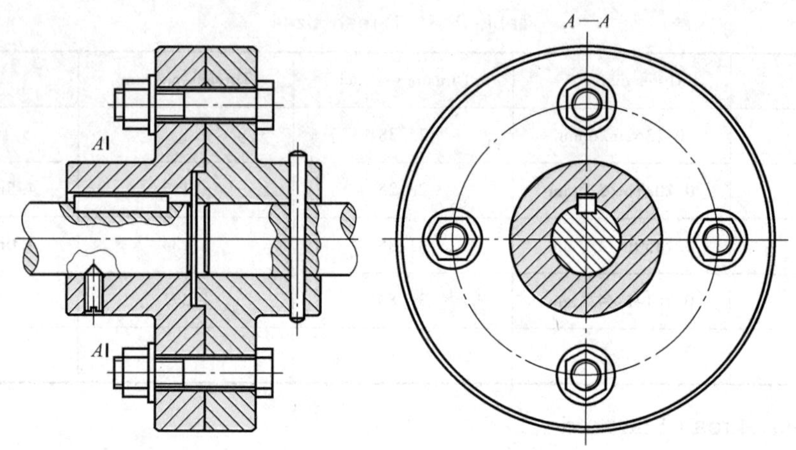

Figure 2. 17 Applications of screwed fasteners

2. 3 Keys and Pins

2. 3. 1 Keys

In mechanical engineering, a key is a machine element used to connect a rotating machine element to a shaft. Through this connection the key prevents relative rotation between the two parts and allows torque to be transmitted through. To function the shaft and rotate machine elements a key must have a keyway, also known as keyseat, which is a slot or pocket for the key to fit in. The whole system is called a keyed joint. A keyed joint still allows relative axial movement between the parts. Commonly keyed components include gears, pulleys and couplings.

There are three commonly used types of keys: parallel, semicircular and tapered keys, as shown in Figure 2. 18 a-c.

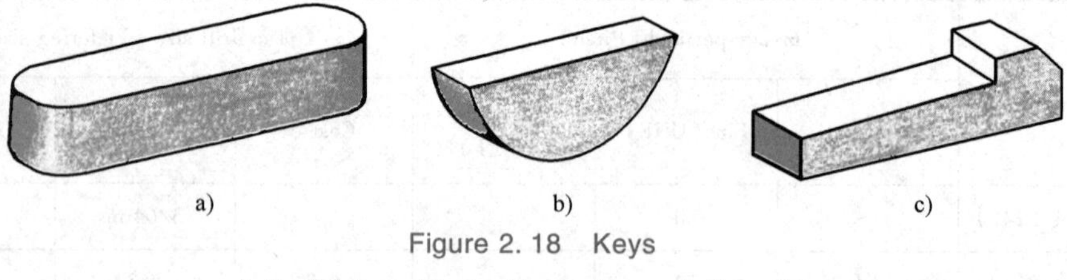

Figure 2. 18 Keys

a) Parallel keys b) Semicircular keys c) Tapered keys

Parallel keys are the most widely used. They have a square or rectangular cross-section. Square keys are used for smaller shafts and rectangular faced keys are used for shaft diameters over 170mm or when the wall thickness of the mating hub is of concern. Set screws often accompany parallel keys to lock the mating parts into place so they do not move. The keyway is a longitudinal slot in both the shaft and mating part (see Figure 2. 19).

Semicircular keys are semicircular-shaped keys that, when installed, leave a protruding tab. The keyway in the shaft is a semicircular pocket and the mating part has a longitudinal slot (see

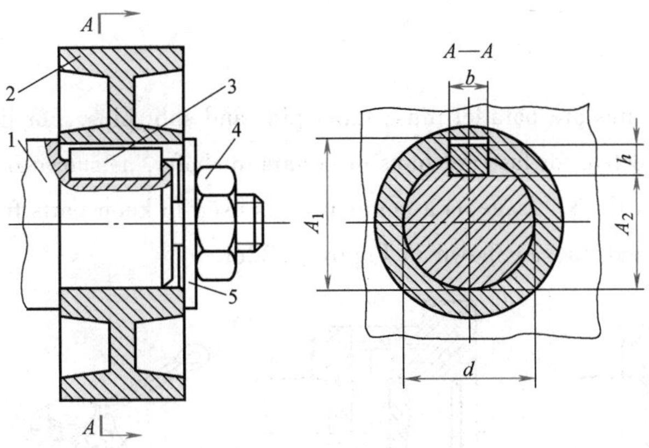

Figure 2. 19 Parallel key joints

1—Shaft 2—Hub 3—Key 4—Nut 5—Washer

Figure 2. 20). They are used to improve the concentricity of the shaft and the mating part, which is critical for high speed operations. The main advantage of the semicircular key is that it avoids the milling of a keyway near shaft shoulders, which already have stress concentrations.

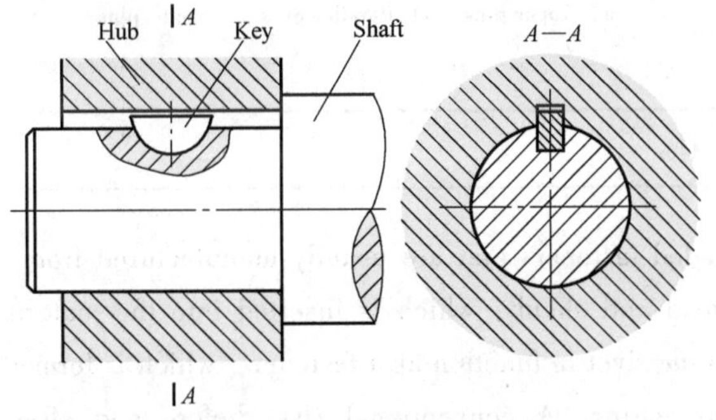

Figure 2. 20 Semicircular keys

The final type of key is the tapered key, which is tapered on one side, the side that engages the hub (see Figure 2. 21). The keyway in the hub is broached with a taper matching that of the tapered key. Some tapered keys have a gib or tab, for easier removal during disassembly. The purpose of the taper is to secure the key itself as well as firmly engage the shaft to the hub without set screw. The problem with taper keys is that they can result the center of rotation of the shaft to be slightly off of the mating part.

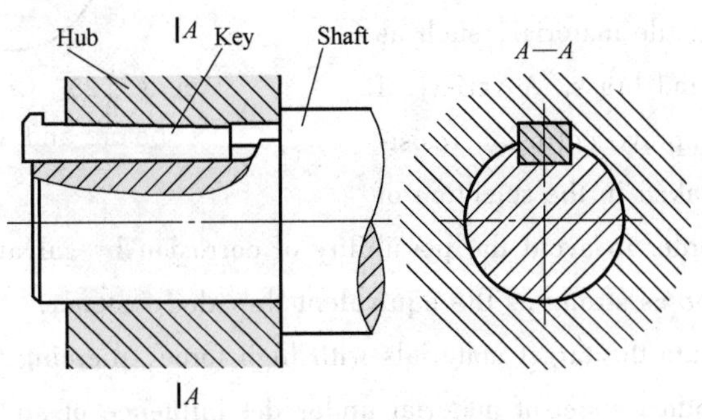

Figure 2. 21 Tapered keys

2.3.2 Pins

The most common pins are parallel pins, taper pins and split pins. For light work, the taper pin or the parallel pin is effective for fastener hubs or collars to shafts, as shown in Figure 2.22a, b. The split pin, also known in U. S. usage as a cotter pin, is used to keep parts from accidentally coming apart. Its ends are opened up, as shown in Figure 2.22c.

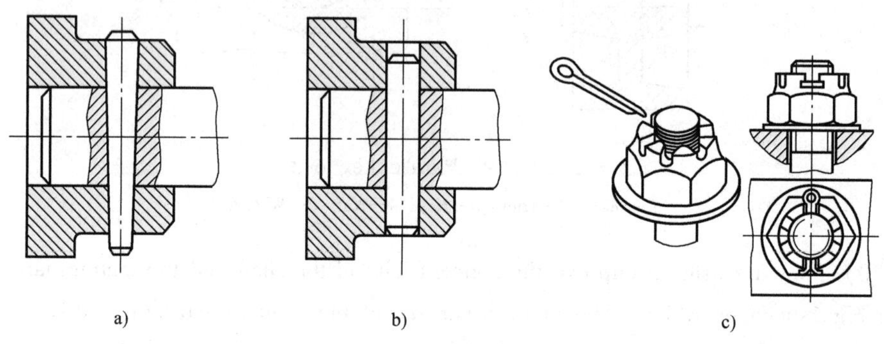

Figure 2.22 Pin joints

a) Taper pins b) Parallel pins c) Split pins

2.4 Riveted Joints

Rivets are non-threaded fasteners that are usually manufactured from steel or aluminum. They consist of a preformed head and shank, which is inserted into the material to be joined, and the second head that enables the rivet to function as a fastener, which is formed on the free end by a variety of means known as setting. A conventional rivet before and after setting is illustrated in Figure 2.23.

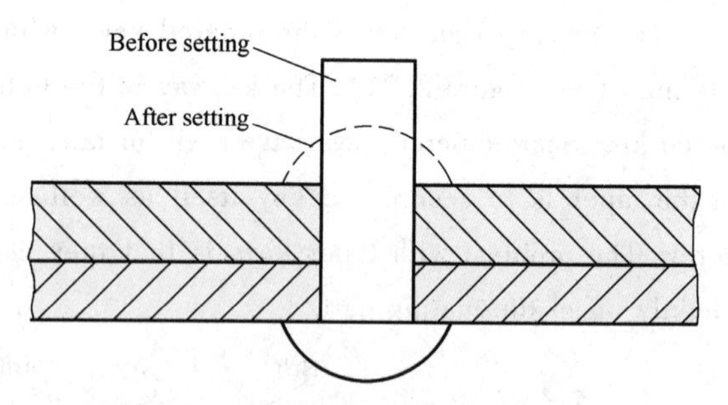

Figure 2.23 Conventional rivet before and after setting

Rivets are widely used to join components in aircraft, boilers, ships, boxes and other enclosures. Rivets tend to be much cheaper to install than bolts and the process can be readily automated with single riveting machines capable of installing thousands of rivets an hour. Rivets can be made from any ductile material, such as carbon steel, aluminum and brass. A variety of coatings is available to improve corrosion resistance. Care needs to be taken in the selection of material and coating in order to avoid the possibility of corrosion by galvanic action. In general, a given size rivet will not be as strong as the equivalent threaded fastener.

The rivet is driven into the target materials with high force, piercing the top sheets and spreading outwards into the bottom sheet of material under the influence of an upsetting die to form the joint.

Factors in the design and specification of rivets include the size, type and material for the rivet, the type of join and the spacing between rivets.

2.5 Welded Joints

Welding can be described as the process of joining material together by raising the temperature of the surfaces to be joined so that they become plastic or molten.

The metal welding involves the metallurgical bonding of components usually by application of heat from an electric arc, gas flame or from resistance heating under heavy pressure. There are numerous types of welding including tungsten inert gas welding (TIG), metal inert gas welding (MIG), manual metal arc welding (MMA), submerged arc welding (SAW), resistance welding and gas welding. Thermoplastics can also be welded. Heat can be applied by means of a hot gas, which is usually inert. Other types of welding for plastics include inertia, ultrasonic and vibration welding. Soldering and brazing involve the joining of heated metals while the components are in their solid state by means of molten filler metals.

2.6 Springs

Springs are used to: ①clamp, ②store elastic energy, and ③ reduce shock. There is a wide range of types of spring that are readily available from specialist suppliers or that can be designed and manufactured fit-for-purpose, as shown in Figure 2.24. The helical springs are most commonly used.

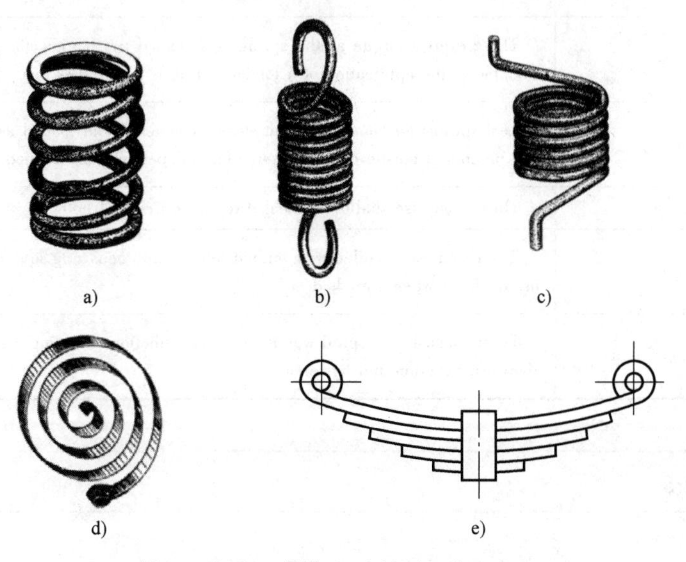

Figure 2.24 Different kinds of springs

a) Helical compression springs b) Helical extension springs

c) Torsion springs d) Spiral torsion springs e) Leaf springs

Springs can be classified by the direction and nature of the force exerted when they are deflected. Several types of spring are listed in Table 2.7.

Table 2.7 Classification of springs

Compressive	Helical compression springs Belleville springs Flat springs, e.g.cantilever or leaf springs
Tensile	Helical extension springs Flat springs, e.g.cantilever or leaf springs Drawbar springs Constant force springs
Radial	Garter springs Elastomeric bands Spring clamps
Torque	Torsion springs Power springs

According to the nature of force or torque exerted, the principal characteristics of the various classes of springs are summarized in Table 2.8.

Table 2.8 Principle of characteristics of a variety of type of spring

Type of spring	Principal characteristics
Helical compression springs	These are usually made from round wire wrapped into straight cylindrical form with a constant pitch between adjacent coils
Helical extension springs	These are usually made from round wire wrapped into straight cylindrical form but with the coils closely spaced in the no-load condition. As an axial load is applied, the spring will extend but resisting the motion
Drawbar springs	A helical spring is incorporated into an assembly with two loops of wire. As a load is applied, the spring is compressed in the assembly resisting the motion
Torsion springs	These exert a torque as the spring is deflected by rotation about their axes. A common example of the application of a torsion spring is the clothes peg
Leaf springs	Leaf springs are made from flat strips of material and loaded as cantilever beams. They can produce a tensile or compressive force depending on the mode of loading applied
Belleville springs	These comprise shallow conical discs with a central hole
Garter springs	These consist of coiled wire formed into a continuous ring so that they can exert a radial inward force when stretched
Volute springs	These consist of a spiral wound strip that functions in compression. They are subject to significant friction and hysteresis

2.7 Bearings

2.7.1 Bearing classification

The term "bearing" typically refers to contacting surfaces through which a load is transmitted.

Bearings may roll or slide or do both simultaneously. The range of bearing types available is extensive, although they can be broadly split into two categories: sliding bearings, also known as plain surface bearings, where the motion is facilitated by a thin layer or film of lubricant, and rolling element bearings, where the motion is aided by a combination of rolling motion and lubrication. Lubrication is often required in a bearing to reduce friction between surfaces and to remove heat. Figure 2.25 illustrates two of the more commonly known bearings: a deep groove ball bearing and a journal bearing. A general classification scheme for the distinction of bearings is given in Figure 2.26.

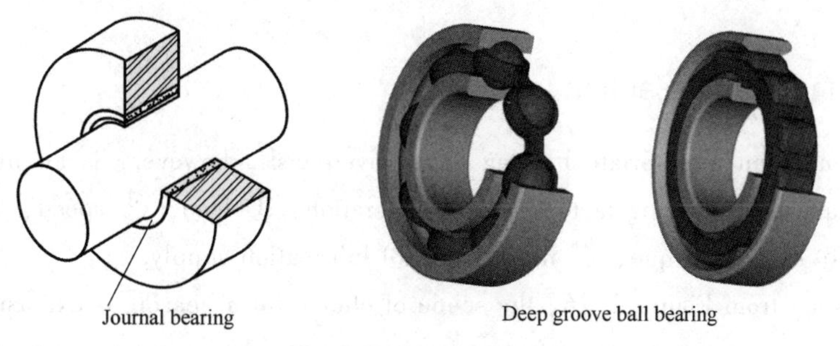

Figure 2.25 bearings

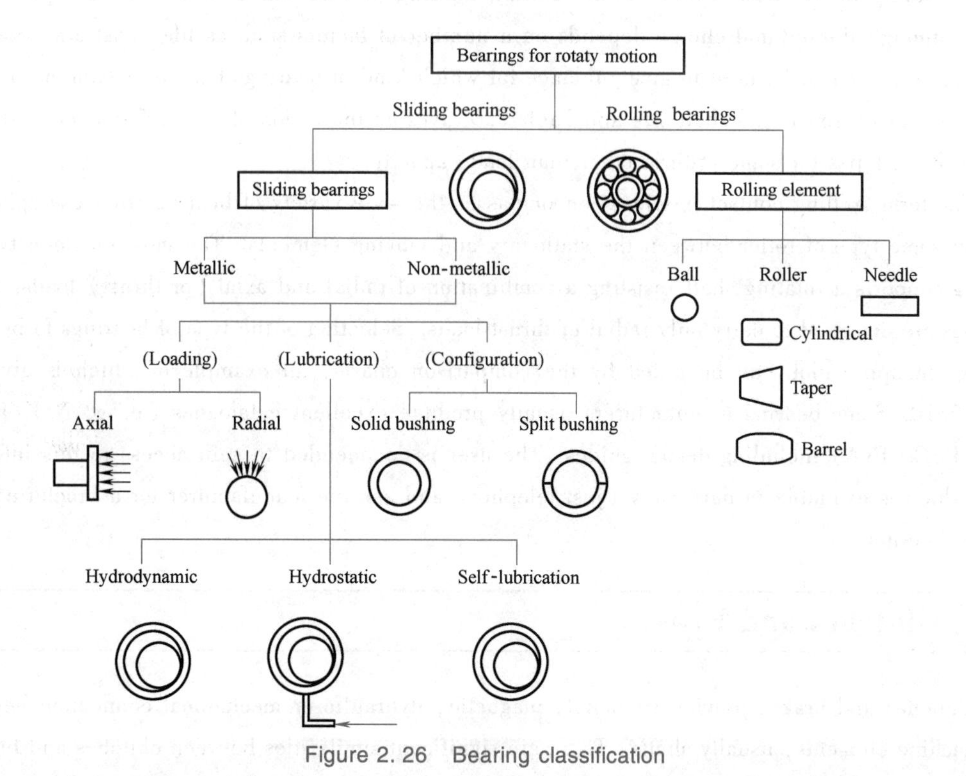

Figure 2.26 Bearing classification

Rolling-contact bearings are designed to support and locate rotating shafts or parts in machines. They transfer loads between rotating and stationary members and permit relatively free rotation with a minimum of friction. They consist of rolling elements (balls or rollers) between an outer and inner ring. Cages are used to space the rolling elements from each other.

Bearing performance is dependent on the type of lubrication occurring and the viscosity of the lubricant. The viscosity is a measure of a fluid's resistance to shear. Lubricants can be solid, liquid or gaseous, although the most commonly known are oils and greases. The principal classes of liquid lubricants are mineral oils and synthetic oils. Their viscosity is highly dependent on temperature.

Journal bearings are also available from specialist bearing suppliers. In principle they can be significantly cheaper than rolling element bearings as they involve fewer moving components.

2.7.2 Bearing type selection

The selection of an appropriate bearing for a given task, however, is an involved activity, which needs to take the following factors into consideration: ① load, ② speed, ③ location, ④ size, ⑤ cost, ⑥ starting torque, ⑦ noise, and ⑧ lubrication supply.

As can be seen from Figure 2.26, the scope of choice for a bearing is extensive. For a given application it may be possible to use different bearing types. For example, in a small gas turbine engine rotating at 50,000 r/min, either rolling bearings or journal bearings could typically be used, although the optimal choice depends on a number of factors such as life, cost and size.

Figure 2.27 can be used to give guidance for which kind of bearings has the maximum load capacity at a given speed and shaft size and Table 2.9 gives an indication of the performance of the various bearing types for some criteria other than load capacity.

The term "rolling contact bearings" encompasses the wide variety of bearings that use spherical balls or some type of roller between the stationary and moving elements. The most common type of bearing supports a rotating shaft resisting a combination of radial and axial (or thrust) loads. Some bearings are designed to carry only radial or thrust loads. Selection of the type of bearings to be used for a given application can be aided by the comparison charts, an example of which is given in Table 2.10. Some bearing manufacturers usually produce excellent catalogues (e.g. NSK/RHP, SKF, FAG, INA) including design guides. The user is commended to gain access to this information, which is available in hard copy (just telephone and ask the manufacturer for a brochure) and via the internet.

2.8 Clutches and Brakes

Clutches and brakes provide frictional, magnetic, hydraulic or mechanical connection between two machine elements, usually shafts. There are significant similarities between clutches and brakes. If both shafts rotate, then the machine element will be classed as a clutch and the usual function is to connect or disconnect a driven load from a driving shaft. If one shaft rotates and the other is stationary, then the machine element is classed as a brake and the likely function is to decelerate a shaft. In reality, however, the same device can function as a brake or clutch by fixing its output element to a shaft that can rotate or to the ground, respectively.

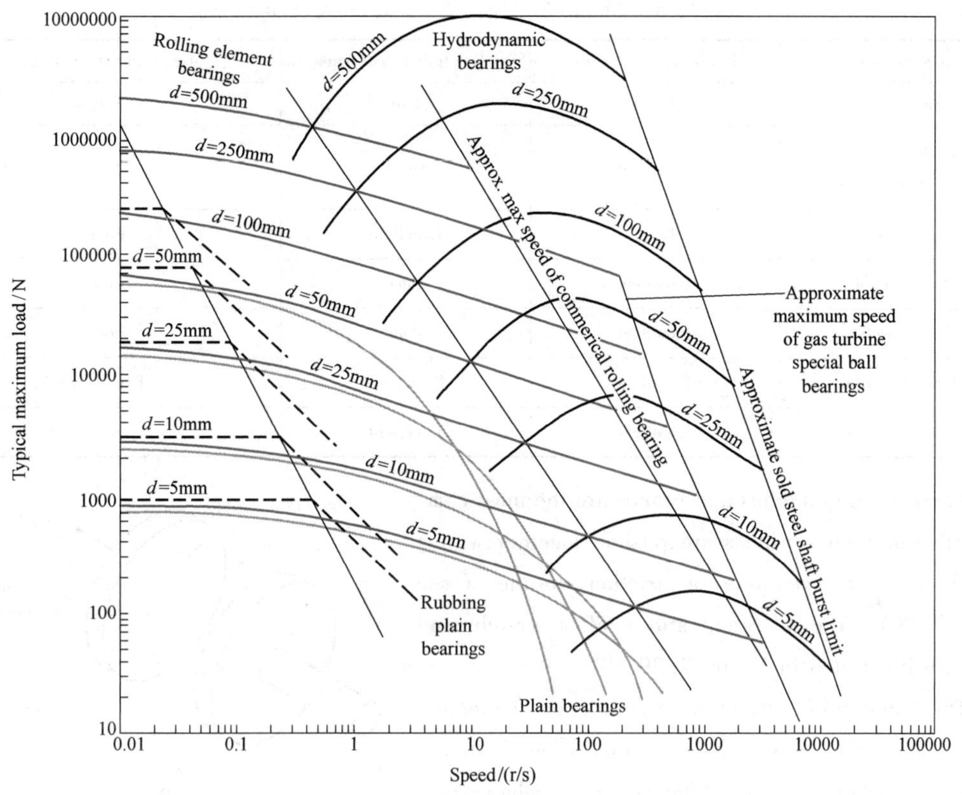

Figure 2. 27 Bearing type selection by load capacity and speed

Table 2. 9 Comparison of bearing performance for continuous rotation

Bearing type	Accurate radial location	Combined axial and radial load capability	Low starting torque capability	Silent running	Standard parts available	Lubrication simplicity
Rubbing plain bearings (non-metallic)	Poor	Some in most cases	Poor	Fair	Some	Excellent
Porous metal plain bearings oil impregnated	Good	Some	Good	Excellent	Yes	Excellent
Fluid film hydrodynamic bearings	Fair	No, separate thrust bearing needed	Good	Excellent	Some	Usually requires a recirculation system
Hydrostatic bearings	Excellent	No, separate thrust bearing needed	Excellent	Excellent	No	Poor special system needed
Rolling bearings	Good	Yes, in most cases	Very good	Usually satisfactory	Yes	Good when grease lubricated

Table 2.10 Merits of different rolling contact bearings

Bearing type	Radial load capacity	Axial or thrust load capacity	Misalignment capability
Single row	Good	Fair	Fair
Double row deep groove ball	Excellent	Good	Fair
Angular contact	Good	Excellent	Poor
Cylindrical roller	Excellent	Poor	Fair
Needle roller	Excellent	Poor	Poor
Spherical roller	Excellent	Fair/good	Excellent
Tapered roller	Excellent	Excellent	Poor

Friction type clutches and brakes are the most common. Two or more surfaces are pushed together with a normal force to generate a friction torque (see Figure 2.28). Typical applications of a clutch and brake are illustrated in Figure 2.29.

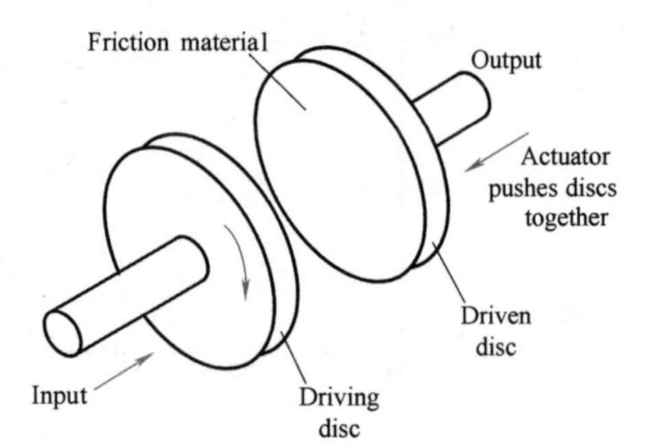

Figure 2.28 Idealized friction disc clutch or brake

The function of a clutch is to permit the connection and disconnection of two shafts, either when both are stationary or when there is a difference in the relative rotational speed of the shafts. Clutch connection can be achieved by a number of techniques from direct mechanical friction, electro-magnetic coupling, hydraulic or pneumatic means or by some combination. There are various types of clutches as outlined in Figure 2.30.

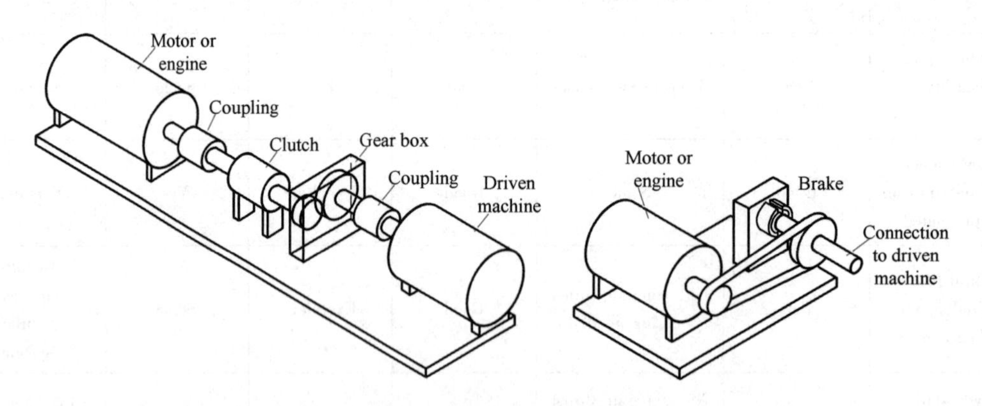

Figure 2.29 Typical applications of clutches and brakes

Complete clutches and brakes are available from specialist manufacturers. In addition key components such as discs, hydraulic and pneumatic actuators are also available enabling the designers to opt, if appropriate, to design the overall configuration of the clutch or brake.

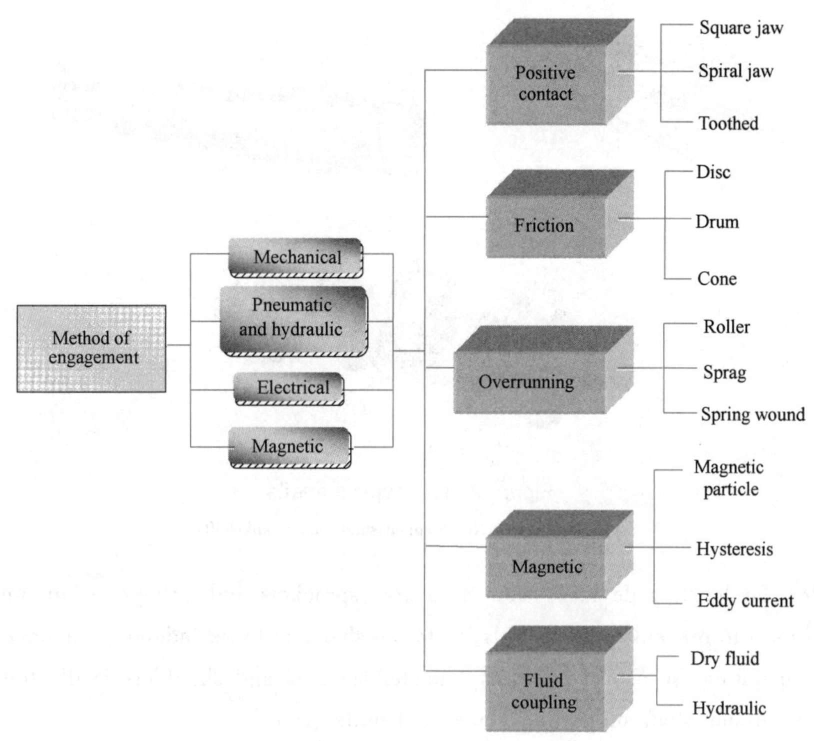

Figure 2. 30 Clutch classification

2. 9 Shafts and Couplings

2. 9. 1 Introduction

The term "shaft" usually refers to a component of circular cross-section that rotates and transmits power from a driving device, such as a motor or engine, through a machine. Shafts can carry gears, pulleys and sprockets to transmit rotary motion and power via mating gears, belts and chains. Alternatively, a shaft may simply connect to another via a coupling. A shaft can be stationary and support a rotating member, such as the short shafts that support the non-driven wheels of automobiles often referred to as spindles. Some common shafts are illustrated in Figure 2. 31.

Shaft design considerations include:

(1) size and spacing of components (as on a general assembly drawing), tolerances.

(2) material selection, material treatments.

(3) deflection and rigidity, such as bending deflection, torsional deflection, slope at bearings and shear deflection.

(4) stress and strength, such as static strength, fatigue and reliability.

(5) frequency response.

(6) manufacturing constraints.

Shafts typically consist of a series of stepped diameters accommodating bearing mounts and pro-

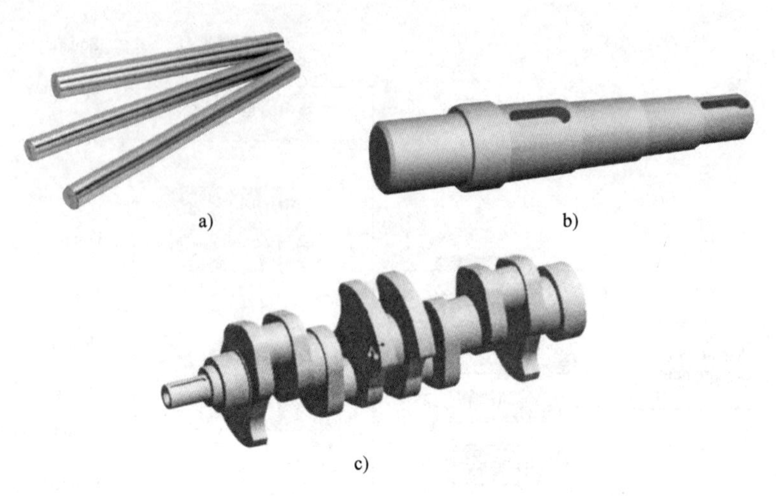

Figure 2. 31 Typical shafts

a) Plain shafts b) Stepped shaft c) Crankshafts

viding shoulders for locating devices, such as gears, sprockets and pulleys to butt up against and keys are often used to prevent rotation, relative to the shaft, of these "added" components. A typical arrangement illustrating the use of constant diameter sections and shoulders is illustrated in Figure 2. 32 for a transmission shaft supporting a gear and pulley wheel.

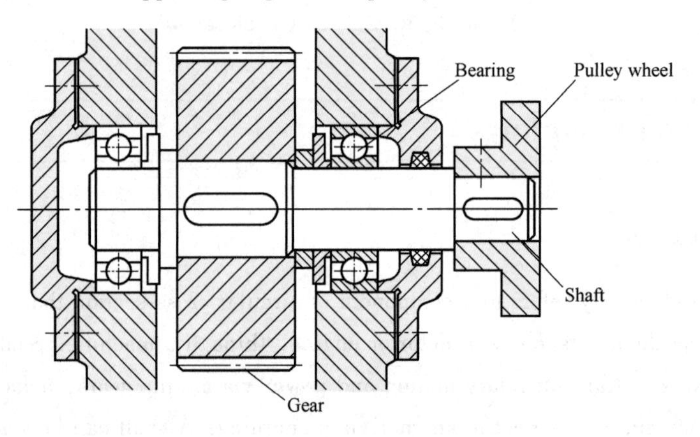

Figure 2. 32 A typical shaft arrangement

2. 9. 2 Shaft-Hub Connection

Power transmitting components such as gears, pulleys and sprockets need to be mounted on shafts securely and located axially with respect to mating components. In addition, a method of transmitting torque between the shaft and the component must be supplied. The portion of the component in contact with the shaft is called the hub and can be attached to or driven by keys, pins and set screws. Alternatively the component can be formed as an integral part of a shaft as, for example, the cam on an automotive cam-shaft.

Figure 2. 33 illustrates the practical implementation of several shaft-hub connection methods.

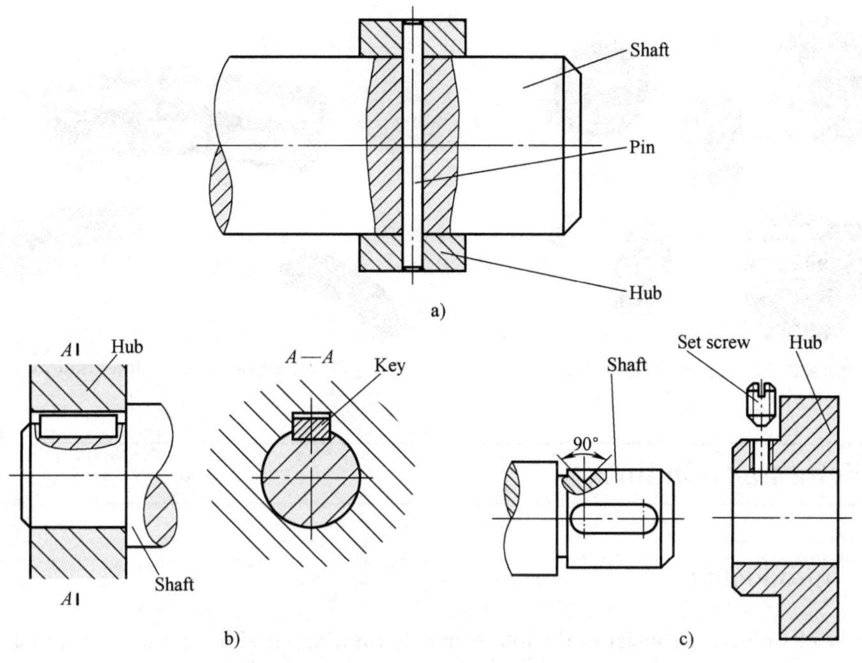

Figure 2. 33 Alternative methods of shaft-hub connection

a) Pin joints b) Key joints c) Set screw

2. 9. 3 Shaft-Shaft Connection—Couplings

In order to transmit power from one shaft to another, a coupling or clutch can be used. There are two general types of coupling: rigid and flexible. Rigid couplings are designed to connect two shafts together so that no relative motion occurs between them (see Figure 2. 34). Rigid couplings are suitable when precise alignment of two shafts is required. If significant radial or axial misalignment occurs, high stresses may result which can lead to early failure. Flexible couplings (see Figure 2. 35) are designed to transmit torque, whilst permitting some axial, radial and angular misalignment. Many forms of flexible couplings are available. Each coupling is designed to transmit a given limiting torque. Generally flexible couplings are able to tolerate up to 3° of angular misalignment and up to 0. 75mm parallel misalignment depending on their design. If more misalignment is required, a universal joint can be used (see Figure 2. 36).

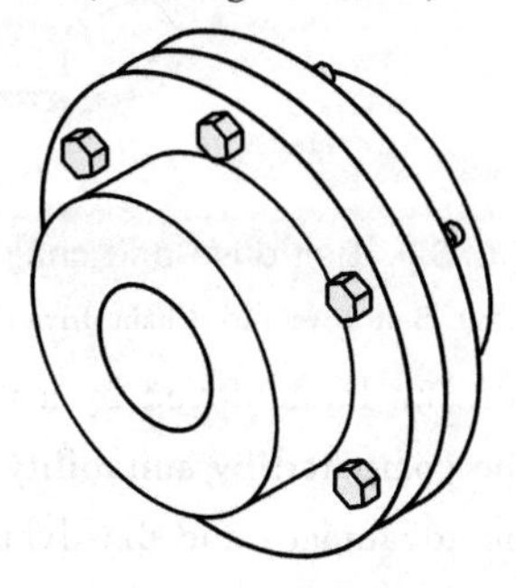

Figure 2. 34 Rigid couplings

Figure 2. 35 Flexible couplings

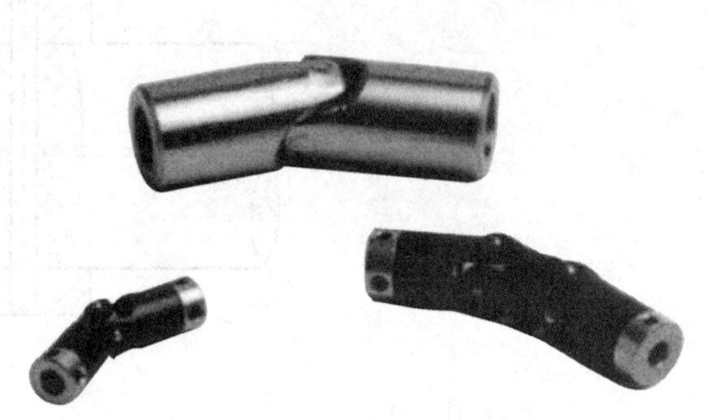

Figure 2. 36 Universal joints

2. 10 Belts and Chains

2. 10. 1 Introduction

Belt and chain drives consist of flexible elements running on either pulleys or sprockets as illustrated in Figure 2. 37. The purpose of a belt or chain drive is to transmit power from one rotating shaft to another. The speed ratio between the driving and driven shaft is dependent on the ratio of the pulley or sprocket diameters as is given by:

$$V_{pitchline} = \omega_1 R_1 = \omega_2 R_2$$

Angular velocity ratio $= \omega_1 / \omega_2 = R_2 / R_1$

Where $V_{pitchline}$——pitch velocity (m/s);

ω_1——angular velocity of driving pulley or sprocket (rad/s);

ω_2——angular velocity of driven pulley or sprocket (rad/s);

R_1——radius of driving pulley or sprocket (m);

R_2——radius of driven pulley or sprocket (m).

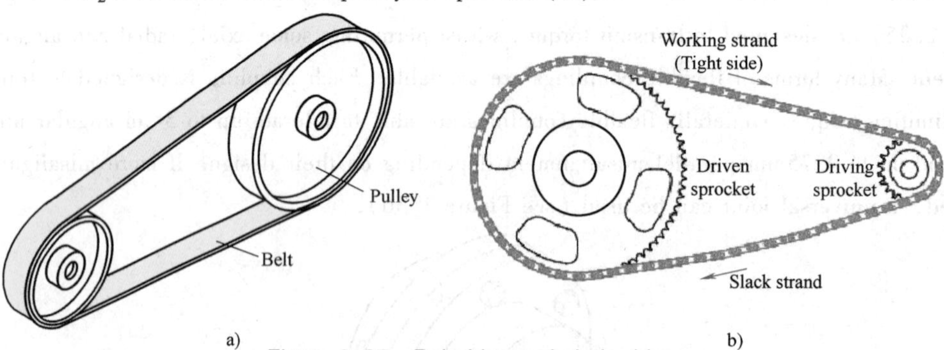

Figure 2. 37 Belt drive and chain drive

a) Belt drive b) Chain drive

Power transmission between shafts can be achieved by a variety of means including belt, chain and gear drives and their use should be compared by suitability and optimization for any given application. The following is to describe the advantages and disadvantages of a belt over gears and chains.

Consequence of Failure: In this consideration, belts have a distinct advantage over chains and gears since a sudden overload will not break a belt. Instead, the belt slips until the overload is ended. However, even momentary overloads can break gear teeth or chain links.

Versatility in Shaft Connection: Since belts are more flexible than chains and gears, they are more versatile in connecting two shafts with unusual geometrical arrangements or large center distances. Gears are the least versatile from a practical point of view, especially if the center distances are large.

Effect on Shaft Bearing Life: Belts are driven by friction, and therefore require initial tension on the slack side of the strand increased bearing loads. Gears and chains are positive drives and do not place this increased load on the bearings.

Speed Ratio: Generally speaking, belt drives do not provide an exact speed ratio as gear systems. The slippage that protects belt drives from damage by sudden over-loads ironically prevents an exact timing between the driving and driven shafts. However, special timing belts that produce positive drive are available.

Cost: Belts are the least expensive of either gears or chains, while chains are less expensive than gears. The required precision of machining and mounting of gears is the principal reason for their higher cost. When using chains, the alignment of the shafts must be more precise than for belts.

Noise and Vibration: Belt drives produce the least amount of noise and vibration, and thus they are used where vibration levels must be low.

Speed and Power: Gears can operate at higher speeds and transmit more power than chains or belts.

Maintenance: Chains and belts require periodic adjustment resulting from wear and stretch, respectively. Chains and gears require lubrication. Putting aside unexpected overloads, properly designed gear systems require the least amount of maintenance.

2. 10. 2 Belt Drives

There are various types of belt drive configurations including flat, round, V and synchronous belt drives, as shown in Figure 2. 38, each with their individual merits. The cross-sections of

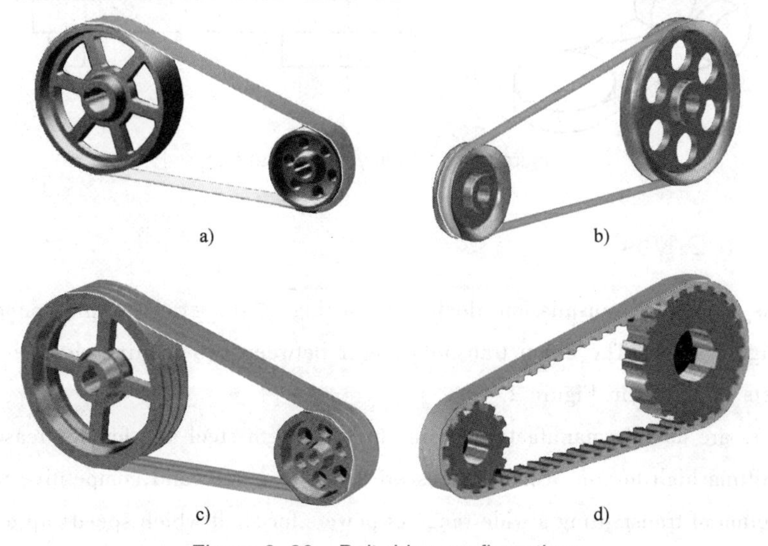

Figure 2. 38 Belt drive configurations

a) Flat belt drives b) Round belt drives c) V belt drives d) Synchronous belt drives

various belts are illustrated in Figure 2. 39. Most belts are manufactured from rubber or polymer-based materials.

A frequent application of belt drives is to reduce the speed output from electric motors which typically run at specific synchronous speeds which are high in comparison to the desired application drive speed. Because of their good "twistability", belt drives are well suited to applications where the rotating shafts are in different planes. Some of the standard layouts are shown in Figure 2. 40. Belts are installed by moving the shafts closer together, slipping the belt over the pulleys and then moving the shafts back into their operating locations.

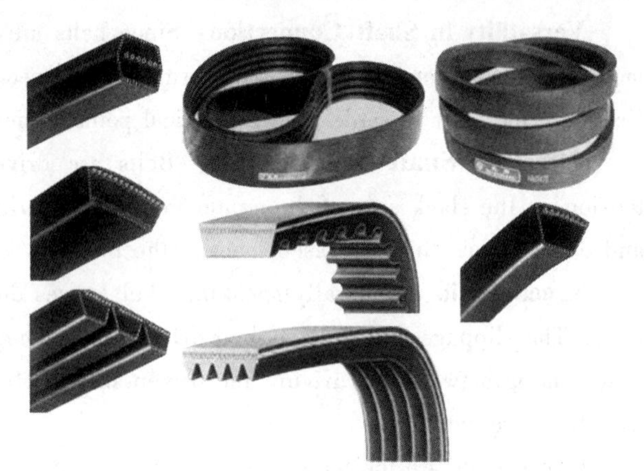

Figure 2. 39 Various belt cross-sections

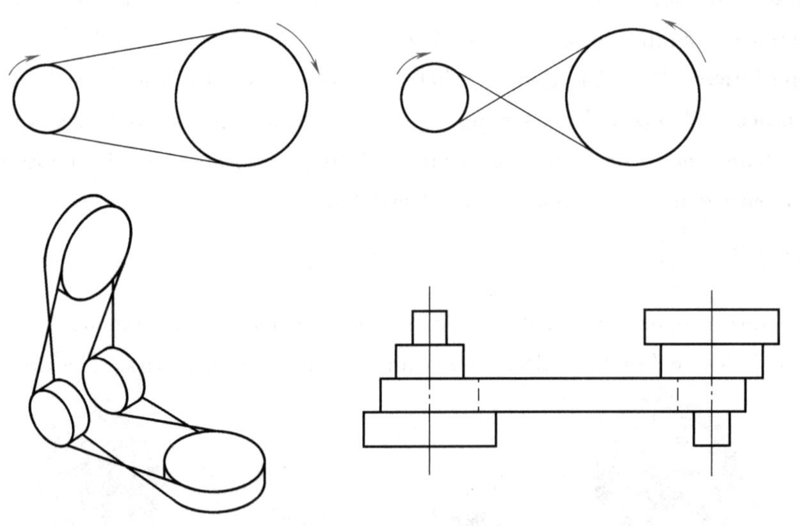

Figure 2. 40 Pulley configurations

2. 10. 3 Chain Drives

A chain is a power transmission device consisting of a series of pin-connected links as illustrated in Figure 2. 41. The chain transmits power between two rotating shafts by meshing with toothed sprockets as shown in Figure 2. 42.

Chain drives are usually manufactured using high strength steel and for this reason they are capable of transmitting high torque. Chain drives are complementary and competitive with belt drives serving the function of transmitting a wide range of powers for shaft which speeds up to about 6000 r/min. At higher speeds the cyclic impact between the chain links and the sprocket teeth, high noise and difficulties in providing lubrication limit the application of chain drives.

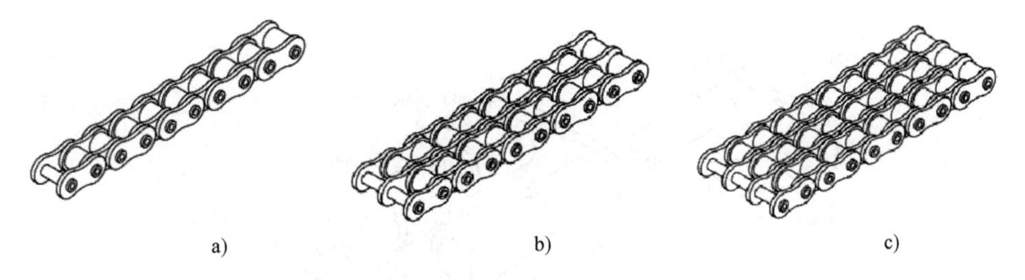

Figure 2.41 Roller chain

a) Simple chain b) Duplex chain c) Triplex chain

The range of chain drives is extensive as illustrated in Figure 2.43. The most common type is roller chain, which is used for high power transmission and conveyor applications. Roller chain is made up of a series of links. Each chain consists of side plates, pins, bushes and rollers as shown in Figure 2.44. The chain runs on toothed sprockets and as the teeth of the sprockets engage with the rollers, rolling motion occurs and the chain articulates on to the sprocket.

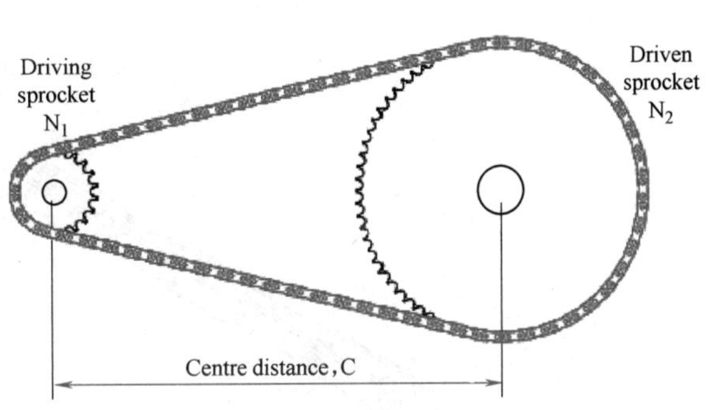

Figure 2.42 Simple chain drives

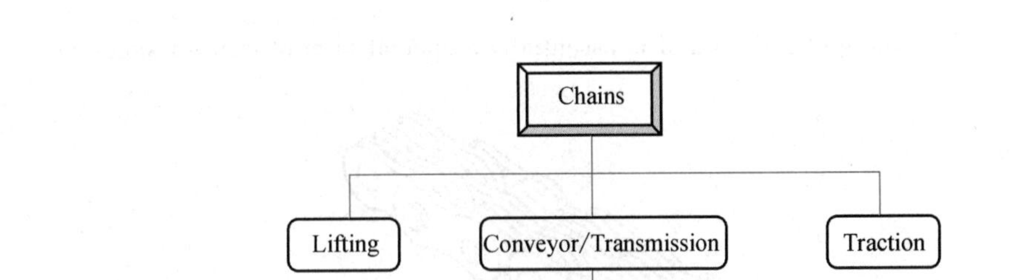

Figure 2.43 Chain types

Conveyor chain is specially designed for use in materials handling and conveyor equipment and is characterized by its long pitch, large roller diameters and high tensile strength. The applications of conveyor chain demand that it is capable of pulling significant loads usually along a straight line at relatively low speeds. The sides of conveyor chains often incorporate special features to aid connection to conveyor components as illustrated in Figure 2.45.

Leaf chain consists of a series of pin connected side plates as illustrated in Figure 2.46. It is

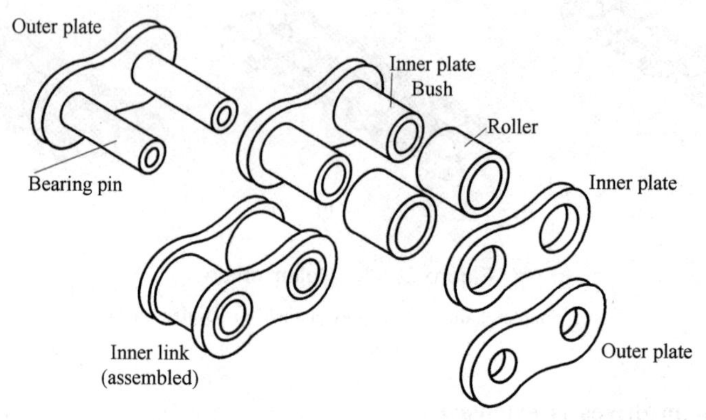

Figure 2. 44　Roller chain components

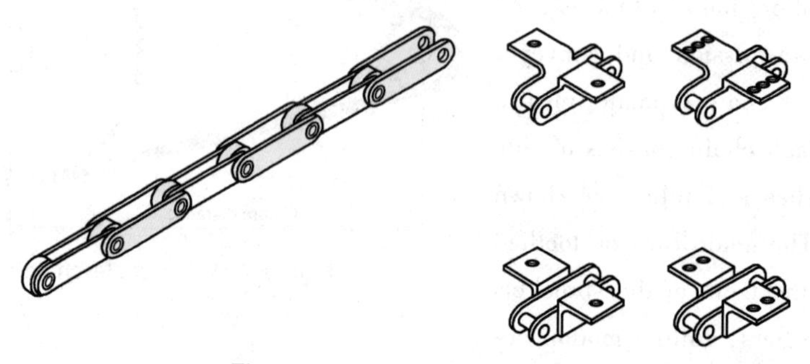

Figure 2. 45　Conveyor chain

generally used for load balancing applications and is essentially a special form of transmission chain.

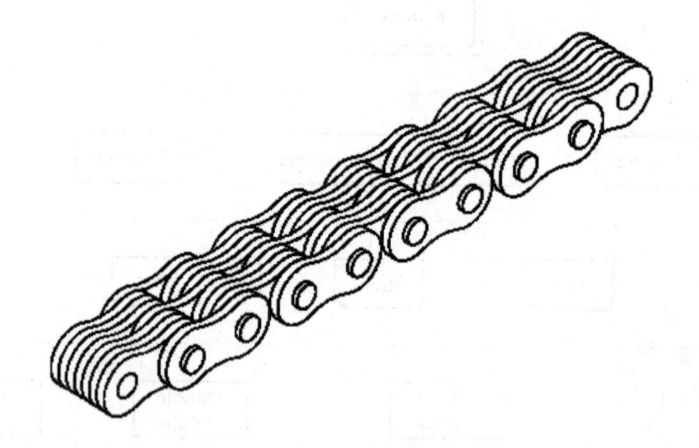

Figure 2. 46　Leaf chain

Silent chain (also called inverted tooth chain) has the teeth formed in the link plates as shown in Figure 2. 47. The chain consists of alternately mounted links so that the chain can articulate on to the mating sprocket teeth. Silent chains can operate at higher speeds than comparatively sized roller chain and as the name suggests more quietly. Figure 2. 48 is an application example of silent chains inside a 1912 Sperber gearbox.

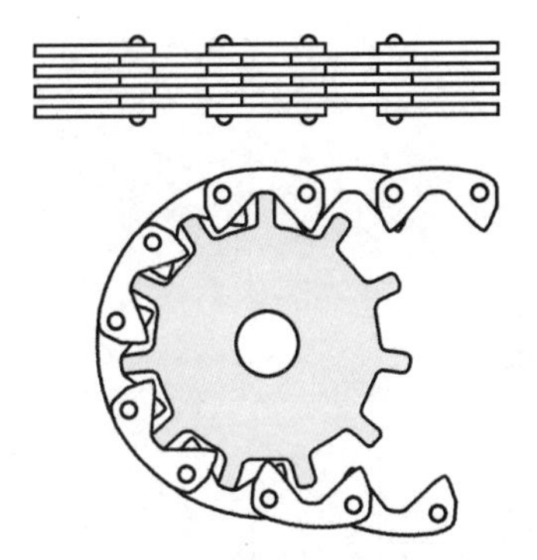

Figure 2. 47 Silent chain

Figure 2. 48 Silent chains inside a 1912 Sperber gearbox

2. 11 Gears

2. 11. 1 Introduction

Gears are toothed cylindrical wheels used for transmitting mechanical power from one rotating shaft to another. Several types of gears are in common use. This part introduces various types of gears and details the design, specification and selection of spur gears in particular.

When transmitting power from a source to the required point of application, a series of devices is available including gears, belts, pulleys, chains, hydraulic and electrical systems. However, when a compact, efficient or high-speed drive is required, gear trains offer a competitive and suitable solution. Additional benefits of gear drives include reversibility, configuration at almost any angle between input and output and their suitability to operate in arduous conditions.

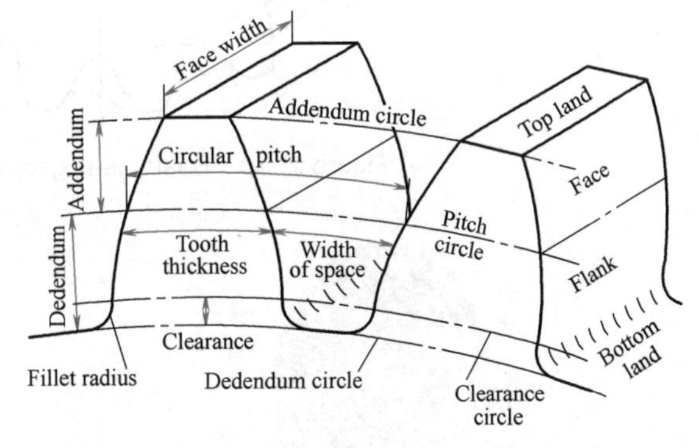

Figure 2. 49 Spur gear schematic showing principle terminology

Various definitions used for describing gear geometry are illustrated in Figure 2. 49. For a pair of meshing gears, the smaller gear is called the "pinion", the larger is called the "gear wheel" or simply the "gear", as shown in Figure 2. 50.

2. 11. 2 Gear Classification

Gears can be divided into several broad classifications.

1. Parallel axis gears

(1) spur gears (see Figure 2. 51).

(2) helical gears (see Figures 2. 52).

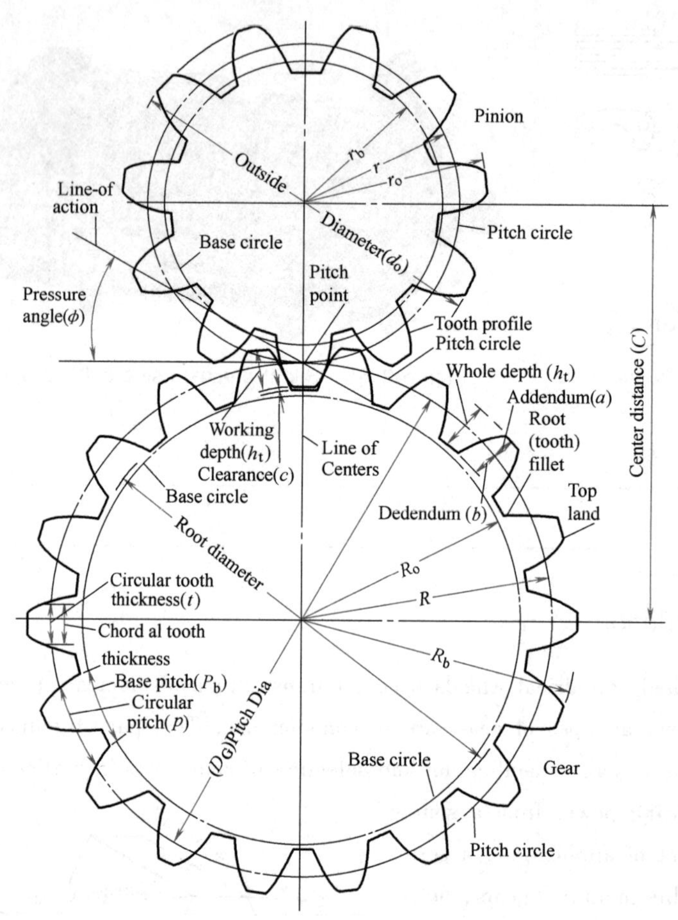

Figure 2. 50 Basic gear geometry and nomenclature

Figure 2. 51 Straight tooth spur gears

Figure 2. 52 Helical gears (Single-helical tooth)

(3) internal gears.

2. Non-parallel, coplanar gears (intersecting axes)

(1) bevel gears (see Figure 2. 53).

(2) conical involute gearing.

3. Non-parallel, non-coplanar gears (nonintersecting axes)

(1) crossed axis helical gears (see Figure 2. 54).

Figure 2. 53 Straight tooth bevel gears

Figure 2. 54 Cross axis helical gears

(2) cylindrical worm gearing (see Figure 2. 55).

(3) single enveloping worm gearing.

(4) double enveloping worm gearing.

(5) hypoid gears.

(6) spiroid and helicon gearing.

4. Special gear types

(1) square and rectangular gears.

(2) elliptical gears.

(3) scroll gears.

(4) multiple sector gears.

Some gear axes can be allowed to rotate about others. In such cases the gear trains are called planetary or epicyclic (see Figure 2. 56). Planetary trains always include a sun gear, a ring gear, a planet carrier or arm, and one or more planet gears.

Figure 2. 55 Worm and worm wheel

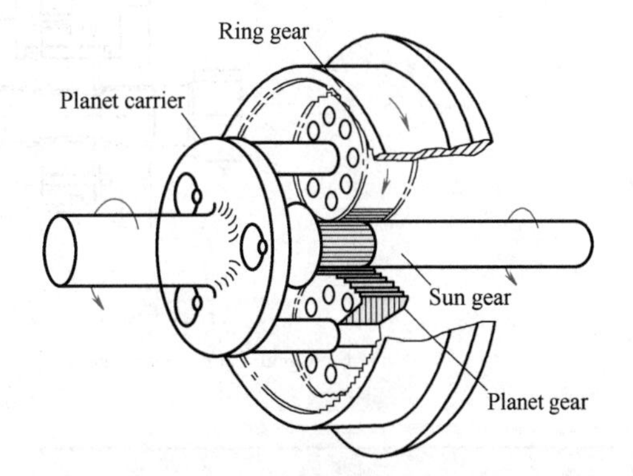

Figure 2. 56 Planetary gears

2. 11. 3 Gear Chains Application

A typical manually shifted gearbox for a passenger car is illustrated in Figure 2. 57. This has five forward speeds and reverse. The basic elements of a manually shifted transmission are a single or multiplate clutch for engaging and disengaging the power from the load, a variable ratio transmission unit with permanent mesh gears and a gear shift mechanism and lever.

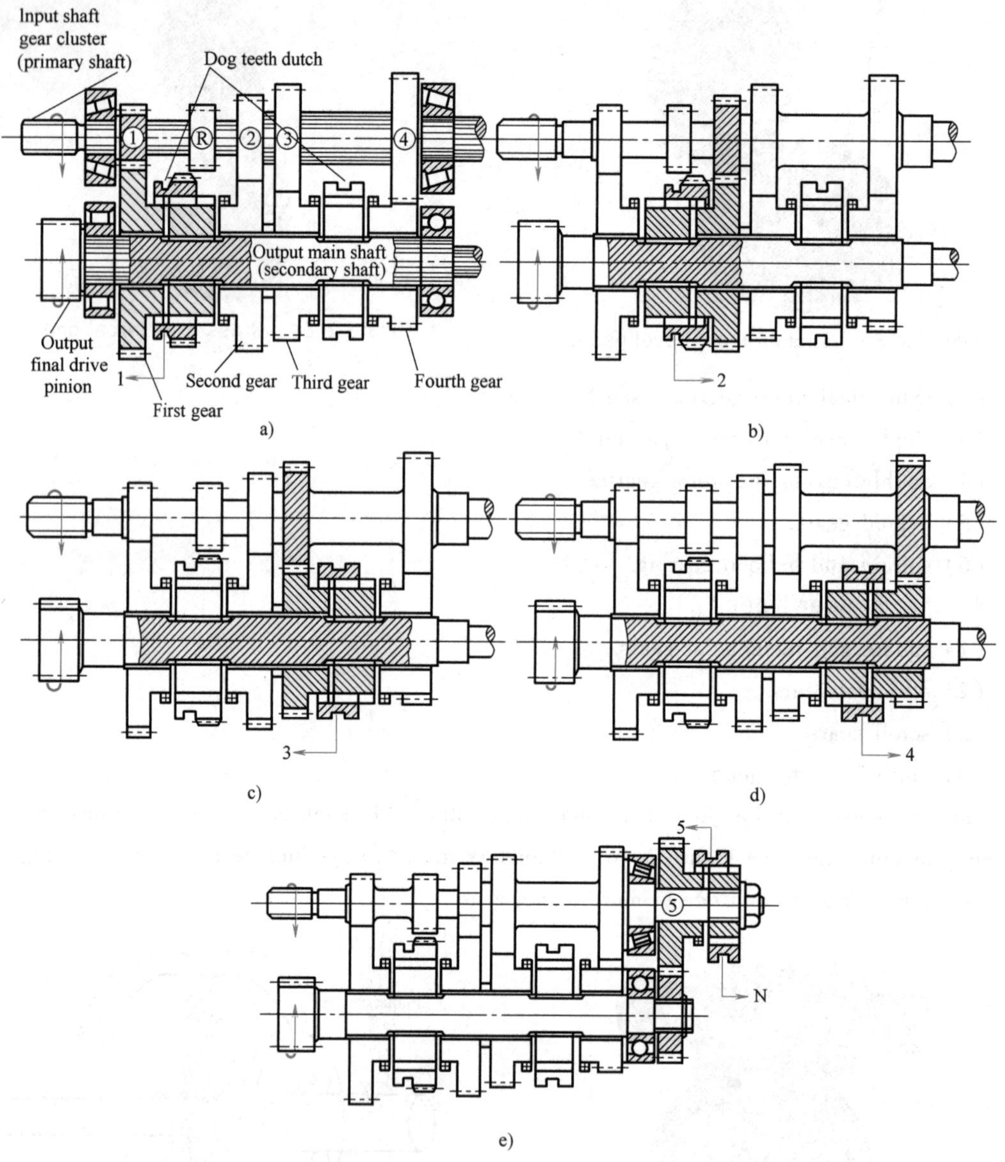

Figure 2.57 A shift gearbox

a) First gear b) Second gear c) Third gear d) Fourth gear e) Fifth gear

2.12 Planar Linkages

2.12.1 Linkage Mechanisms and Terminology

Linkages are perhaps the most fundamental class of machines that humans employ to turn thought into action. From the first lever and fulcrum to the complex shutter mechanism, linkages translate one type of motion into another.

A linkage, or kinematic chain, is an assembly of links and joints that provide a desired output motion in response to a specified input motion. An example of a mechanism is the windshield wiper mechanism shown in Figure 2.58. The motion is transferred from the crank driven by a motor through the coupler to the left rocker. The two rockers, left and right, are connected by the rocker coupler, which synchronizes their motion. The mechanism comprising links ① frame, ② crank, ③ coupler, and ④ rocker is called a 4-bar mechanism. In this example, revolute joints connect all links.

A link is a rigid body that possesses at least 2 nodes. A node is an attachment point to other links via joints. The order of a link indicates the number of joints to which the link is connected (or the number of nodes per link). There are binary (2 nodes), ternary (3 nodes) and quaternary (4 nodes) links. A joint is a connection between two or more links at their nodes, which allows motion to occur between the links. A pivot is a joint that allows rotary motion, and a slider is a joint that allows linear motion. A mechanism is a kinematic chain in which at least one link is connected to a frame of reference (ground), where the ground is also counted as a link. A planar mechanism is the one in which all points move in parallel planes.

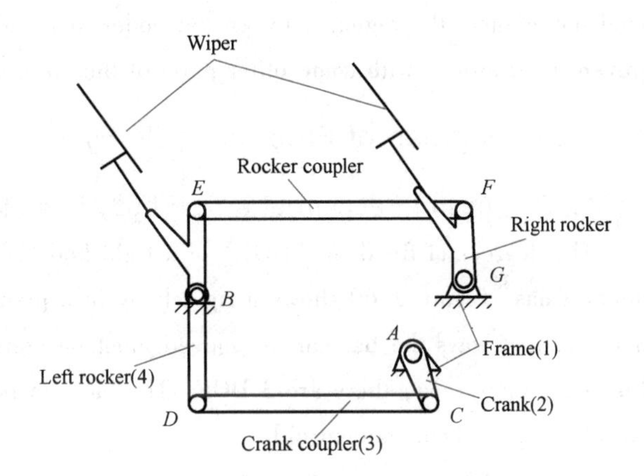

Figure 2.58 Windshield wiper mechanism

A joint between two links restricts the relative motion between these links, and thus imposes a constraining condition on the mechanism motion. The type of constraining condition determines the number of degrees of freedom (DOF) a mechanism has. If the constraining condition allows only one DOF between the two links, the corresponding joint is called a lower-pair joint. The examples are a revolute joint between links 1 and 2, 2 and 3, 3 and 4, left rocker and rocker coupler, right rocker and rocker coupler in Figure 2.58. If the constraint allows two DOF between the two links, the corresponding joint is called a high-pair joint. An example of a high-pair joint is a connection between the cam and the roller.

Even a lever with some sort of means to apply an input force is a linkage. One of the most common types of linkages is the 4-bar linkage, which is comprised of four links and four joints as shown in Figure 2.59. A ground link acts as the reference for all motions of the other three links and the power input device, usually a motor, and another joint are attached to the ground link. The motor output shaft is connected to the link called rocker in the case of oscillating input motion, while the same link is called crank in the case of continuous input motion. The follower is connected to the

ground link through a joint at one end. The coupler link couples the ends of the crank (or rocker) and the follower link. These four links are thus geometrically constrained to each other. However, their motion may not be deterministic, for there are link lengths and ground joint locations that can lead to instability in the linkage. Even though two points define a line, a straight line structure need not connect the region between the nodes of a link. A link may be curved or have a notch to prevent interference with some other parts of the structure or linkage as it moves.

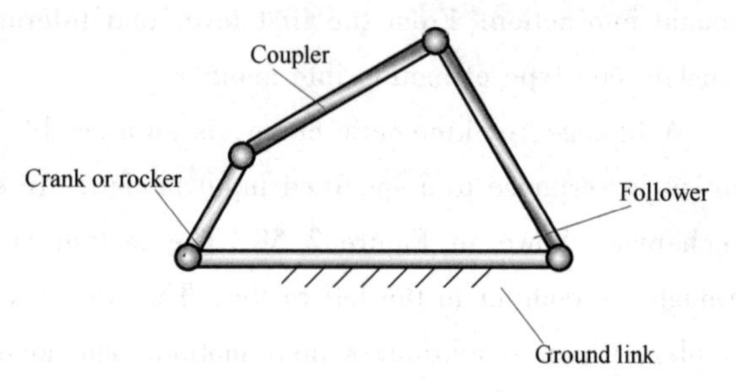

Figure 2. 59 Diagram of 4-bar linkage

2. 12. 2 Degrees of Freedom (DOF)

1. Degrees of freedom of a rigid body in a plane

The degrees of freedom (DOF) of a rigid body is defined as the number of independent movements it has. Figure 2. 60 shows a rigid body in a plane. To determine the DOF of this body, how many distinct ways the bar can be moved must be considered. In a two-dimensional plane such as this computer screen, there are 3 DOF. The bar can be translated along the x axis the y axis and it can also rotate about its centroid.

2. Degrees of freedom of a rigid body in space

An unrestrained rigid body in space has six degrees of freedom: three translation motions along the x, y and z axes and three rotary motions around the x, y and z axes respectively, as shown in Figure 2. 61.

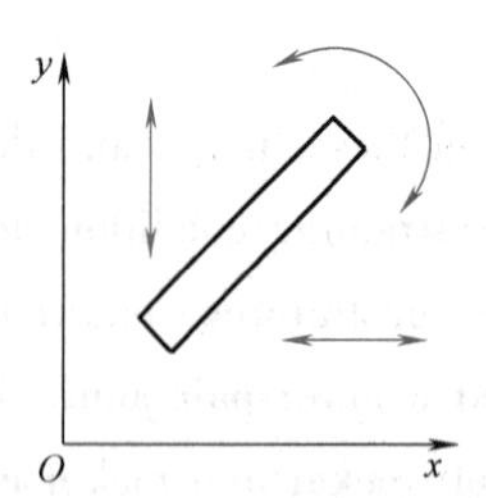

Figure 2. 60 Degrees of freedom of a rigid body in a plane

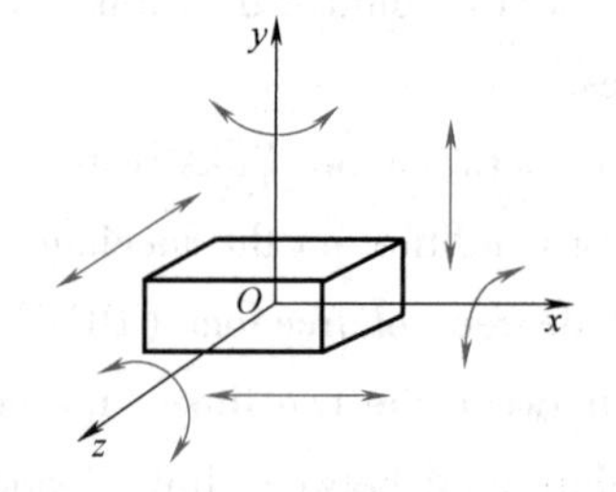

Figure 2. 61 Degrees of freedom of a rigid body in space

3. Kinematic constraints

Two or more rigid bodies in space are collectively called a rigid body system. The motion of these independent rigid bodies with kinematic constraints can be hindered. Kinematic constraints are constraints between rigid bodies that result in the decrease of the degrees of freedom of rigid body system.

The term kinematic pairs actually refers to kinematic constraints between rigid bodies. The kinematic pairs are divided into lower pairs and higher pairs, depending on how the two bodies are in contact.

2.12.3 Lower Pairs in Planar Mechanisms

There are two kinds of lower pairs in planar mechanisms (Figure 2.62): revolute pairs and prismatic pairs.

A rigid body in a plane has only three independent motions: two translation and one rotary, so introducing either a revolute pair or a prismatic pair between two rigid bodies removes two degrees of freedom.

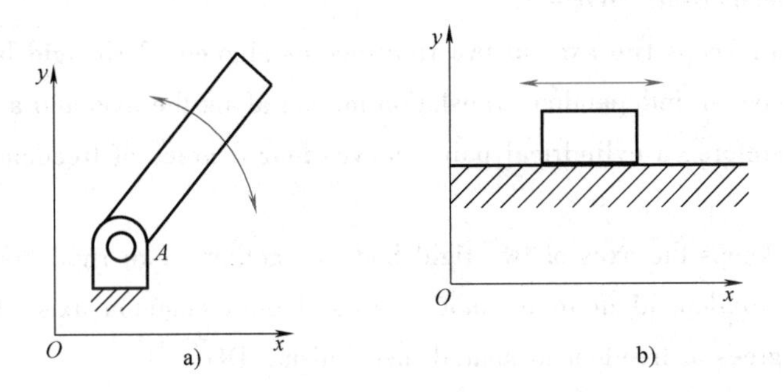

Figure 2.62 Lower pairs in planar mechanisms

a) A planar revolute pair (R-pair) b) A planar prismatic pair (P-pair)

2.12.4 Lower Pairs in Spatial Mechanisms

There are six kinds of lower pairs under the category of spatial mechanisms. The types are: spherical pair, planar pair, cylindrical pair, revolute pair, prismatic pair and screw pair, as demonstrated in Figure 2.63.

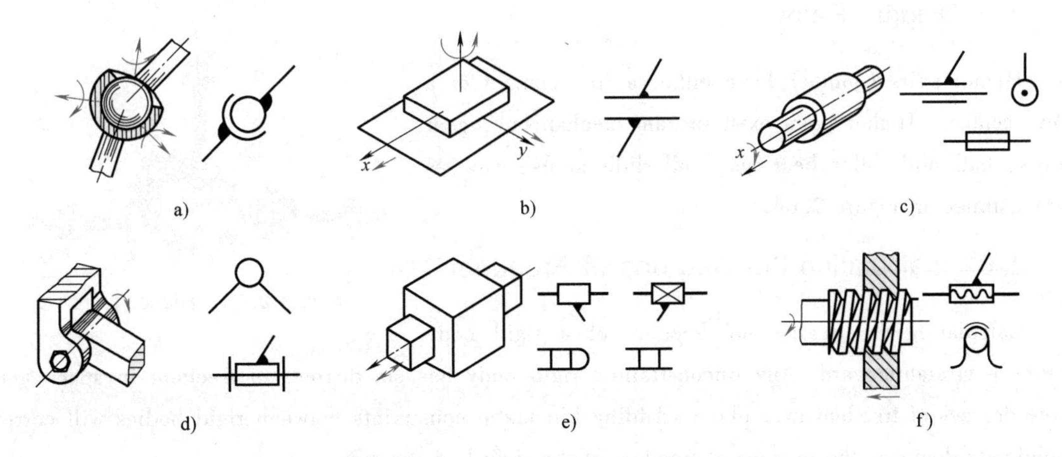

Figure 2.63 Lower pairs in spatial mechanisms

a) A spherical pair (S-pair) b) A planar pair (E-pair) c) A cylindrical pair (C-pair)

d) A revolute pair (R-pair) e) A prismatic pair (P-pair) f) A screw pair (H-pair)

A spherical pair keeps two spherical centers together. Two rigid bodies connected by this constraint are able to rotate relatively around x, y and z axes, but there is no relative translation along any of these axes. Therefore, a spherical pair removes three degrees of freedom in spatial mechanism. DOF = 3.

A plane pair keeps the surfaces of two rigid bodies together. To visualize this, imagine a book lying on a table where it can move in any direction except off the table. Two rigid bodies connected by this kind of pair have two independent translation motions in the plane, and a rotary motion around the axis that is perpendicular to the plane. Therefore, a plane pair removes three degrees of freedom in spatial mechanism. DOF = 3.

A cylindrical pair keeps two axes of two rigid bodies aligned. Two rigid bodies that are part of this kind of system have an independent translation motion along the axis and a relative rotary motion around the axis. Therefore, a cylindrical pair removes four degrees of freedom from spatial mechanism. DOF = 2.

A revolute pair keeps the axes of two rigid bodies together. Two rigid bodies constrained by a revolute pair have an independent rotary motion around their common axis. Therefore, a revolute pair removes five degrees of freedom in spatial mechanism. DOF = 1.

A prismatic pair keeps two axes of two rigid bodies align and allow no relative rotation. Two rigid bodies constrained by this kind of constraint will be able to have an independent translational motion along the axis. Therefore, a prismatic pair removes five degrees of freedom in spatial mechanism. DOF = 1.

A screw pair keeps two axes of two rigid bodies aligned and allows a relative screw motion. Two rigid bodies constrained by a screw pair have a motion which is a composition of a translation motion along the axis and a corresponding rotary motion around the axis. Therefore, a screw pair removes five degrees of freedom in spatial mechanism. DOF = 1.

2. 12. 5 Higher Pairs

Higher pairs (joints) have either a line contact or a point contact. Higher pairs exist in cam mechanisms, gear trains, ball and roller bearings, roll-slide joints, etc. as demonstrated in Figure 2. 64.

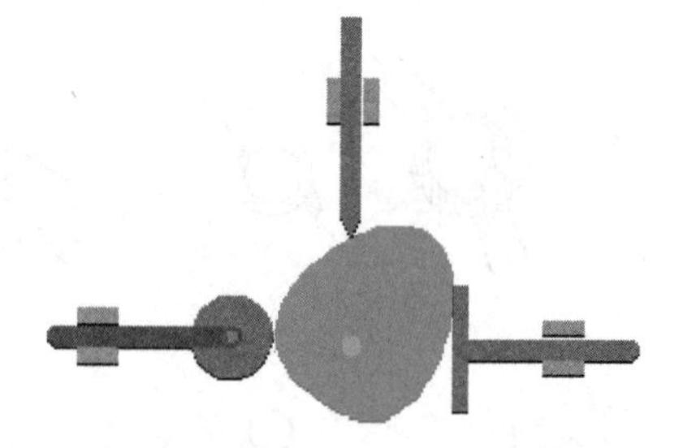

Figure 2. 64 Higher pairs

2. 12. 6 Calculating the Degrees of Freedom

Calculating the degrees of freedom of a rigid body system is straightforward. Any unconstrained rigid body has six degrees of freedom in space and three degrees of freedom in a plane. Adding kinematic constraints between rigid bodies will correspondingly decrease the degrees of freedom of the rigid body system.

The definition of the degrees of freedom of a mechanism is the number of independent relative motions among the rigid bodies. For example, Figure 2. 65 shows several cases of rigid bodies constrained by different kinds of planar pairs.

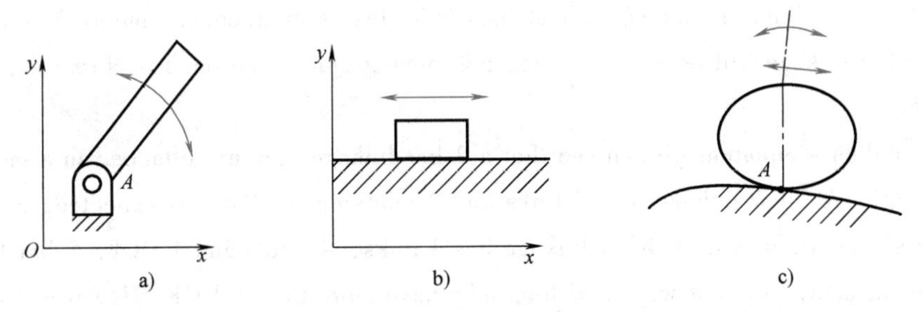

Figure 2. 65 Rigid bodies constrained by different kinds of planar pairs

In Figure 2. 65a, a rigid body is constrained by a revolute pair which allows only rotational movement around an axis. It has one degree of freedom, turning around point A. The two lost degrees of freedom are translation movements along the x and y axes. The only way the rigid body can move is to rotate about the fixed point A.

In Figure 2. 65b, a rigid body is constrained by a prismatic pair which allows only translation motion. In two dimensions, it has one degree of freedom, translating along the x axis. In this example, the body has lost the ability to rotate about any axis and it can't move along the y axis.

In Figure 2. 65c, a rigid body is constrained by a higher pair. It has two degrees of freedom: translating along the curved surface and turning about the instantaneous contact point.

In general, a rigid body in a plane has three degrees of freedom. Kinematic pairs are constraints on rigid bodies that reduce the degrees of freedom of a mechanism. Figure 2. 65 shows three kinds of pairs in planar mechanisms. These pairs reduce the number of the degrees of freedom. If a lower pair is created (Figure 2. 65a, b), the degrees of freedom are reduced by 2. Similarly, if a higher pair is created (Figure 2. 65c), the degrees of freedom are reduced by 1.

Therefore, the following equation can be written:

$$F=3(n-1)-2l-h \tag{4.1}$$

Where F——total degrees of freedom in the mechanism;

n——number of links (including the frame);

l——number of lower pairs (one degree of freedom);

h——number of higher pairs (two degrees of freedom).

This equation is also known as Gruebler's equation.

Examples:

Slider-crank: $n=4$, $f=4$, $F=1$ (Figure 2. 66)

4-bar linkage: $n=4$, $f=4$, $F=1$ (Figure 2. 67)

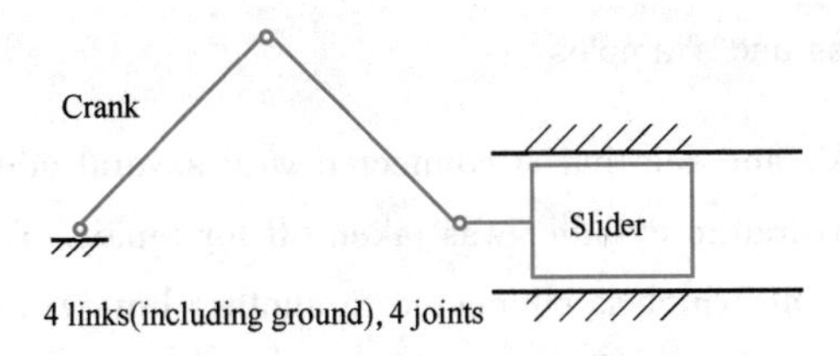

Figure 2. 66 Slider-crank mechanism

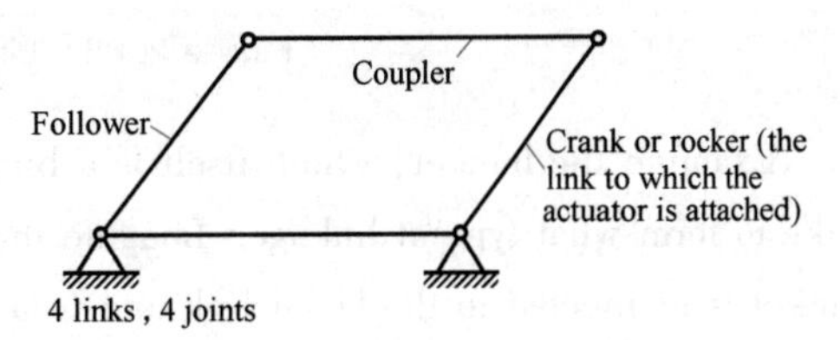

Figure 2. 67 4-bar linkage

Notice that the simplest linkage with at least one degree of freedom (motion) is thus a 4-bar linkage. A 3-bar linkage will be rigid, stable, not moving unless you bend it, break it, or throw it, it has 0 DOF.

From Gruebler's equation we can see that a 2-bar linkage, an arm attached to a motor's output shaft has 1 DOF. A 3-bar linkage with 3 links and 3 joints has 0 DOF, as expected, and hence triangles make stable structures. A 4-bar linkage has 4 links, 4 joints and 1 DOF. 5-bar linkages can be configured in many different ways and thus may have more than 1 DOF. However, these are not generally stable unless multiple input power sources are used. 6-bar linkages can have 1 DOF and they can be extraordinarily useful.

2.12.7 Types of Links

The four most common links are known as binary, ternary, quaternary and pentanary links and they have two, three, four and five joints (nodes) respectively on their structures, as shown in Figure 2.68. Look closely at the picture of the excavator and try to identify each of these types of links. What types of links represent the hydraulic cylinders? The hydraulic cylinders have pivot joints at each end and the rod slides inside the cylinder, thus a hydraulic cylinder is comprised of two binary links, each with a pivot joint and a slider in between. Note the first link which has a letter F on it, what type of link is it ? This link has a pivot at its base, which can't be seen but obviously it must be present, a pivot at its end for the second link, and two other pivots to which hydraulic cylinders are attached; thus it is a quaternary link. How about the second major link? How many joints are on it and what type of link is it? Look closely and you can see it is a pentanary link.

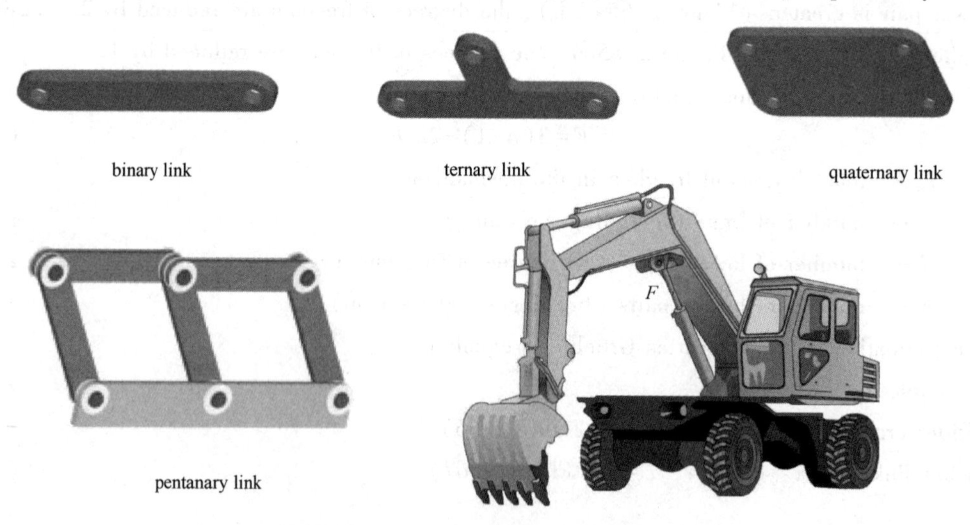

Figure 2.68 Links types and examples

Examine the bucket, which itself is a binary link, and see that is connected with several other links to form what type of linkage. Imagine that the hydraulic cylinder was taken off for repair. The bucket is connected to the boom link and to a binary link which is connected to another binary link that is also connected to the boom link. The bucket could be said to be a follower, and the binary link opposite it is a rocker link. Thus the bucket linkage is a 4-bar linkage. The rocker is driven by the hydraulic cylinder which is connected to the boom link. Recall from above that the hydraulic cylinder is modeled as two binary links with pivots at their ends, but they happen to share a slider joint. Thus the bucket system is comprised of two 4-bar linkages that share a common link, the follower for one (the hydraulic cylinder side) and the rocker for the other (the bucket side). Together, they actually form a 6-bar linkage.

2.12.8 4-Bar Linkages

A 4-bar linkage has four binary links and 4 revolute joints; hence from Gruebler's equation there are $3\times(4-1)-2\times4=1$ degree of freedom. This means that only one input is required to make the linkage move. The planar 4-bar linkage is probably the simplest and most common linkage. A 4-bar linkage is the simplest closed-loop kinematic linkage, as shown in Figure 2.69. They perform a wide variety of motions with a few simple parts. They were also popular in the past due to the ease of calculations without computers, compared to more complicated mechanisms.

Examples of 4-bar linkages are:

1) The crank-rocker linkage (Figure 2.70), in which the input crank fully rotates and the output link rocks back and forth.

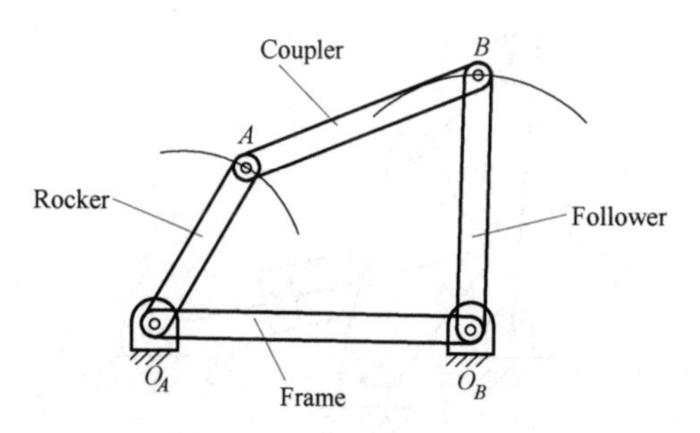

Figure 2.69 A 4-bar linkage

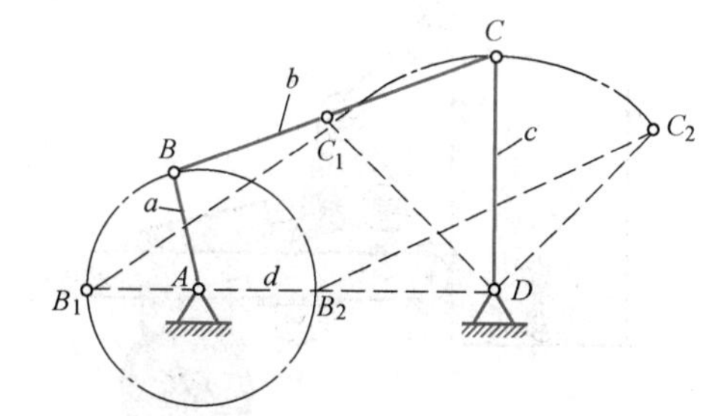

Figure 2.70 A crank-rocker linkage

2) The slider-crank linkage (Figure 2.71), in which the input crank rotates and the output slide moves back and forth. If the eccentricity is zero ($e=0$), the slider crank mechanism is called an in-line/a centric slider-crank. If the eccentricity is not zero ($e\neq0$, it is usually called an offset slider-crank mechanism.

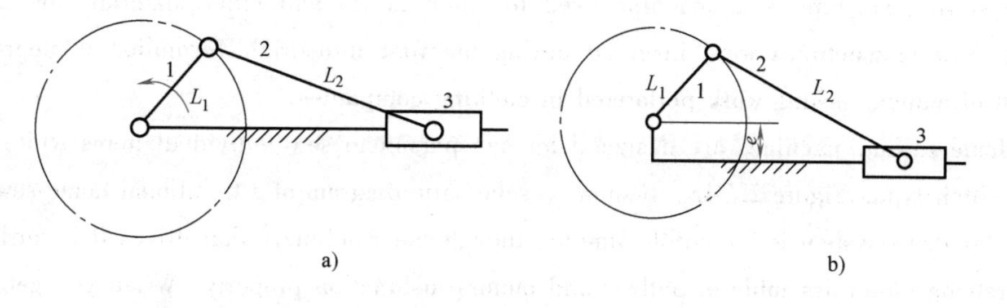

Figure 2.71 A slider-crank linkage

a) A centric slider-crank linkage b) An offset slider-crank linkage ($e\neq0$)

3) The double-rocker linkage, in which the input crank fully rotates and drags the output crank in a fully rotational movement, as shown in Figure 2.72.

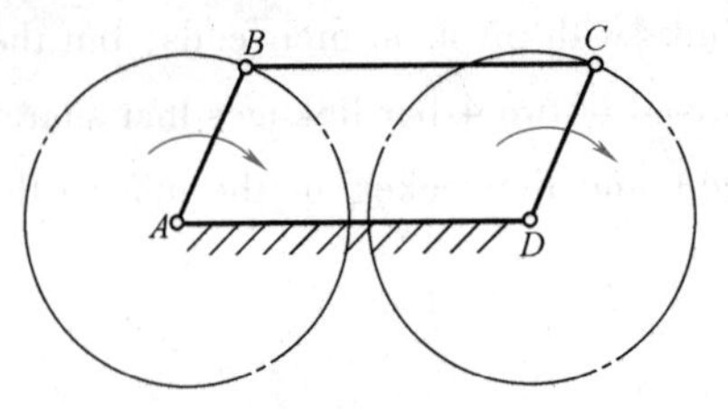

Figure 2.72 A double-rocker linkage

2.12.9 Applications of 4-Bar Linkages

Example 2.1 A Dump Truck Mechanism

A dump truck mechanism is shown in Figure 2.73a, and its skeleton diagram in Figure 2.73b. This is an example of a compound mechanism comprising two simple bar linkage: the first, links 1-2-3, is the slider-crank mechanism and the second, links 1-3-5-6, is 4-bar linkage. The two mechanisms work in sequence (or they are functionally in series): the input is the displacement of the piston in the hydraulic cylinder, and the output is the tipping of the dump bed.

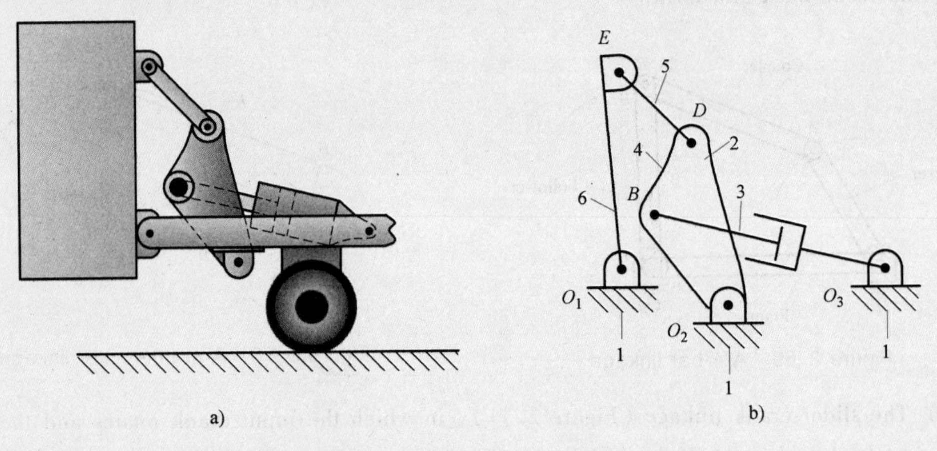

Figure 2.73 A dump truck mechanism

Example 2.2 Home Sewing Machine

A sewing machine is a machine used to stitch fabric and other materials together with thread. Sewing machines were invented during the first Industrial Revolution to decrease the amount of manual sewing work performed in clothing companies.

Home sewing machines are designed for one person to sew individual items while using a single stitch type. Figure 2.74a, b shows a schematic diagram of a traditional home sewing machine; its drive system is incredibly simple, though the machinery that drives it is fairly elaborate, relying on an assembly of pulleys and motors to function properly. When you get down to it, the drive system of the sewing machine is actually a crank-rocker linkage mechanism, as demonstrated in Figure 2.74c.

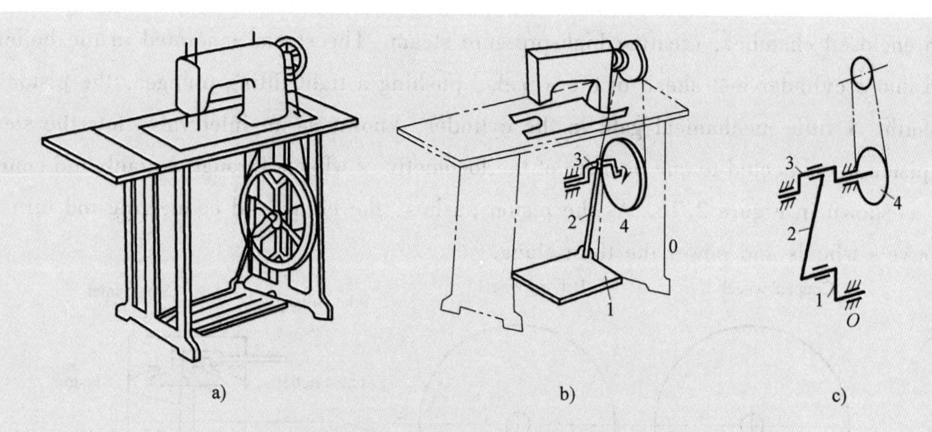

Figure 2. 74 A home sewing machine

The linkage consists of frame 0, foot pedal 1 (rocker), connecting rod 2 and crank 3. Press the foot pedal, with the input rocker 1 rocking back and forth, the output crank 3 fully rotating to drive sewing machine by pulley 4.

Example 2. 3 Car Engine

The engine is the heart of a car and it is a very complicated machine. However, the working principle of the engine is pretty simple. A four cylinder four-stroke engine is consisted of four cylinders and each cylinder makes its power by endlessly repeating a series of four strokes. This work is done by a slider-crank linkage (Figure 2. 75).

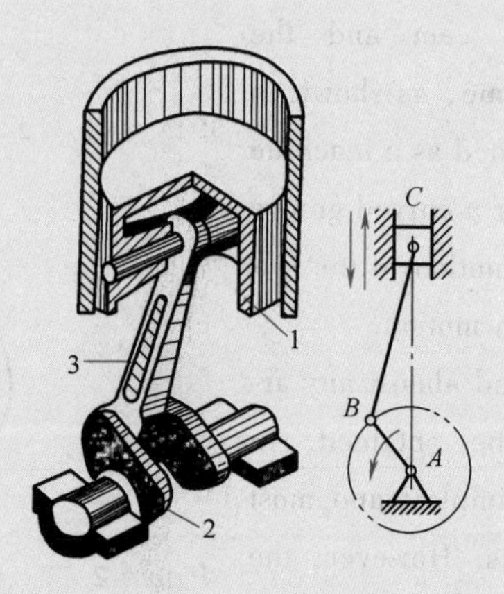

Figure 2. 75 A slider-crank linkage in a car engine

1—piston (slider) 2—crank 3—coupler/connecting rod

Example 2. 4 Wheel Linkage in Steam Locomotive

Steam engines powered most trains from the early 1800s to the 1950s. Though the engines varied in size and complexity, their fundamental operation remained essentially as illustrated in Figure 2. 76. In a steam engine, the boiler (fueled by wood, oil or coal) continuously boils water

in an enclosed chamber, creating high-pressure steam. The steam generated in the boiler flows down into a cylinder just ahead of the wheels, pushing a tight-fitting plunger, the piston, back and forth. A little mechanical gate in the cylinder, known as an inlet valve lets the steam in. The piston is connected to one or more of the locomotive's wheels through a crank and connecting rod, as shown in Figure 2.76. As the piston pushes, the crank and connecting rod turn the locomotive's wheels and power the train along.

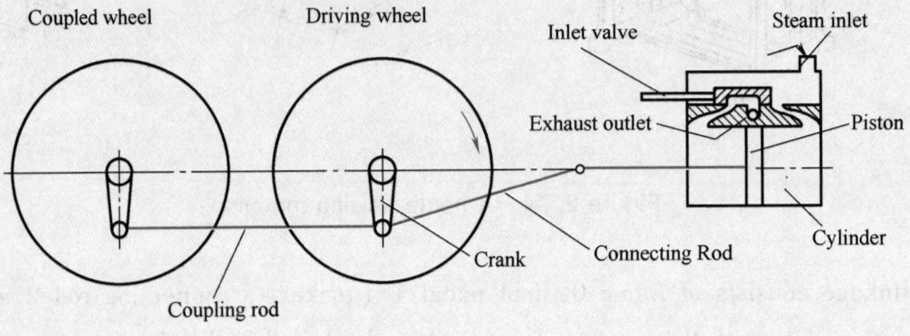

Figure 2.76 The mechanism drives the locomotive's wheels

2.13 Cam Mechanisms

2.13.1 Cam Mechanism and Terminology

The transformation of one of the simple motions, such as rotation, into any other motions is often conveniently accomplished by means of a cam mechanism. A cam mechanism usually consists of two moving elements, the cam and the follower, mounted on a fixed frame, as shown in Figure 2.77. A cam may be defined as a machine element which a curved outline or a curved groove gives a predetermined specified motion to the follower by its oscillation or rotation motion.

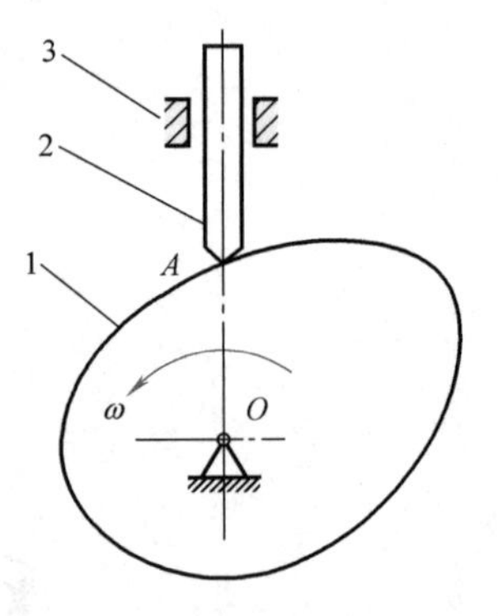

Figure 2.77 Composition of a cam mechanism

1—Cam 2—Follower 3—Frame

Cam devices are versatile and almost any arbitrarily-specified motion can be obtained. In some instances, they offer the simplest and most compact way to transform motions. However, the manufacturing of cams is expensive and the wear effect due to the contact stress is a disadvantage. On the other hand, cams are not proper for the systems with high speed and heavy load.

2.13.2 Classification of Cam Mechanisms

Cam mechanisms can be classified by the modes of input/output motion, the configuration and arrangement of the follower and the shape of the cam, or by different types of motion events of the follower and by means of a great variety of the motion characteristics of the cam profile, as shown in Figure 2. 78.

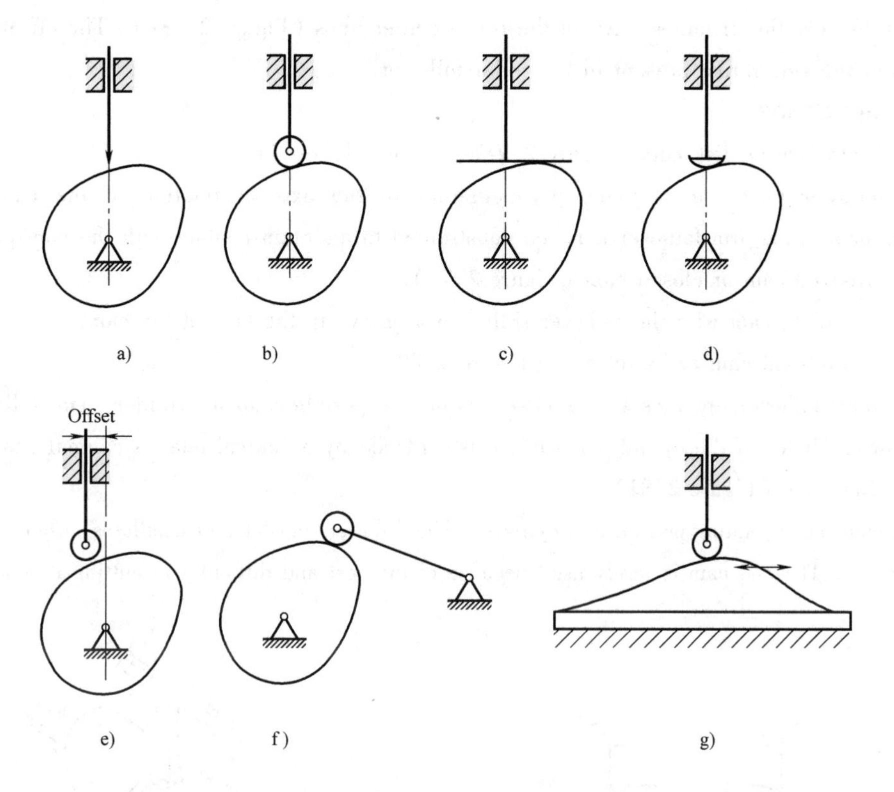

Figure 2. 78 Classification of Cam Mechanisms

1. Modes of Input/Output Motion

(1) Rotating cam with translating follower (Figure 2-78a, b, c, d, e).

(2) Rotating follower (Figure 2-78f).

The follower arm swings or oscillates in a circular arc with respect to the follower pivot.

(3) Translating cam with translating follower (Figure 2-78g).

(4) Stationary cam-rotating follower.

The follower system revolves with respect to the center line of the vertical shaft.

2. Follower Configuration

(1) Knife-edge follower (Figure 2. 78a).

(2) Roller follower (Figure 2. 78b, e, f).

(3) Flat-faced follower (Figure 2. 78c).

(4) Oblique flat-faced follower.

(5) Spherical-faced follower (Figure 2. 78d).

3. Follower Arrangement

(1) In-line follower (Figure 2. 78a, b, c, d).

The center line of the follower passes through the center line of the camshaft.

(2) Offset follower.

The center line of the follower does not pass through the center line of the camshaft. The amount of offset is the distance between these two center lines (Figure 2.78e). The offset causes a reduction of the side thrust present in the roller follower.

4. Cam Shape

(1) Plate cam or disk cam (Figure 2.78a, b, c, d, e, f).

The follower moves in a plane perpendicular to the axis of rotation of the camshaft. A translating or a swing arm follower must be constrained to maintain contact with the cam profile.

(2) Grooved cam or closed cam (Figure 2.79).

This is a plate cam with the follower riding in a groove in the face of the cam.

(3) Cylindrical cam or barrel cam (Figure 2.80).

The roller follower operates in a groove cut on the periphery of a cylinder. The follower may translate or oscillate. If the cylindrical surface is replaced by a conical one, a conical cam results.

(4) End cam (Figure 2.81).

This cam has a rotating portion of a cylinder. The follower translates or oscillates, whereas the cam usually rotates. The end cam is rarely used because of the cost and difficulty in cutting its contour.

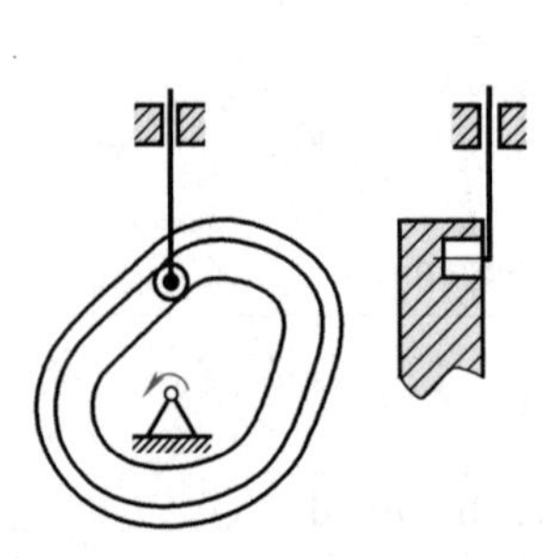

Figure 2.79 Grooved cam

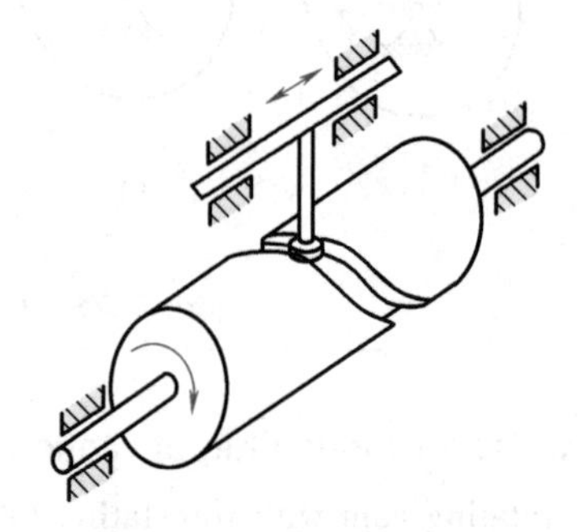

Figure 2.80 Cylindrical cam

5. Constraints on the Follower

(1) Gravity constraint (Figure 2.82a).

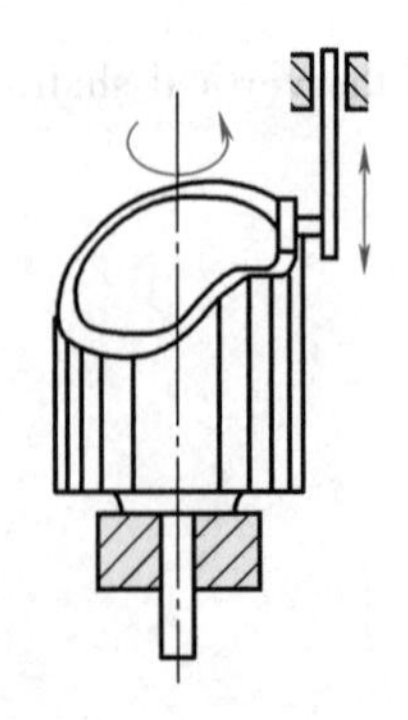

Figure 2.81 End cam

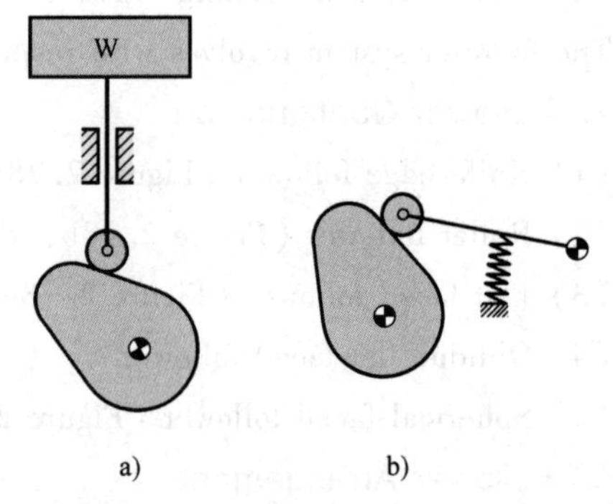

Figure 2.82 Constraints on the follower

a) Weight b) Spring force

The weight of the follower system is sufficient to maintain contact.

(2) Spring constraint (Figure 2. 82b).

The spring must be properly designed to maintain contact.

(3) Positive mechanical constraint.

A groove maintains positive action as shown in Figure 2. 79 and Figure 2. 80. For the cam in Figure 2. 83, the follower has two rollers, separated by a fixed distance, which act as the constraint; the mating cam in such an arrangement is often called a constant-diameter cam.

(4) A dual or conjugate cam.

A mechanical constraint cam can also be introduced by employing a dual or conjugate cam in arrangement similar to what is shown in Figure 2. 84. Each cam has its own roller, but the rollers are mounted on the same reciprocating or oscillating follower.

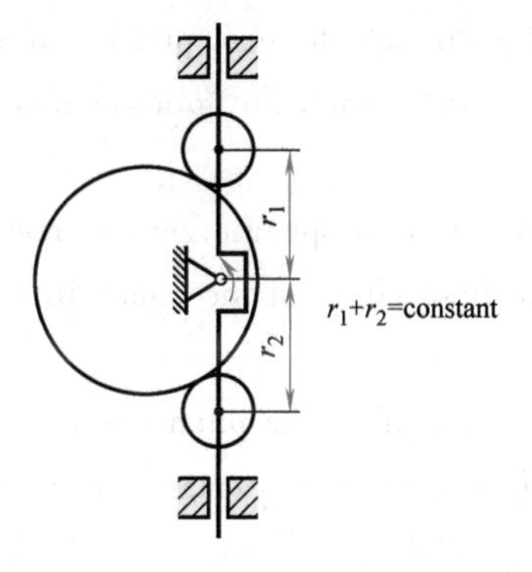

Figure 2. 83 Constant-diameter cam

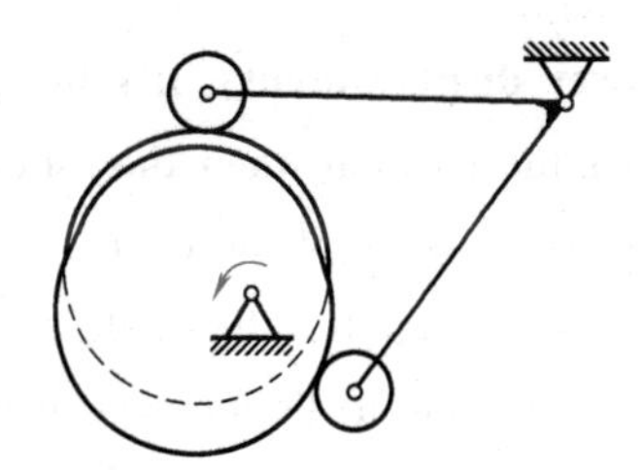

Figure 2. 84 A dual or conjugate cam

2. 13. 3 Cam Nomenclature

Figure 2. 85 illustrates some cam nomenclatures.

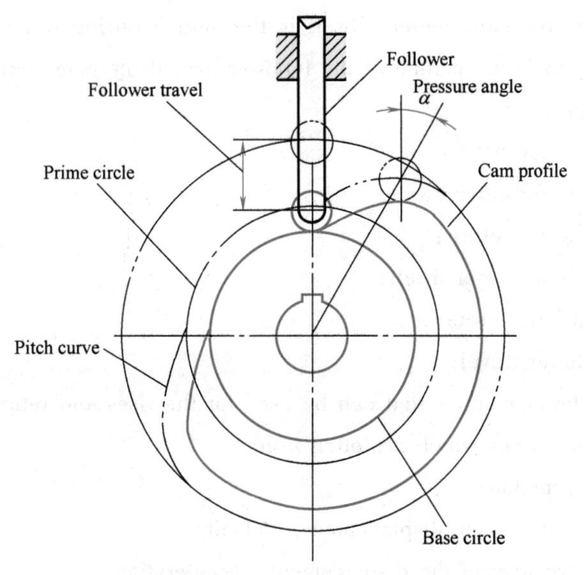

Figure 2. 85 Cam nomenclatures

Trace point: A theoretical point on the follower, corresponding to the point of a fictitious knife-edge follower. It is used to generate the pitch curve. In the case of a roller follower, the trace point is at the center of the roller.

Pitch curve: It's The path generated by the trace point at the follower is rotated about a stationary cam.

Working curve: It's The working surface of a cam in contact with the follower.For the knife-edge follower of the plate cam, the pitch curve and the working curve coincide. In a close or grooved cam there is an inner profile and an outer working curve.

Pitch circle: It's a circle from the cam center through the pitch point.The pitch circle radius is used to calculate a cam of minimum size for a given pressure angle.

Prime circle (reference circle): The smallest circle from the cam center through the pitch curve.

Base circle: It's the smallest circle from the cam center through the cam profile curve.

Stroke or throw: It's the greatest distance or angle through which the follower moves or rotates.

Follower displacement: It's the position of the follower from a specific zero or rest position (usually it's the position when the follower contacts with the base circle of the cam) in relation to time or the rotary angle of the cam.

Pressure angle: It's the angle at any point between the normal to the pitch curve and the instantaneous direction of the follower motion.This angle is important in cam design because it represents the steepness of the cam profile.

2.13.4 Motion Events

When the cam turns through one motion cycle, the follower executes a series of events consisting of rises, dwells and returns, as shown in Figure 2.86. Rise (from A to B) is the motion of the follower away from the cam center; dwell is the motion during which the follower is at rest (from B to C); and return is the motion of the follower toward the cam center (from D to A).

Where φ_t——cam angle;

φ——cam angle for rise;

φ_s——cam angle for outer dwell;

φ'——cam angle for return;

φ'_s——cam angle for inner dwell;

r_0——radius of base circle;

h——total follower travel.

There are many follower motions that can be used for the rises and returns. Figure 2.87 shows a number of basic motion curves which are often used.

s——displacement function;

v——the first derivative of the displacement, velocity;

a——the second derivative of the displacement, acceleration.

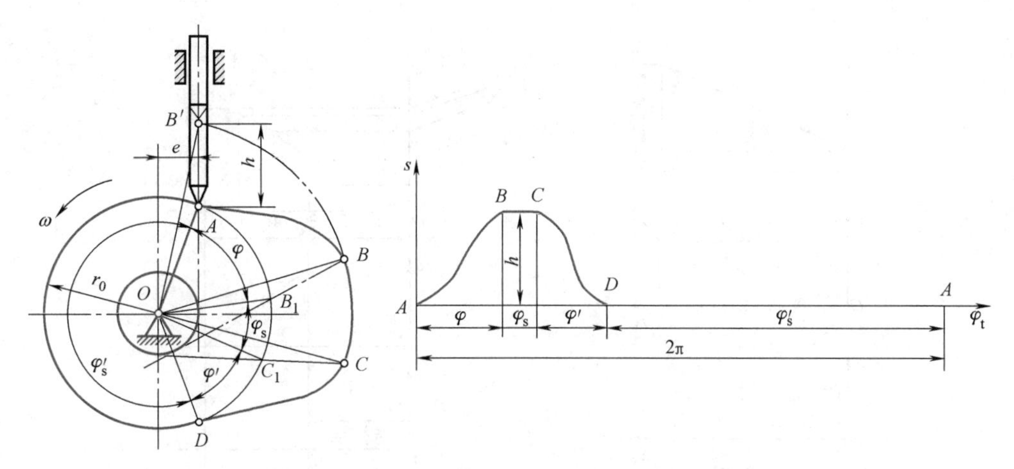

Figure 2. 86 One motion cycle of the follower

1. Constant velocity motion curve

The motion curve and velocity and acceleration curves are as shown in Figure 2. 87a. Note that the acceleration is zero for the entire motion ($a=0$) but it is infinite at the end. Due to infinite acceleration, high inertia forces will be created at the start and at the end even at moderate speed. The cam profile will be discontinuous. These forces are undesirable, especially when the cam rotates at high velocity. The constant velocity motion is therefore only of theoretical interest.

Note that one basic rule in cam design is that this motion curve must be continuous and the first and second derivatives (corresponding to the velocity and acceleration of the follower) must be finite even at the transition points.

2. Constant acceleration and deceleration motion curve

The motion curve and velocity and acceleration curves are as shown in Figure 2. 87b. As it can be seen that the velocity increases at a uniform rate during the first half of the motion and decreases at a uniform rate during the second half of the motion. The acceleration is constant and positive throughout the first half of the motion, and is constant and negative throughout the second half. This type of motion gives the follower the smallest value of maximum acceleration along the path of motion. In high-speed machinery this is particularly important because of the forces that are required to produce the accelerations.

Note that the acceleration curve shows there are acceleration changes at the start, the middle and the end. Due to the changes, soft inertia forces will be created at the start, the end and the middle at high speed. So this curve is suitable for low and moderate speed cams.

3. Cosine acceleration curve (simple harmonic motion curve)

Simple harmonic motion curve is widely used since it is simple to design. The curve is the projection of a circle about the cam rotation axis as shown in Figure 2. 87c .

Note that even though the velocity and acceleration is finite, the maximum acceleration is discontinuous at the start and end of the rise period. Hence the third derivative, jerk, will be infinite at the start and end of the rise portion. This curve isn't suitable for high or moderate speed cams.

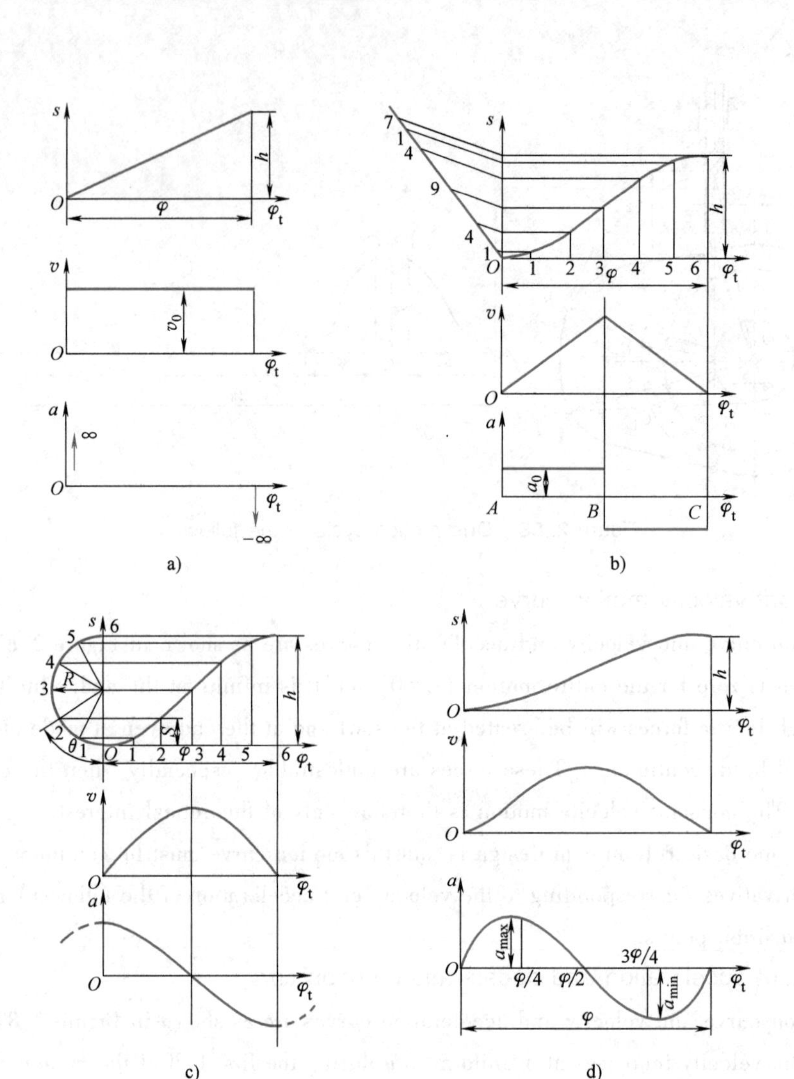

Figure 2. 87 Basic motion curves of the follower

a) Constant velocity motion curve b) Constant acceleration and deceleration motion curve

c) Cosine acceleration curve d) Cycloidal motion curve

4. Cycloidal Motion Curve (Sine acceleration curve)

Cycloidal motion curve has the best dynamic characteristics. The acceleration is finite at all times and the starting and ending acceleration is zero, as shown in Figure 2. 87d . It will yield a cam mechanism with the lowest vibration, stress, noise and shock characteristics. Hence for high speed applications this motion curve is recommended.

2. 13. 5 Applications of Cam Mechanism

Cams are used in many machines. They are numerous in automatic packaging, shoemaking, typesetting machines and so on, but they are often found in machine tools, reciprocating engines and compressors. They are occasionally used in rotating machinery.

A familiar application of cam mechanism is in the opening and closing of valves in an automotive engine, as shown in Figure 2.88.

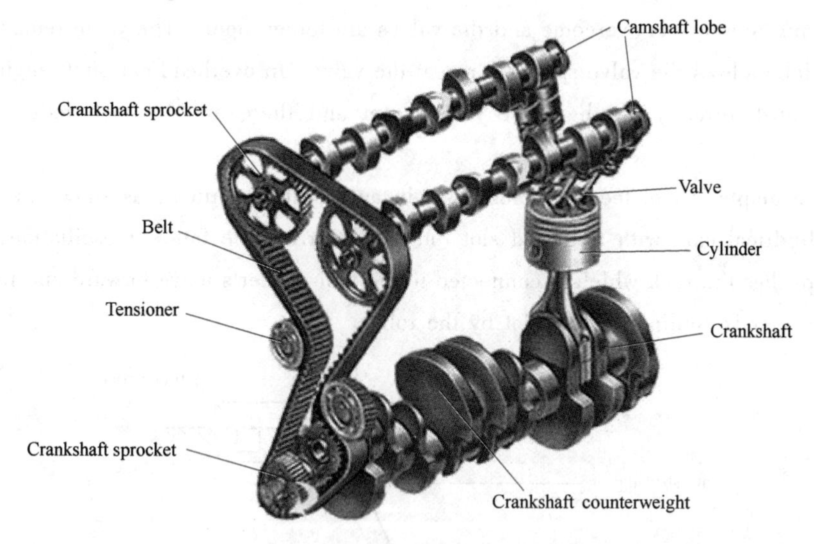

a)

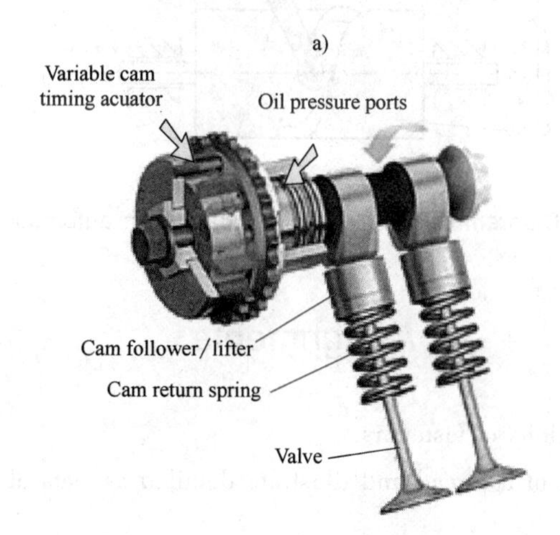

b)

Figure 2.88 Cams in the opening and closing of the valves in an the automotive engine

One or more intake valves are used to control the flow of the air into each cylinder. One or more exhaust valves are used to control the flow of exhaust gases out of each cylinder. Valves are designed to open and close at precise moments to allow the engine to run efficiently at all speeds. Valves also seal the cylinder during the compression and power strokes. Intake and exhaust valves are identical in shape, but intake valves are usually larger. Opening and closing of the valves are controlled by the valve train. Valve train operation is similar in both overhead camshaft and cam-in-block engines. The engine crankshaft turns camshaft via a chain, belt or gear set. The camshaft controls the distance the valves open and the duration of time over when they are open. There is one camshaft lobe for each valve.

In a cam-in-block engine, the camshaft lobes push on valve lifters installed into bores which are machined into the block. The motion is transferred through push rods and rocker arms to the valves. The valve spring pressure is overcome and the valves are forced open. The valve remains open until the camshaft lobe allows the valve spring to reseat the valve. In overhead camshaft engines, the cam lobes usually push directly on the valve rocker arm and there are no push rods, as shown in Figure 2. 88b.

Another example is tool feeding mechanism in automatic machines, as shown in Figure 2. 89. When the cylindrical cam with a curved slot rotates, it drives the follower oscillation, so the fan-shaped gear pushes the rack which is connected to the tool carrier's move forward and backward.One end of the follower is confined in the slot by the roller.

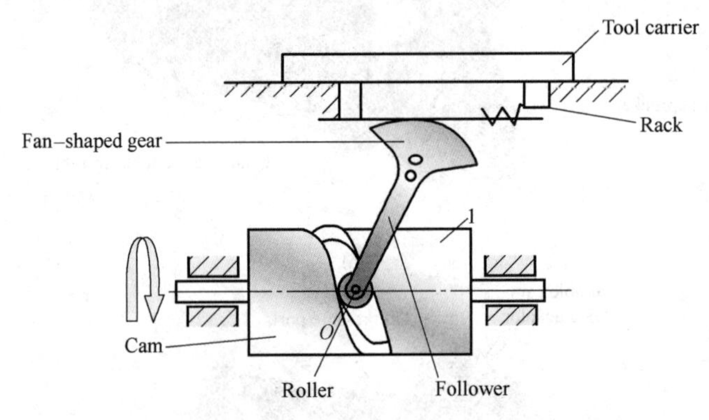

Figure 2. 89 Diagram of tool feeding mechanism in automatic machines

Assignments

2. 1 Describe the functions of fasteners.

2. 2 Prepare a drawing of a thread and illustrate detailed aspects of the thread as well as the special terminology.

2. 3 Which of the bearings in Figure 2. 90:

(a) can support axial loads?

(b) can support combined axial and radial loads?

(c) has the highest load capacity?

(d) has the highest speed capacity?

(e) is the most common?

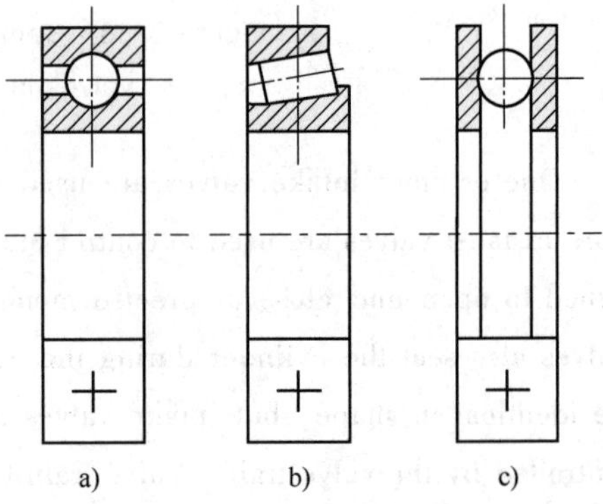

Figure 2. 90 Rolling element bearings

2. 4 A bearing is required for the floating end of a heavy-duty lathe to carry a radial load of up to 9 kN. The shaft diameter is 50 mm and rotates at 3000 r/min. A life of 7500 hours for the bearings is desired. Select and specify an appropriate bearing.

2. 5 Based on your personal experience or observation, describe the functions of belt drives,

chain drives and gear drives, and list their advantages and disadvantages in application.

2.6 The standard bearing number or bearing code indicates the type of the bearing as well as its specification. Write down the typical format of the basic code for bearings, and give two examples to explain the meaning of codes.

2.7 Observe those simple machines around you, select one of them as an example, and try to figure out the mechanism which drives machine working. Make a drawing and share it with your classmates.

2.8 The following demonstration of a differential applies to a "traditional" rear-wheel-drive car (see Figure 2.91). Try to explain how the differential mechanism works using whatever English you know. Write your explanation down, and then work in pairs, A and B. Each of you has to describe your explanation to your partner.

Hints: The following texts will help you with the vocabulary you need.

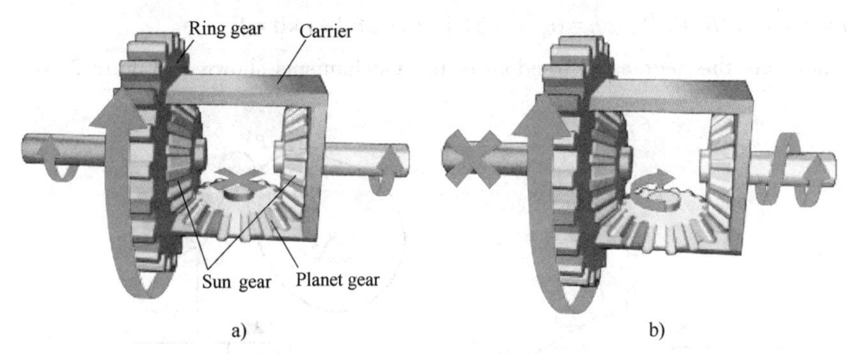

Figure 2.91 A differential mechanism

Input torque is applied to the ring gear, which turns the entire carrier. The carrier is connected to both the sun gears only through the planet gear. Torque is transmitted to the sun gears through the planet gear. The planet gear revolves around the axis of the carrier, driving the sun gears. If the resistance at both wheels is equal, the planet gear revolves without spinning about its own axis, and both wheels turn at the same rate, as shown in Figure 2.91a.

If the left sun gear encounters resistance, the planet gear spins as well as revolving, allowing the left sun gear to slow down, with an equal speeding up of the right sun gear, as shown in Figure 2.91b.

2.9 What are gear ratios? How to calculate gear ratios?

2.10 Figure 2.92 shows compound gears in which two gears are on the middle shaft. Gears B and D rotate at the same speed since they are keyed (fixed) to the same shaft. The number of teeth on each gear is given in the figure. Given these numbers, if gear A rotates at 100 r/min clockwise, figure out the rotating speed of gear B and gear C.

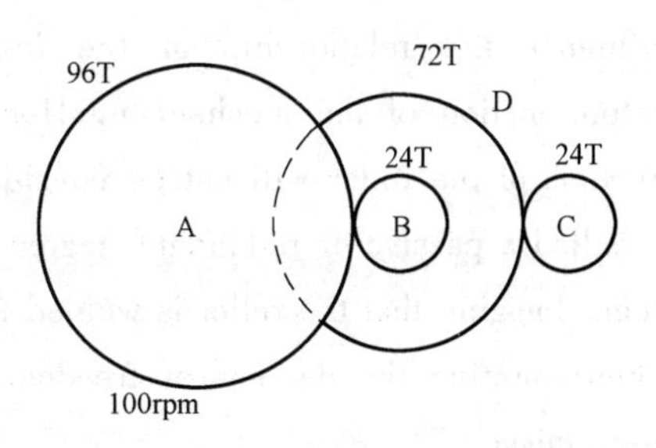

Figure 2.92 Compound gears

2.11 Calculate the degrees of freedom of the mechanisms shown in Figure 2.93.

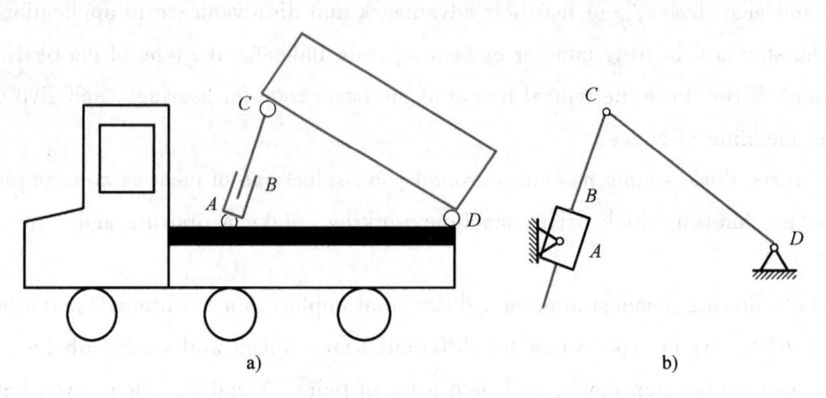

Figure 2. 93 Dump truck

Hints:

$n=4$, $l=4$ (at A,B,C,D), $h=0$; $F=3(4-1)-2\times4-1\times0=1$.

2. 12 Calculate the degrees of freedom of the mechanisms shown in Figure 2. 94.

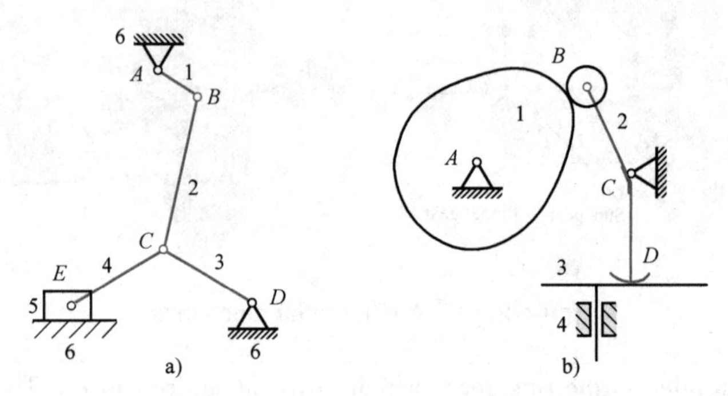

Figure 2. 94 Degrees of freedom calculation

Hints:

a) $n=6$, $l=7$, $h=0$; $F=3\ (6-1)-2\times7-1\times0=1$.

b) $n=4$, $l=3$, $h=2$; $F=3\ (4-1)-2\times3-1\times2=1$.

Note: The rotation of the roller does not influence the relationship of the input and output motion of the mechanism. Hence, the freedom of the roller will not be considered. It is called a passive or redundant degree of freedom. Imagine that the roller is welded to link 2 when counting the degrees of freedom for the mechanism.

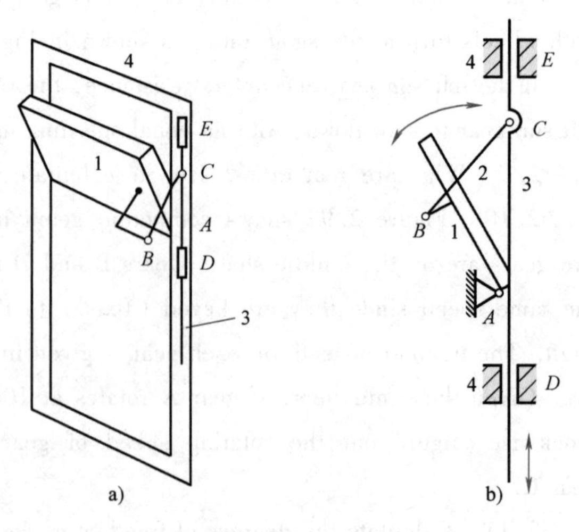

Figure 2. 95 Transom mechanism

2. 13 Look at the transom above the door in Figure 2. 95a. The opening and closing

mechanism is shown in Figure 2.95b. Calculate its degrees of freedom.

$n=4$ (link1, 3, 3 and frame 4), $l=4$ (at A, B, C, D), $h=0$;

$F=3(4-1)-2\times4-1\times0=1$.

Note: D and E function as a same prismatic pair, so they only count as one lower pair.

2.14 What is the dead point in a linkage? Is it useful or not? Give examples and discuss how to overcome the dead point? Work in pairs while answering the above questions.

Hints:

When a side link such as AB in Figure 2.96, becomes aligned with the coupler link BC, it can only be compressed or extended by the coupler. In this configuration, a torque applied to the link on the other side, CD, can't induce rotation in link AB. This link is therefore said to be at a dead point (sometimes called a toggle point).

In Figure 2.96, if AB is a crank, it can become aligned with BC in full extension along the line AB_1C_1 or in flexion with AB_2 folded over B_2C_2. The angle ADC is denoted by ϕ and the angle DAB is denoted by θ. The subscript 1 is to denote the extended state and 2 is to denote the flexed state of links AB and BC. In the extended state, link CD can't rotate clockwise without stretching or compressing the theoretically rigid line AC_1. Therefore, link CD can't move into the forbidden zone below C_1D, and ϕ must be at one of its two extreme positions; in other words, link CD is at an extremum. A second extremum of link CD occurs with $\phi=\phi_1$.

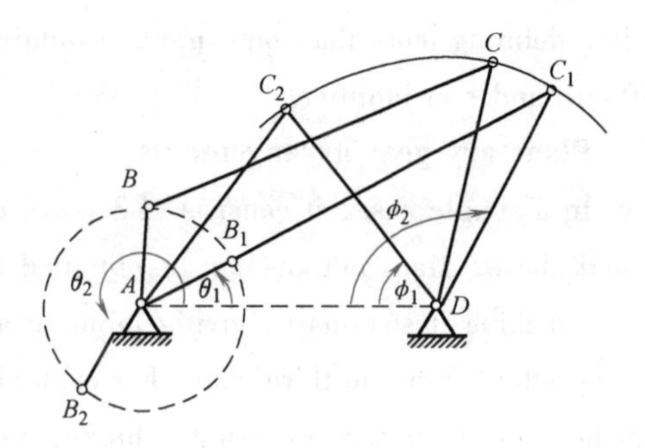

Figure 2.96 Dead point

Note that the extreme positions of a side link occur simultaneously with the dead points of the opposite link.

In some cases, the dead point can be useful for tasks such as work fixturing (Figure 2.97). The dead point can be used to lock the mechanism.

In other cases, dead point can be overcome with the moment of inertia of links (such as the linkage mechanism in a sewing machine) or with the asymmetrical deployment of the mechanism like in V engines as shown in Figure 2.98.

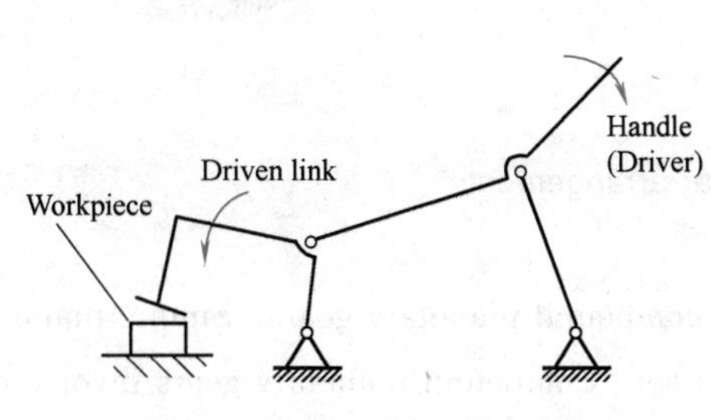

Figure 2.97 Work fixturing

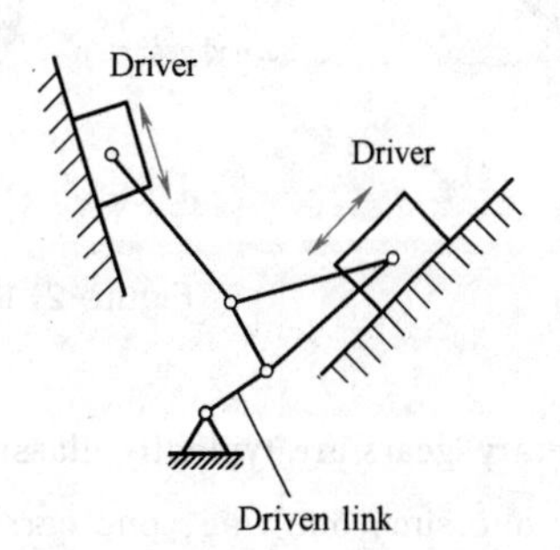

Figure 2.98 Overcoming the dead point by asymmetrical deployment (V engine)

2.15 According to your observation, what products use a cam mechanism If you know one of them, draft a diagram to show how it works and write down your description.

2.16 How does a planetary gear train work and what is the mechanism of transmission? Discuss advantages and disadvantages of planetary gear train over parallel axis gears with your partner.

Hints:

What is a planetary gear train?

A planetary gear train is often suitable when a large torque/speed ratio is required in a compact envelope. It is made up of a number of elements which are interconnected to form the train. Basically, it involves 3 gears: a sun gear, a planet gear and a ring gear, as shown in Figure 2.99. The underlying concept of many gear ratios can be obtained from a small volume as compared to other types of gear trains which take up more space. Unlike simple gear trains, a planetary gear train requires defining more than one input to obtain a specific output, hence making the analysis a little difficult and non-intuitive.

Planetary gear arrangements

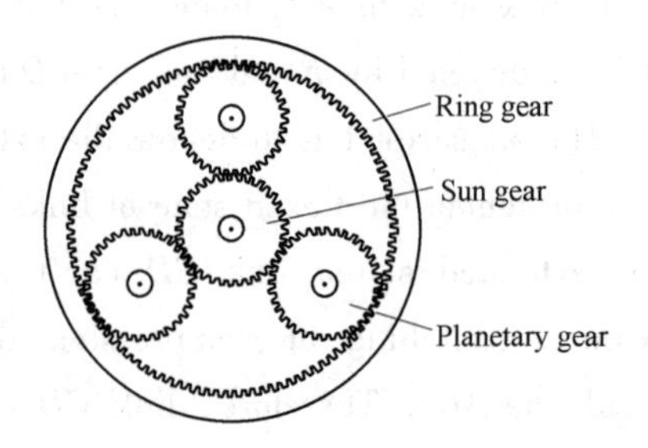

Figure 2.99 A planetary gear train

In a simple case, it consists of 3 types of gears as mentioned above. Many outputs can be obtained by fixing a gear, i.e., making it stationary, giving input to another gear and taking output from the third gear. For example, the ring gear can be made stationary by using a brake, input given to the sun gear, and output taken from the planetary gear by using a spider arrangement. This arrangement is shown in Figure 2.100a. Various arrangements of the sun and planetary gear are possible and used across the industry, as the examples shown in Figure 2.100b, c.

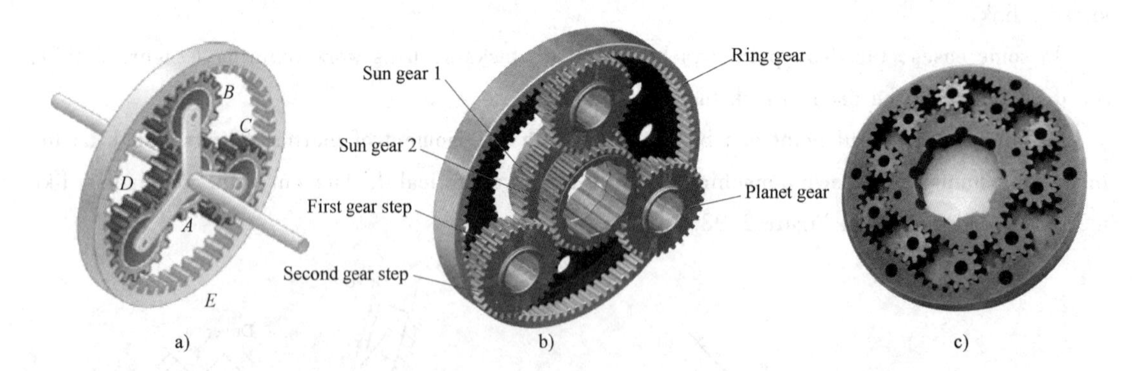

Figure 2.100 Planetary gear arrangements

Planetary gears are typically classified as simple or compound planetary gears. Simple planetary gears have one sun, one ring, one carrier and one planet set. Compound planetary gears involve one or more of the following three types of structures: meshed-planet (there are at least two more planets in mesh with each other in each planet train), stepped-planet (there exists a shaft connection between two planets in each planet train), and multi-stage structures (the system contains two or more planet sets). Compared to simple planetary gears, compound planetary gears have the advantages of larger reduction ratio, higher torque-to-weight ratio and more flexible con figurations.

Typically, the planet gears are mounted on a movable arm or carrier which itself may rotate relative to the sun gear. In many planetary gearing systems, one of these three basic components is held stationary; one of the two remaining components is an input, providing power to the system, while the last component is an output, receiving power from the system. The possible combinations are as follows:

1) The ring gear is held stationary while the sun gear is used as input (Figure 2. 101a). The planet gears are driven gears and they turn in a ratio determined by the number of teeth in each gear in the same direction as the sun gear. This configuration will produce an increase in gear ratio and reduction in speed.

2) The ring gear is held stationary while the planet gear carrier is used as input (Figure 2. 101b). The sun gear is the driven gear and turns in a ratio determined by the number of teeth in each gear in the same direction as the planet gears. This configuration will produce a decrease in gear ratio and increase in speed.

3) The sun gear is held stationary while the ring gear is used as input (Figure 2. 101c). The carrier is the driven part; it turns in a ratio determined by the number of teeth in each gear in the same direction as the ring gear. This configuration will produce an increase in gear ratio and reduction in speed.

4) The sun gear is held stationary while the planet gear carrier is used as input (Figure 2. 101d). The ring gear is the driven gear and turns in a ratio determined by the number of teeth in each gear in the same direction as the planetary gears. This configuration will produce a decrease in gear ratio and increase in speed.

5) The carrier is held stationary while the sun gear is used as input (Figure 2. 101e). The ring gear is the driven gear; it turns in a ratio determined by the number of teeth in each gear in the opposite direction as the sun gear. This configuration will produce an increase in gear ratio and reduction in speed.

6) The carrier is held stationary while the ring gear is used as input (Figure 2. 101f). The sun gear is the driven gear; it turns in a ratio determined by the number of teeth in each gear in the opposite direction as the ring gear. This configuration will produce a decrease in gear ratio and increase in speed.

In addition, two of these three basic components can be fixed together and used as an input, while the last component is an output.

Advantages and disadvantages of a planetary gear train

Advantages of planetary gears over parallel axis gears include high power density, large reduction in a small volume, multiple kinematic combinations, pure torsional reactions and coaxial shafting.

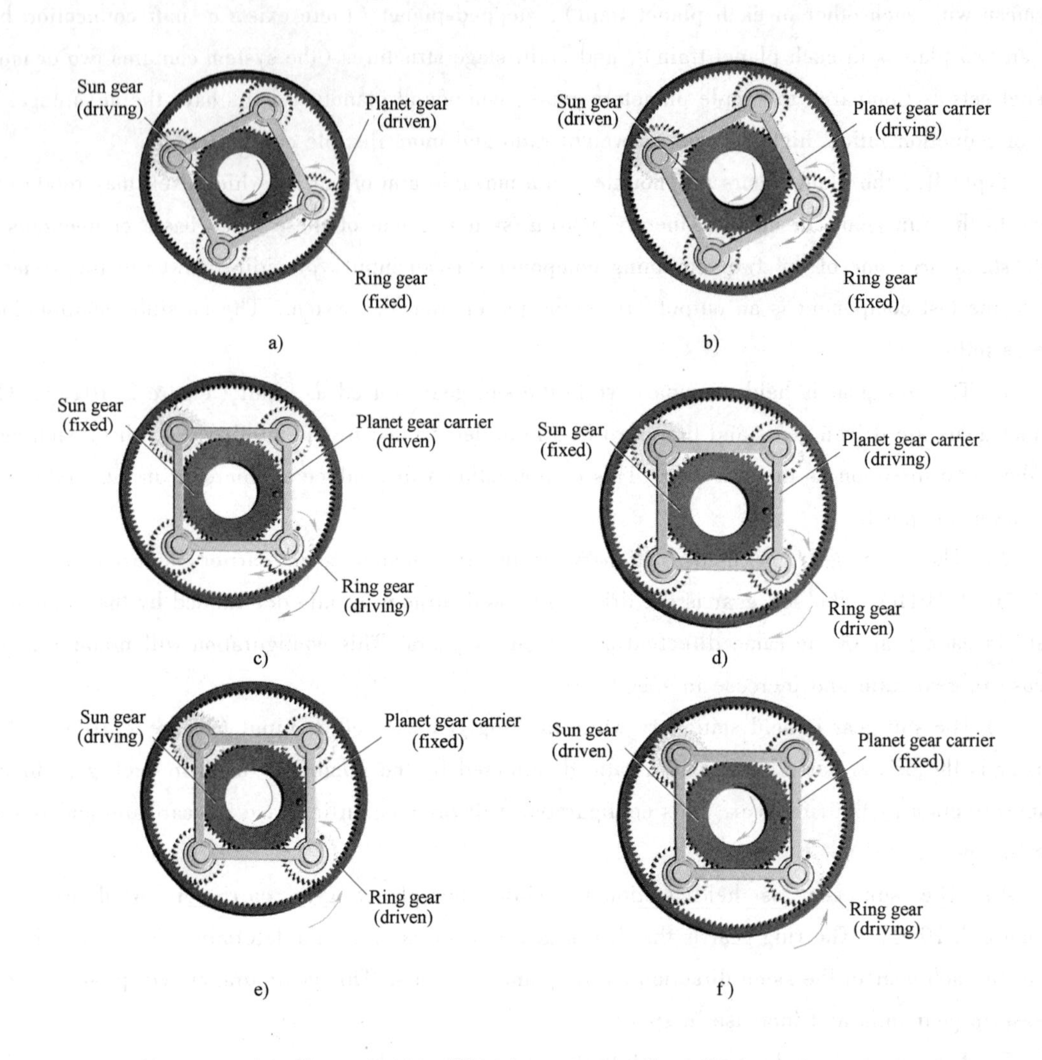

Figure 2. 101 Planetary gear configurations

The planetary gearbox arrangement is an engineering design that offers many advantages over traditional gearbox arrangements. One advantage is its unique combination of both compactness and outstanding power transmission efficiency. A typical efficiency loss in a planetary gearbox arrangement is only 3% per stage. This type of efficiency ensures that a high proportion of the energy as the input into the gearbox is multiplied and transmitted into torque, rather than being wasted on mechanical loss inside the gearbox. Another advantage of the planetary gear box arrangement is load distribution. Because the load transmitted is shared between multiple planets, torque capability is greatly increased. The more planets there are in the system, the greater load ability and the higher the torque density the system has. The planetary gearbox arrangement also creates great stability (because it's a balanced system) and increased rotational stiffness.

Disadvantages of planetary gear train include high bearing loads, inaccessibility and design complexity.

相关词汇注解

1 bearing *n*. 轴承
2 axle *n*. 轴
3 key *n*. 键
4 fastener *n*. 连接件，坚固件
5 seal *n*. 密封（材料）
6 lubricant *n*. 润滑油（剂）
7 involute gear 渐开线齿轮
8 plain bearing 滑动轴承
9 leadscrew/powerscrew/lead screw/power screw 丝杠，导向螺栓
10 inclined plane 斜面
11 bushing *n*. 衬套/轴套
12 clutch *n*. 离合器
13 brake *n*. 制动
14 belt drive 带传动
15 thread *n*. 螺纹
16 terminology *n*. 术语
17 metric thread 米制螺纹
18 acme thread 梯形螺纹
19 square thread 矩形螺纹
20 buttress thread 锯齿形螺纹
21 non-equilateral trapezium 非等边梯形
22 pipe thread 管螺纹
23 friction *n*. 摩擦
24 stiction *n*. 静摩擦
25 self-locking *adj*. 自动锁定的
26 ball screw 滚珠丝杠
27 trapezoidal thread 梯形螺纹
28 high-efficiency feed screw 高效进给丝杠
29 preload *n*. 预紧力
30 stud *n*. 双头螺柱
31 nut *n*. 螺母
32 hexagonal *adj*. 六边的，六角形的
33 washer *n*. 垫片
34 screwdriver *n*. 螺钉旋具
35 wrench *n*. 扳手
36 socket *n*. 插座，凹槽

37 stainless steel 不锈钢
38 titanium *n.* 钛
39 brass *n.* 黄铜
40 bronze *n.* 青铜
41 monel *n.* 铜-镍合金
42 silicon bronze 硅铜
43 electrical insulation 绝缘体（材料）
44 corrosion *n.* 腐蚀
45 coarse pitch 粗牙
46 fine pitch 细牙
47 pin *n.* 销
48 rectangular cross-section 矩形断面（截面）
49 longitudinal *adj.* 长度的，纵向的，沿长度方向的
50 slot *n.* 槽
51 hub *n.* 毂
52 parallel key 平键
53 semicircular key 半圆键
54 tapered key 钩头型楔键
55 gib *n.* 拉紧销
56 tab *n.* 伴扣
57 taper pin 圆锥销
58 parallel pin 圆柱销
59 split pin 开口销
60 riveted joint 铆钉连接
61 carbon steel 碳钢
62 aluminum *n.* 铝
63 tungsten inert gas welding（TIG） 钨极惰性气体保护焊
64 metal inert gas welding（MIG） 熔化极惰性气体保护焊
65 manual metal arc welding（MMA） 焊条电弧焊
66 submerged arc welding（SAW） 埋弧焊
67 resistance welding 电阻焊
68 gas welding 气焊
69 thermoplastics 热塑性塑料
70 spring *n.* 弹簧
71 helical compression spring 螺旋压缩弹簧
72 helical extension spring 螺旋拉伸弹簧
73 torsion spring 扭转弹簧
74 spiral torsion spring 平面涡卷弹簧
75 leaf spring 钢板弹簧

76 belleville spring 碟形弹簧
77 flat spring 平板弹簧
78 journal bearing 滑动轴承
79 deep groove ball bearing 深沟球轴承
80 bearing performance 轴承性能
81 rolling contact bearing 滚动轴承
82 hydraulic and pneumatic actuator 液压气动传动装置（执行机构）
83 coupling *n.* 联轴器
84 plain shaft 光轴
85 stepped shaft 阶梯轴
86 crankshaft *n.* 曲轴
87 pulley wheel 滑轮
88 shaft-hub connection 轴-毂连接
89 cam-shaft *n.* 凸轮轴
90 rigid coupling 刚性联轴器
91 flexible coupling 弹性联轴器
92 universal joint 万向联轴器
93 angular velocity 角速度
94 sprocket *n.* 链轮
95 flat belt drive 平带传动
96 round belt drive 圆带传动
97 V belt drive V 带传动
98 synchronous belt drive 同步带传动
99 chain drive 链传动
100 conveyor chain 输送链
101 roller chain 滚子链
102 leaf chain 板式链
103 silent chain 无声链
104 spur gear 直齿圆柱齿轮
105 helical gear 斜齿圆柱齿轮
106 internal gear 内齿轮
107 bevel gear 锥齿轮
108 cylindrical worm gearing 圆柱蜗杆
109 worm and worm wheel 蜗杆蜗轮
110 planetary/epicyclic train 行星轮系
111 sun gear 太阳轮
112 a planet carrier 行星齿轮架
113 planet gear 行星齿轮
114 ring gear 环形齿轮，齿圈

115　polytetrafluoroethylene（PTFE）　*n*. 聚四氟乙烯

116　lockwire　*n*. 止推垫圈

117　Overhead camshaft　顶置式凸轮轴（指发动机中配气机构的配置形式，将往复式内燃机气缸盖的凸轮轴放在气缸顶上，即置于活塞和燃烧室之上，从而直接对气阀进行打开和关闭）

118　Cam-in-block engine　指凸轮轴置于气缸体内的发动机

119　viscosity　*n*. 黏度，黏性

Part3

Communication Skills in Mechanical Engineering

Objective

After completing this chapter, you will be able to:

- understand that it is very important for engineers to know how to communicate well with others both orally and written. Be familiar with various ways of giving written technical reports, oral reports and an engineering presentation;
- understand the field of mechanical engineering from an international perspective;
- expand your cross-cultural communication and problem-solving skills;
- better prepare yourself to work in an increasingly diverse and international workplace;
- improve your English language skills.

3.1 Background of Cross-Cultural Communication

In the 20th century, "culture" emerged as a concept central to anthropology, encompassing all human phenomena that are not purely the results of human genetics. Specifically, the term "culture" in American anthropology has two meanings: ① the evolved human capacity to classify and represent experiences with symbols, and to act imaginatively and creatively, and ② the distinct ways that people living in different parts of the world classified and represented their experiences, and acted creatively. It can be seen that culture is the full range of learned human behavior. Culture includes arts, beliefs, customs, inventions, language, technology and traditions.

Cross-cultural communication (also frequently referred to as intercultural communication) is a field of study that looks at how people from differing cultural backgrounds communicate, in similar and different ways among themselves, and how they endeavor to communicate across cultures.

We communicate with others all the time—in our homes, in our workplaces, in the groups we belong to and in the community. No matter how well we think and understand each other, communication is hard. Just think, for example, how often we hear things like, "He doesn't get it," or "She didn't really hear what I meant to say." "Culture" is often at the root of communication challenges. Our culture influences how we deal with problems, and how we participate in groups and communities. When we participate in groups, we are often surprised at how differently people approach their work.

As anyone who works internationally can tell you, intercultural communication is not always a smooth ride. When different cultures come together in a business setting, their differences can often cause confusion, misunderstanding, mistake and the like. These intercultural differences can be anything from contrasting approaches to communication, etiquette, meeting styles or body language.

The modern business world demands that people from all corners of the earth communicate with one another. A manager in the USA may have staff in Germany, India and China; the importer in France may have associates in Turkey, Italy and Japan. Now people are increasingly communicating across intercultural lines.

The connection between culture and language has been noted as far back as the classical period and probably long before. Some anthropologists maintain that the shared language of a community is the most essential carrier of their common culture. Franz Boas is a German-American anthropologist and a pioneer of modern anthropology. Boas is the first anthropologist who considers it unimaginable to study the culture of a foreign people without also becoming acquainted with their language. For Boas, the fact that the culture of a group is largely constructed, shared and maintained through the use of language, means that understanding the language of a group is the key to understand its culture.

In the past decade, there has been an increasing pressure for universities across the world to incorporate intercultural and international understanding and knowledge into the education of their students. International literacy and cross-cultural understanding have become critical to a country's cultural, technological, economic and political health. Students must possess a certain level of global competence to understand the world they live in and how they fit into this world.

3.2 Cross-Cultural Communication for Mechanical Engineers

3.2.1 Challenges for Mechanical Engineers

Engineers are responsible for some of the greatest inventions and technologies the world depends on. Mechanical engineers specialize in the design and planning of machines, tools and other mechanical devices. Many products from bridges to air conditioning systems to space shuttles require the work of mechanical engineers.

In a global economy environment, many employers compete for business overseas, have multinational operations, and work through overseas partners. Mechanical product realization is often an international team effort, in which a manufacturing company might design a product in the U. S., modify it for assembly in Europe, use overseas contractors and suppliers, or set up and run a plant in China. Even if you do not work overseas, it's entirely possible that you will someday deal with international clients.

One might think that a mechanical engineer does not need strong communication skills. As a matter of fact every mechanical engineer needs to have a strong grip over his communication skills in order to meet the following challenges in modern mechanical engineering.

(1) Mechanical engineers are problem solvers. These problems are not purely technical problems—mechanical engineers deal with management requirements, unique customer needs, budgetary and legal constraints, environmental and social issues, as well as changes in technology. Engineers have to stay abreast of emerging technologies. Therefore, aside from being mechanically adept, mechanical engineers need to have a number of effective communication skills to perform their job correctly and to be successful in the field of engineering.

(2) Currently, the Global Engineering Services industry is becoming one of the professional service industries in China. The global market for engineering services consists of the work performed by consultant engineering firms along with the in-house services undertaken by construction contractors, designers, manufacturers, government agencies and utility owners. Among them, the mechanical engineering services occupy a large proportion.

(3) Nowadays there are many multinational companies in China. Engineers in multinational companies need to acquire cross-cultural communication skills to perform successfully in overseas assignments. Today companies are doing business in a global environment. The people that count in any business from the suppliers to clients to employees are increasingly based in remote locations in foreign countries. The need for effective and clear intercultural communication is becoming vital in securing success in today's global workplace.

There is no doubt about the fact that technical knowledge about machines is an absolute essential for an engineer to perform his job diligently and efficiently. Communication skills are also essential for the growth of an engineer in the organization. A blend of both is what determines your level of success as a professional.

3.2.2 Challenges for Students

It is critical that engineering students develop and enhance basic communication skills, i. e. reading, writing, speaking as well as cross-cultural communication skills. These skills build a necessary foundation that influences and shapes the types of engineers that institutions of higher learning produce. The importance of these communication skills, often referred to as "soft skills", are downplayed in higher education in favor of an emphasis on the technical "hard skills". This attitude and practice engenders beliefs and attitudes that the "soft skills" are secondary, even unimportant. As a result, engineering students are often equipped with technical knowledge, but lack of the "soft skills", leaves them under-prepared for the real world.

Engineers are problem solvers. Once they have obtained a solution to a problem, they need to communicate effectively their solution to various people inside or outside their organization. Presentations are an integral part of any engineering project. As an engineering student, you would be asked to present your solution to assigned homework problems, write a technical report, or give an oral presentation to your class, student organization, or the audience at a student conference. Later, as an engineer, you could be asked to give presentations to your boss, colleagues in your design group, sales and marketing people, the public, or to an outside customer. Depending on the size of the project, the presentations might be brief, lengthy, frequent, and may follow a certain format. You might be asked to give detailed technical reports or presentations containing graphs, charts and engineering drawings, or they may take the form of brief project updates.

The above mentioned challenges will accompany with the students all the time, even when they apply for a job, these "invisible" challenges still test their ability. The importance of "soft skills" in business and industry is also emphasized in the checklist of the characteristics which the company wants in their employees when considering new applicants. The following is a checklist example of the Boeing Company. It can be seen that 60% of items in the checklist are related to "soft skills", and good communication skills (induding written, verbal, graphic, listening) are one of highlights in the checklist.

Employer's Checklist—Boeing Company

√ A good grasp of these engineering fundamentals:

Mathematics (including statistics)

Physical and life sciences

Information technology

√A good understanding of the design and manufacturing process (i. e. and understanding of engineering)

√A basic understanding of the context in which engineering is practiced, including:

Economics and business practice

History

The environment

Customer and social needs

√ A multidisciplinary systems perspective

√ Good communication skills

Written

Verbal

Graphic

Listening

√ High ethical standards

√ An ability to think critically and creatively as well as independently and cooperatively

√ Flexibility—an ability and the self-confidence to adapt to rapid/major change

√ Curiosity and a lifelong desire to learn

√ A profound understanding of the importance of teamwork

√ An awareness of the boundaries of one's knowledge, along with an appreciation for other areas of knowledge and their interrelatedness with one's expertise

√ An awareness and strong appreciation for other cultures and their diversity, their distinctiveness, and their inherent value

√ A strong commitment to teamwork, including extensive experience and understanding of team dynamics

√ An ability to impart knowledge to others

It can be seen that communication skills could become an item on your list of "lifelong learning" objectives. A number of U. S. engineering schools participate in exchange programs with universities in America, Europe, Asia, Africa, etc. Students who participate in these programs find that language communication skills and international experiences distinguish them from other engineering graduates and job candidates. Later on, engineers with this background have a wider choice of assignments.

3.2.3 Technical Communication Skills for Mechanical Engineers

As an engineering student, you need to develop good written and oral communication skills. During the 4 years study in university, you will learn how to express your thoughts, present a concept for a product or a service, analyze a problem and its solution, or show your findings from experimental work. Moreover, you will learn how to communicate design ideas by means of engineering drawings or computer-aided modeling techniques. Starting right now, it is important to understand that the ability to communicate your solution to a problem is as important as the solution itself. You may spend several months on a project, but if you can't effectively communicate to others, the results of all your efforts may not be understood or appreciated. Most engineers are required to write reports. These reports might be lengthy and detailed technical reports containing charts, graphs and engineering drawings, or they may take the form of a brief memorandum or executive summary.

Under the background of globalization, the term "cross-cultural communication" is often used to refer to the wide range of communication issues that inevitably arise within an organization composed of individuals from a variety of religious, social, ethnic and technical backgrounds. Each of these individuals brings a unique set of experiences and values to the workplace and many of the experiences and values can be traced to the culture in which they grew up and now operate. Therefore a large number of communication skills should be involved in cross-cultural communication.

It's impossible to cover all topics related to cross-cultural communication in one section. The aim of this section focuses on providing a guide of technical communication skills for students in mechanical engineering. Four general types of communication skills are considered and discussed:

1. Problem finding

Problem finding means problem discovery. Problem finding is the creative ability to define or identify and frame a problem. It is a part of the problem process that includes problem shaping and problem solving. Building solutions for most problems is the easy part; the hard part is finding the right problem. For students homework is the backbone of higher education in technical communication. Students should have a better understanding of problems in assignments. A problem is half-solved if properly interpreted and stated.

2. Written technical reports

Skills for writing good reports are essential for effective communication. In today's information overload world, it's vital to communicate clearly, concisely and effectively. Readers don't have time to read book-length reports and they don't have the patience to scour badly constructed reports for finding main points. Therefore, the better your professional writing skills are, the better impression you'll make on the people around you—including your teachers, classmates, bosses, colleagues and clients. Students should cultivate writing skills by writing formal reports based on writing assignments given in many classes. These reports may vary in content and format; they may be involved in design reports in project classes, lab reports in experimental classes, and term project or research reports in other subjects.

3. Oral reports

If you are a non-native speaker of English, you may find it more challenging to speak extempore in English than in your native language. Still, even imperfect extemporaneous English is more likely to engage the audience than reciting a more polished, less spontaneous written text if you make oral reports in conjunction with visual media (such as a PowerPoint presentation) in a professional way. To improve your delivery and overall presentation as a non-native speaker, practice more and support your spoken discourse with appropriate slides.

4. Slide presentation of technical material

To make a good graphical presentation which involves techniques of graphing data, measurements, and equations: choose coordinates, labels and symbols as well as engineering drawings of components or systems that make research work clear.

3.3 Problem Finding

The following example of homework is cited from the MIT Open Courseware. Imagine you were a student at MIT, how would you solve the following problems clearly and correctly, think about this and then attempt to give solutions. Meanwhile, by studying the following problems, learn some skills to find, shape, formulate and describe problems in mechanical engineering. In addition, reference solutions to the following problems are available on the multimedia tutorial CD.

2007 Design and Manufacturing 1

Homework#1 Design Process & Machine Elements

NAME: ______________________________

Date Issued: Tuesday 10 FEB, 11AM

Date Due: Tuesday 24 FEB, 11AM

Please answer the following 3 questions showing your work to the extent possible within the allotted time. Point allocations are listed for each question. The points sum up to 100.

As described in the course policies document, this is one of four assignments you will complete in this course. Each assignment counts as 5% of your total grade. You will submit your work via the course web site.

(1) (50 total) It is proposed to develop a bicycle including a new feature.

A) (10 points) Make an annotated sketch of a standard bicycle. A side view could suffice. Ensure that the ways in which parts are interconnected are accurate, the proportions are roughly correct, and key dimensions are labeled so that your functional understanding of the bicycle is evident.

B) (20 points) Now consider a new feature. The bicycle should enable the rider to store energy from pedaling or going down hill and later use that energy to aid in either acceleration from a stand still or else in hill climbing. Develop, at the conceptual level, three very different ways of implementing this function. Simple abstract descriptions and sketches will be enough.

C) (20 points) Begin to set up a Pugh chart for selection among your alternatives from B. List at least three criteria by which you might choose among them. Choose a datum concept. Then fill out a single row, rating the two concepts against the datum ("+" indicates superior to, "-" indicates inferior to, "S" indicates essentially the same as the datum). That would imply you need to make just two relative assessments. Below your Pugh chart, provide a few sentences of text, sketches, equations and so on that defend each of these ratings. Be quantitative in your exposition to the extent possible.

(2) (20 total) Consider a capstan such as the one as depicted below (see Figure 3.1).

A) (5 points) How much tension, T, must be applied to raise the weight at a constant rate?

B) (5 points) How much tension, T, must be applied to lower the weight at a constant rate?

C) (5points) How much tension, T, must be applied to hold the weight at a constant position?

D) (5 points) Explain the differences in scenarios A, B and C in a style that a bright ten-year-old would understand.

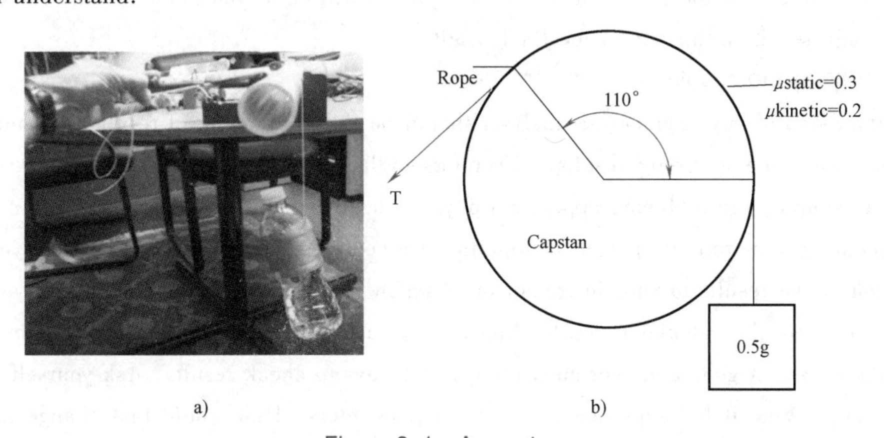

Figure 3.1 A capstan

From the above study of the case, it can be seen there are four basic steps that must be followed when finding and solving a mechanical engineering problem: ① defining the problem, ② simplifying the problem by assumptions and estimations, ③ performing the solution or analysis, and ④ verifying the results.

1. Defining the problem

Before you can obtain an appropriate solution to a problem, you must first thoroughly understand the problem itself. There are many questions that you need to ask before proceeding to determine a solution. What is it exactly that you want to design or analyze? What do you really know about the problem, or what are some of the things known about the problem? What are you looking for? What exactly are you trying to find a solution to?

Taking time to understand the problem completely at the beginning will save lots of time later and help to avoid a great deal of frustration. Once you understand the problem, you should be able to divide any given problem into two basic questions: What is known? And what is to be found?

2. Simplifying the problem

Before you can proceed with the analysis of the problem, you may first need to simplify it and make some assumptions and estimations.

Once you have a good understanding of the problem, you should then ask yourself this question: Can I simplify the problem by making some reasonable and logical assumptions and yet obtain an appropriate solution? Understanding the physical and fundamental concepts, as well as where and when to apply them and their limitations, will benefit you greatly in making assumptions and solving the problem.

3. Performing the solution or analysis

Once you have carefully studied the problem, you can proceed with obtaining an appropriate solution. You will begin by applying the physical laws and fundamental concepts that govern the behavior of mechanical engineering systems to solve the problem. Among the engineering tools in your toolbox you will find mathematical tools and CAD tools. It is always a good practice to set up the problem in parametric form, that is, in terms of the variables involved. You should wait until the very end to substitute for the given values. This approach will allow you to change the value of a given variable and see its influence on the final result.

4. Verifying the results

The final step of any engineering analysis should be the verification of results. Various sources of error can contribute to wrong results. Misunderstanding a given problem, making incorrect assumptions to simplify the problem, applying a physical law that does not truly fit the given problem, and incorporating inappropriate physical properties are common sources of error. Before you present your solution or the results to your instructor or, later in your career, to your manager, you need to learn to think about the calculated result. You need to ask yourself the following question: Do the results make sense? A good engineer must always find ways to check results. Ask yourself this additional question: What if I change one of the given parameters? How would that change the result? Then consider if the outcome seems reasonable. If you formulate the problem such that the final result is left in parametric form, then you can experiment by substituting different values for various parameters and look at the final result.

3.4 Written Technical Reports

Contrary to current layman thinking that engineering works only involve working with machines, many tasks performed by an engineer involve writing. Written communication, in fact, is an integral part of mechanical engineering tasks. The ability to write a technical report in a clear and concise manner is a mark of a good engineer. An engineer must be able to translate the formulae, numbers and other engineering abstractions into an understandable written form.

There is no fixed format with written technical reports. Each group, institution or company may have its own standard format to follow. There are certain elements common in most engineering writings. These elements can be seen in any typical engineering report. A technical report must inform readers of the problems, reasons, means, results and conclusions of the subject matter being reported.

For students many mechanical engineering writing is centered on reporting of experimental works and reporting of a project. In such cases, lab reports and project reports are necessary to assess students'academic performance.

3. 4. 1 Lab Reports

Experimental work is fundamental in developing the understanding of the theoretical knowledge in mechanical engineering students. It also gives students practical experience in the use of equipment and the experimental techniques in the field.

A lab report is a detailed account of an experiment, its methods, results and conclusions which answer a question. Although there are different kinds of lab report formats, there are certain elements common in most lab writings.

Lab reports should follow a set format and structure. This section explains the tasks involved in doing a lab report and covers some important points students should keep in mind. It outlines the structure of this kind of report and what is required in each section.

The following template, including the title page, is a guideline to students for writing good experimental lab reports in mechanical engineering lab courses.

1. Title page

Title Page of Lab Report

(2 points)

Course Number and Name:	
Semester and Year:	
Name of Team Leader:	Name of Lab Instructor:
Lab Section and Meeting Time:	Experiment Number:
Title of Experiment:	
Date of Experiment Performed:	Instructor Comments:
Date of Report Submitted:	
Names of Group Members:	Grade:

2. Abstract (10 points)

[Primary Contributor: ________; Secondary Contributors: ________]

The experiment is usually performed by teamwork, therefore the name of contributors should be placed here in order to value their performances based on their contributions.

Place your abstract here.

An abstract is a brief summary of the report. It appears as a single paragraph. The Abstract usually includes the following key elements:

- What is the objective of the experiment?
- Why is this objective significant?
- What type of experiment is performed to achieve the objective?
- What are the major results of the experiment?
- Conclusion, including brief evaluation and/or recommendation.

From the abstract, the reader can make an informed decision as to whether or not the lab report applies to what they are researching or what they are interested in. The reader can then decide whether or not to devote more time to reading the full report.

3. Objective and Introduction (5 points)

[Primary Contributor: ________; Secondary Contributors: ________]

Place your objective and introduction here.

State the objective clearly in a complete sentence. A few explanatory sentences may be included, if needed. The objective should answer the question: What is the experiment designed to determine? The introduction must start in a separate paragraph; it provides explanation of the engineering problem. It explains the significance as well as any significant background information of the problem.

4. Theory and Experimental Methods (20 points)

[Primary Contributor: ________; Secondary Contributors: ________]

Place your theory and experimental methods here.

The theory should explain all equations, theoretical principles and assumptions that are used in the experiment and analysis. The primary purpose of the theory section is to show how the raw data is manipulated to become results. Relevant equations used are to be presented and described to illustrate their basis and origin. This section should include diagrams where needed. Define all variables used in the equations.

The experimental methods should give a detailed description in your own words of how you accomplish the experimentation. This should include equipment used in the experiment as well as how it is used. The description should have sufficient detail so that another experimenter could duplicate your efforts.

Use sketches, diagrams or photos, to describe the experimental set-up. Label the main components. Provide dimensions and materials of test samples where applicable.

The equipment listing in the appendices is the appropriate place for model numbers and serial numbers. In this section, it is adequate to use generic names for the equipment, e.g. the fluid network apparatus.

5. Results and Discussion (40 points)

[Primary Contributor: ________; Secondary Contributors: ________]

Place your results and discussion here.

This is the heart of your report. Summarize your results in the introductory sentence. Relate your results to your objective. Present the results in the easiest way for your reader to understand: graphs, tables, figures, etc. Spreadsheets are often an ideal tool for organizing and analyzing the data, and generating graphs and tables. All tables and figures should be accompanied by comments or discussions in the text of report; use a numbering system for identification of each one. All figures and tables must have numbers and captions. While the table captions should be placed over the table, figure captions should be placed below the figure.

Calculations and formulas are not presented in this section. Your calculations should be detailed in the appendices under sample calculations. Formulas should be discussed in the theory section.

Explain the results of the experiment, for instance, comment on the shapes of the curves, compare obtained results with expected results, give probable reasons for discrepancies from the theory, answer any questions outlined in the instructions and solve any problems that may have been presented. Experimental errors should be discussed here. When discussing sources of error in experimental measurement, be quantitative: estimate or calculate percentage of error explained by measurement.

6. Conclusion (10 points)

[Primary Contributor: ________; Secondary Contributors: ________]

Place your conclusions here.

State your discoveries, judgments and opinions from the results of this experiment. Make recommendations for further study. Suggest ways to improve the results of this experiment.

7. References (3 Points)

Place your references here.

Itemize any books, publication or websites that you referenced in compiling your report. Provide authors, publisher, date of publication, page number, etc.

Refer to the following format for typing a reference.

[1] Little P and Cardenas M. Use of Studio Methods in the Introductory Engineering Design Curriculum, Journal of Engineering Education, 2001, 90 (3): 309-318.

[2] Nunally J. Psychometric Theory 2nd edition. New York: McGraw-Hill, 1978.

[3] Lister B. Next Generation Studio: A New Model for Interactive Learning. www.ciue.rpi.edu/pdfs/nextGenStudio.pdf.

8. Appendices (5 points)

[Primary Contributor: ________; Secondary Contributors: ________]

A. Data Tables

Place your data tables here. Data tables are for the convenience of the extremely interested reader. These tables may contain any additional comparisons or calculations that you have prepared. Results may contain only summaries of your work. Data tables are the place to show everything that you have done.

B. Sample Calculations

Place your sample calculations here. Demonstrate how you have performed the calculations made in the experiment. Include tabular results of computations where such are made. Show the generic calculations to support all your work. Provide any computer or calculator program listings, along with sample input and output. Use equation writer in Microsoft Word or neatly write the equations by hand.

C. Equipment List

Place your equipment list here. Every piece of equipment used in the experiment should be given. Specimens are not equipment.

D. Raw Data Sheets

Attach your raw data sheets to the end of the report. Data sheets must be completed in ink and signed by the instructor at the completion of the laboratory period.

In the case of an error, simply run a line through the mistake and continue. The name of the recorder and the group members should be indicated on the raw data sheets.

9. Participation Information (5 points)

In this section information on the contributions of each team member must be summarized.

This section and section 10 "Peer and Self Evaluations" are optional sections; some lab reports may not include these two items. It depends on different requirements from lab instructors.

10. Peer and Self Evaluations

For all group reports, each student must fill out the peer and self evaluation and submit it to the lab instructor separately from the report. The following is an example of the peer and self evaluation form for lab experiments.

Note: This form must be filled out by each student and submitted to the lab instructor separately from the lab report.

Student Name and Signature: ________________

Course Name: ________________

Date of Experiment: ________________

A. Self Evaluation

a. Rate your overall contribution to this project (5 = key contributions, 1 = little contributions)

5 4 3 2 1

b. Explain briefly what you contributed to this project:

____

____

____

____

B. Peer Evaluation

Group member	Student's name	Rate the contribution (5 = key contributions, 1 = little contributions)	Describe the contribution made during the experiment and in writing the lab report
1		5 4 3 2 1	
2		5 4 3 2 1	
3		5 4 3 2 1	

Comments: ________________

3.4.2 Project Reports

In America, each student will participate in a teamwork on the project professionally, and this is the most important element of the class. The projects are usually paid for by an outside group that is truly committed to the project. This commitment gives the project financial resources as well as a customer. Even if there is no financial commitment, each project should have a customer that is external to the class. It means if the final project is successful, a potential customer will pay for the project.

Typical projects are design task, where the team may be asked to develop a new product, design and build a portion of a new manufacturing process cell, or fabricate a special machine designed for a specific task.

Therefore, a project usually involves in a sponsor who pays for the project, an advisor who guides students to do the project and students who are responsible for the project. Before asking the sponsor to commit to the project, it is important to remind sponsors that not all student projects are successful, and sponsors must be willing to take the sponsorship as a potential investment.

Such projects encompass the entire design process — ideation through to functional prototype build and evaluation. Teams of four or more students work on one sponsored project. The project usually lasts two semesters. These projects are designed as training modules for students, and because the projects involve designing real products, students are treated as engineers, not students.

First semester ends with a formal project review, in which each team demonstrates their design progress to date and gains approval for their plan to complete the project during the second semester. Second semester concludes with a project-ending presentation before industry representatives, faculty members and peers.

Except for a project-ending presentation, students must write up a report of the project. The report includes the progress report and the final report. The progress report is presented by the end of first semester; it focuses on gaining approval for their plan of project. Then students can move onto implementation of the plan during the second semester. By the end of second semester students are expected to finish their project. Each sponsor and advisor will expect a finished, documented project completed to their expectations with a final report.

The project report is the written record of the entire project from start to finish. When read by a person unfamiliar with the project, the report should be clear and detailed enough for the reader to know exactly what you have done, why you do it, what the results are, whether or not the results support your hypothesis, and where you have got your research information. This written document is your "spokesperson" when you are not present to explain your project, but more than that, it documents all your work.

Therefore, such kind of report is a formal technical report. The final written report usually includes the following sections, as shown in Table 3.1.

As we know, all students have to write written reports during 4 year university study in China, for example, writing a graduation thesis (or a graduation report), a design report in a course or a

written report presented for a certain competition, etc. Although the example in this section is about how to write a lab or a project report in regard to the situation in American university, its format, method, contents, and so forth, can be used as a reference when writing a technical written report in English.

Table 3.1 The contents of a written report

Report sections		Explanation
Title page: the title, the names of investigators and institutional affiliation, the name and number of the course, a date and the name of sponsor and advisor as well as their titles.		
Abstract:		
Table of contents		
Introduction	1. Background of the project 2. Design objectives and constraints 3. Weight user requirements	1. Clearly define and document the design goals to make sure that the team is working towards a common goal 2. Establish the various requirements: functional, performance, interface, environmental, etc 3. Refine requirements by conducting trade 4. Capture quantitative constraints which can be used to validate product design
Methods	State methods for functional descriptions and concept designs as well as detailed design	This section would typically answer the following questions: 1. What assumptions are needed? 2. How is the problem modeled? 3. What analysis is used?
Validation and verification	This is the part which proves that the design will work as it should	1. Assure design meets the objectives 2. Verify the design against the requirements
Results and discussion	Briefly restate the problem, your approach and results as well as the result evaluation	1. Evaluate your results: be quantitative 2. State how your results apply to your original objective 3. Suggest how the model might be improved; make recommendations for further work
References: shows where you have got information that is not your own.		
Appendices—Raw data, sample calculations, etc.		

3.5 Oral Reports

An oral report involves face-to-face, more informal spoken communication where you only have time to inform the audience of the key aspects of your work. Your oral presentation may show the results of all your efforts regarding the project that you may have spent months or a year to develop. If the listener can't follow how a product is designed, or how the analysis is performed, then all your efforts become insignificant. It is very important that all information should be conveyed in a manner easily understood by the listener.

As we know, during the 4 years of studying in university, students have to give oral reports on different occasions. For example, each undergraduate student or postgraduate student needs to give a graduation report. It is important to develop skills and confidence in giving oral presentations.

The format of an oral technical presentation is similar to that of a written report. The key features of a good formal oral presentation are: ① a clear logical structure, ② effective visual aids, and ③ good delivery techniques. Many students feel nervous about speaking inpublic, but oral presentations can be successful with good planning, systematic structuring and preparing as well as good presenting, as suggested in the following tables.

(1) Planning

What should you plan?	How do you plan for it?
Analyze your audience	1.What do they know about the subject? What terminology would they know? 2.What do they want or need to know? What is their motivation for listening to you? 3.What aspects of your subject would they be interested in? 4.How much information can they absorb? If they are new to the topic, their level of absorption may be lower than for an expert audience
Determine primary purpose	1.What is your main point? (Can you put it into one sentence?) 2.What do you want your listeners to do or think? (Are you trying to inform/convince/guide/entertain them?)
Select effective supporting information	1.What kind of information will best support your presentation? 2.What kind of information will appeal to your listeners? Provide examples 3.Listeners may only remember two or three supporting points

(2) Structuring and preparing

Structuring	Preparing
Introduction — state briefly and clearly your purpose to give presentation	When your introduction is over, your audience should be interested, know what your main point is, and know how you're going to explain it.Therefore, your introduction should: 1.arouse interest in the topic 2.provide context, i.e.background and definitions 3.clearly state the main point of the topic 4.describe the structure of the topic
Body—give the full details	The body of your presentation is where you provide the actual information, details and evidence to support your main idea or topic.Therefore, the following general rules should be observed 1.Usually there will be several sections in the body, each corresponding to one of the main points in your outline 2.This is where you develop more deeply into your argument, providing clear evidence and relevant examples
Conclusion — give a short summary or conclusion regarding the significance of the work	1.State your conclusion as clearly as possible 2.Summarize your evidence for each conclusion 3.Discuss the theoretical implications of your work, as well as any possible practical applications

(3) Presenting

Practice before delivery	On presentation day	Present
1. Practice your presentation aloud (it will less help if you just say it in your head) 2. Don't read it. Use cue cards with simple dot points on which you elaborate orally 3. Let the audience know what's coming: let them know the structure of your talk, use linking words between sections 4. Audience attention span is short, so break up long sections of information with questions, feedback, activities, and repeat important points 5. Do not rush. Speak more slowly and clearly than you normally would. Provide extra emphasis through intonation and body language 6. Be aware of body language: avoid annoying habits such as talking with your hands in your pockets, slouching, etc	1. Arrive early 2. Check projection and voice equipment 3. Bring a backup of your presentation 4. Expect the unexpected: How will you deal with audience questions? What if you can't answer the question? How will you respond to criticism? What if the audience misunderstand what you say? 5. Time your presentation using the equipment	1. Nervousness could be overcome by knowing your content and practicing it! 2. Pause between points. Emphasize key ideas/information 3. Establish contact with the audience — talk with them before your presentation 4. Walk purposefully and confidently to the front of the lecture room 5. Remember, the purpose of oral presentations is to communicate a topic as interestingly and succinctly as possible, so be expressive and concise

3.6 Slide Presentation of Technical Materials

Engineers use engineering drawings to convey their ideas and design information about products. Therefore, technical presentations in mechanical engineering are different from other presentations in regard to humanities. Mechanical presentations usually contain charts, graphs and different kinds of engineering drawings. These technical materials should be organized, well prepared and get right to the point. Microsoft PowerPoint presentation is used commonly to give an attractive and effective presentation. Engineers have to be familiar with creating a PowerPoint presentation.

This section is primarily intended for improving presentation skills and providing helpful hints on the visual display of data. It's important to note that there is no easy shortcut to make a good presentation. Generally, if you spend time in dedicated preparation and keeping in mind any unforeseen circumstances that may arise, you will be a good presenter.

A skillful presenter usually organizes the presentation into three parts: introduction (background), main body and conclusions. In most cases, graphical presentations are always performed through oral reports. Therefore, the format of a graphical presentation is similar to that of an oral report. What you say is what you should present to the audience.

When making PowerPoint presentation, the following guidelines should be heeded.

(1) Use visual aids to support the presentation, wherever relevant and feasible.

(2) Minimize text: Don't crowd your slides with a lot of text. Especially, avoid using complete sentences or worse, complete paragraphs. Either the audience will become engrossed in

trying to read the text, and will stop paying attention to you, or else they'll wonder why you don't just give them a handout already and save yourself the trouble of reading to them. Therefore, try to use keywords to represent a sentence or the main idea of a paragraph.

(3) Avoid setting potentially annoying animation when slides are played, or else it will distract audience's attention. Keep them simple.

(4) Title all charts, tables and diagrams in order to clarify the purpose of your slides.

(5) A picture is worth a thousand words; try to use pictures and engineering drawings when making slides.

(6) Avoid using more complicated 2D drawings to convey information. If it's unavoidable, try to use local enlarged views or 3D CAD drawings, or extract some useful information related to your main points to design simple and effective graphics.

(7) Use contrasting colors: If you want your audience to be able to see what you have on the slide, a lot of contrast is needed between the text color and the background color. A dark background with light text or light background with dark colors is appropriate. Most projectors make colors duller than they appear on a screen, and it should be checked how the colors look when projected to make sure there is still enough contrast. Always prepare simple slides with suitable color combination.

(8) Use a big enough font: Font size should be big enough so that the audience can read it. The normal text should be at least 22 font. Some prefer the text at a 28 or 32 size, with titles at 36 to 44 point size.

(9) At the end of your presentation, summarize your main points and give a strong concluding remark that reinforces why your information is of value.

3.7 Case Study-Comprehensive Training

践行“六个必须坚持”，以问题为导向，本节给出了37个小型案例分析。

In previous sections, some guidelines related to communication skills in mechanical engineering are provided. In this section, 37 cases are given in order to guide you to have a series of hands-on comprehensive training in improving your communication skills. These cases demonstrate how to express ideas when you meet with different kinds of topics in mechanical engineering. The cases have a wide coverage; they include basic skills in writing, ways of linking ideas, skills in regard to reading and speaking, ways of describing shapes, features, product quality, safety issues at work, working principles of machines or mechanisms etc., and engineering materials are also included. Finally several cases related to patent, trademark, copyright as well as standards and codes are presented.

Case 1. Basic Language Study

What is the link between column A and column B?

A	B
mechanical	Machines
electrical	electricity

Column A lists a branch of engineering or a type of engineer. Column B lists things they are concerned with. We can show the link between them in a number of ways:

1. Mechanical engineering deals with machines.
2. Mechanical engineers deal with machines.
3. Mechanical engineering is concerned with machines
4. Mechanical engineers are concerned with machines.
5. Machines are the concern of mechanical engineers.

Match each item in column A with an appropriate item from column B and link the two in a sentence.

A	B
1. marine	a. air-conditioning
2. aeronautical	b. roads and bridges
3. heating and ventilating	c. body scanners
4. electricity generating	d. cables and switch gear
5. automobile	e. communications and equipment
6. civil	f. ships
7. electronic	g. planes
8. electrical installation	h. cars and trucks
9. medical	i. power stations

Match the verbs from the text (1~10) to the definitions (a~j).

1. adjust	a. carry out planned maintenance
2. drain	b. change an old or damaged part
3. disconnect	c. set up carefully by making small change
4. dismantle	d. empty a liquid
5. examine	e. add more fluid to fit a tank to the recommended level
6. replace	f. check carefully
7. reconnect	g. take apart assembled component
8. service	h. apply the correct torque, for example to loose bolt
9. tighten	I. establish a connection again
10. top up	j. remove or isolate from a circuit or network

Case 2. Ways to Link a Cause and an Effect

Task 1

Study these actions. What is the relationship between them?

1. A load is placed on the platform.
2. The load cell bends very slightly.
3. The strain gauge is stretched.
4. The electrical resistance increases.

In each case, the first action is the cause and the second action is the effect. We can link a cause and effect like this:

1+2 A load is placed on the platform, which causes the load cell to bend very slightly.

3+4 The strain gauge is stretched, which causes the electrical resistance to increase.

In these examples, both the cause and the effect are clauses—they contain a subject and a verb. Study the following example.

Cause: The strain gauge is stretched.

Effect: An increase in electrical resistance.

The effect is a noun phrase. We can link cause and effect like this:

The strain gauge is stretched, which causes an increase in electrical resistance.

The diagram below (see Figure 3.2) is a cause and effect chain which explains how a strain gauge works. Each arrow shows a cause and effect link. Match these actions with the correct boxes in the diagram.

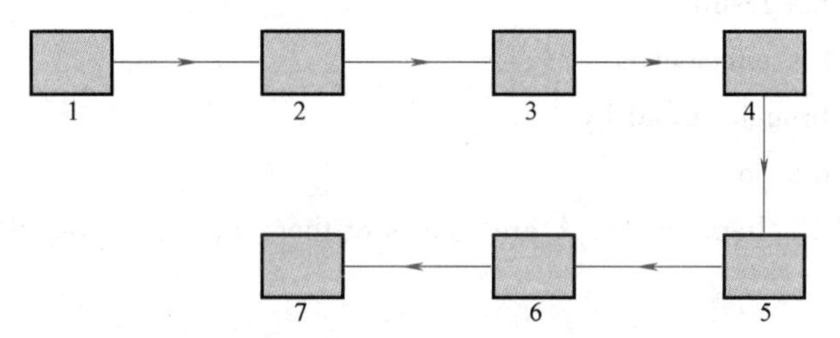

Figure 3.2 A cause and effect chain

1. An increase in resistance.
2. A load is placed on the scale.
3. A drop in voltage across the gauge.
4. The load cell bends very slightly.
5. They become longer and thinner.
6. The strain gauge conductors stretch.
7. The strain gauge bends.

Now practise linking each pair of actions, i. e. 1+2, 2+3, and so on.

Task 2

In this exercise we will study other ways to link a cause and an effect.

What connection can you see between the followings?

corrosion

loss of strength

dampness

reduction in cross-section

Put them in the correct order to show this connection.

Cause and effect links like these are common in engineering explanations. You can link a cause and effect when both are nouns or noun phrases, like this:

1. If you want to put the cause first.

Cause	Effect
Dampness	corrosion

causes

results in
gives rise to
brings about
leads to

2. If you want to put the effect first.

Effect	Cause
Dampness	corrosion

is caused by
results from
is the result of
is the effect of is
is brought about by
is due to

Study lists A and B. Items in list A are causes of those in list B, but the items are mixed up. Link the related items.

A	B
1. reduction in cross-section	a. corrosion
2. insulation breakdown	b. bearing failure
3. over tightening	c. excessive heat
4. overloading a circuit	d. shearing in metal
5. carelessness	e. loss of strength
6. impurities	f. shearing in bolts
7. lack of lubrication	g. blown fuses
8. friction	h. short circuits
9. repeated bending	i. accidents
10. overrunning an electric motor	j. wear and tear in machinery

Now write sentences to show the link. For example:

A	B
reduction in cross-section	loss of strength

The link: Loss of strength results from reduction in cross-section.

Case 3. Ways of Linking Ideas

When we write, we may have to describe, explain, argue, persuade, complain, etc. In all these forms of writing, we use ideas. To make our writing effective, we have to make sure our readers can follow our ideas. One way of helping our readers is to make the links between the ideas in our writing.

What are the links between these pairs of ideas? What words can we use to mark the links?

1. Mechanisms are important to us.
2. They allow us to travel.

3. Mechanisms deliver the power to do work.
4. They play a vital role in industry.
5. Friction is sometimes a help.
6. It is often a hindrance.

Sentence 2 is a reason for sentence 1. We can link 1 and 2 like this:

Mechanisms are important to us **because/since/as** they allow us to travel.

Sentence 4 is the result of sentence 3. We can link 3 and 4 like this:

Mechanisms deliver the power to do work **so** they play a vital role in industry.

Mechanisms deliver thepower to do work; **therefore** they play a vital role in industry.

Sentence 6 contrasts with sentence 5. We can link 5 and 6 like this:

Friction is sometimes a help **but** it is often a hindrance.

Show the links between these sets of ideas using appropriate linking words.

1. Copper is highly conductive.
   It is used for electric wiring.
2. Weight is measured in Newton.
   Mass is measured in kilogram.
3. Nylon is used for bearings.
   It is self-lubricating.
4. ABS has high impact strength.
   It is used for safety helmets.
5. The foot pump is a class 2 lever.
   The load is between the effort and the fulcrum.
6. Friction is essential in brakes.
   Friction is a nuisance in an engine.

Case 4. Reading and Scanning a Text

Hints: Scanning is the best strategy for searching for specific information in a text. Move your eyes up and down the text until you find the word or words you want. Again, try to ignore any information which will not help you with your task.

Now read the following text to find the answers to the questions.

Mechanisms

Mechanisms are an important part of everyday life. They allow us to do simple things like switch on lights, turn taps and open doors. They also make it possible to use escalators and lifts, travel in cars and fly from continent to continent.

Mechanisms play a vital role in industry. While many industrial processes have electronic control systems, it is still mechanisms that deliver the power to do the work. They provide the forces to press steel sheets into car body panels, to lift large components from place to place, to force plastic through dies to make pipes.

All mechanisms involve some kind of motion. The four basic kinds of motion are as follows.

Rotary: Wheels, gears and rollers involve rotary movement.

Oscillating: The pendulum of a clock oscillates—it swings backwards and forwards.

Linear: The linear movement of a paper trimmer is used to cut the edge of the paper.

Reciprocating: The piston in a combustion engine reciprocates.

Many mechanisms involve changing one kind of motion into another type. For example, the reciprocating motion of a piston is changed into a rotary motion by the crankshaft, while a cam converts the rotary motion of the engine into the reciprocating motion required to operate the valves.

1. What does a cam do?
2. What does oscillating mean?
3. How are plastic pipes formed?
4. What simple mechanisms in the home are mentioned directly or indirectly?
5. What is the function of a crankshaft?
6. Give an example of a device which can produce a linear movement.
7. How are car body panels formed?
8. What do mechanisms provide in industry?

Case 5. Describing Jobs in Engineering

Task 1

List some of the jobs in engineering. Combine your list with others in your group.

Task 2

Work in groups of three, A, B and C. Scan your section of this text, A, B or C. How many of the jobs in the combined list you made in Task 1 are mentioned in your section?

Hints: The following text will help you with the vocabulary you need.

Jobs in engineering

Part A

Professional engineers may work as follows.

Design engineers: They work as part of a team to create new products and extend the life of old products by updating them and finding new applications for them. Their aim is to build quality and reliability into the design and to introduce new components and materials to make the product cheaper, lighter or stronger.

Installation engineers: They work on the customer's premises to install equipment produced by their company.

Production engineers: They ensure that the production process is efficient, materials are handled safely and correctly, and faults in the production are corrected. The design and development department consult with them to ensure that any innovations proposed are practicable and cost-effective.

Part B

Just below the professional engineers are the *technician engineers*. They require a detailed knowledge of a particular technology—electrical, mechanical, electronic, etc. They may lead teams of engineering technicians. Technician engineers and engineering technicians may work as follows.

Test/Laboratory technicians: They test samples of the materials and products to ensure that the quality is maintained.

Installation and service technicians: They ensure that equipment sold by the company is installed correctly and carry out preventative maintenance and essential repairs.

Production planning and control technicians: They produce the manufacturing instructions and organize the work of production so that it can be done as quickly, cheaply and efficiently as possible.

Inspection technicians: They check and ensure that incoming and outgoing components and products meet specifications.

Debug technicians: They find fault, repair and test equipment and products down to component level.

Part C

Draftsmen/women and designers: They produce the drawings and design documents from which the product is manufactured. The next grade is craftsmen/women. Their work is highly skilled and practical. Craftsmen and women may work as:

Toolmakers: They make dies and moulding tools which are used to punch and form metal components and produce plastic components such as car bumpers.

Fitters: They assemble components into larger products.

Maintenance fitters: They repair machinery.

Welders: They do specialized joining, fabricating and repair work.

Electricians: They wire and install electrical equipment.

Operators: require fewer skills. Many operator jobs consist mainly of minding a machine, especially now that more and more processes are automated. However, some operators may have to check components produced by their machines to ensure they are accurate. They may require training in the use of instruments such as micrometers, verniers, or simple "go/no go" gauges.

Task 3

Combine answers with the others in your group. How many of the jobs listed in Task 1 are mentioned in the whole text in task 2?

Who would be employed to:

1. test completed motors from a production line?
2. find out why a new electronics assembly does not work?
3. produce a mould for a car body part?
4. see that the correct test equipment is available on a production line?
5. find a cheaper way of manufacturing a crankshaft?
6. repair heating systems installed by their company?
7. see that a new product is safe to use?
8. commission a turbine in a power station?

Task 4

Speaking practice: Role play

Work in pairs, A and B. Each of you has profiles of three workers in a light engineering plant which supplies car electrical components such as starter motors, fuel pumps and alternators.

Play the part of one of these workers and be prepared to answer questions from your partner about your work. Your partner must try to identify your job from your replies.

In turn, find out about your partner. Don't give your partner your job title until he or she has found out as much information as possible and made a guess at your occupation. Try to find out:

Age

Education

Qualifications

Nature of work

Who he/she is responsible to

What he/she feels about his/her work

Before you start, work out useful questions with your partner to obtain this information.

Hints: The following text will help you with this speaking practice.

Student A: Your profile may be organized into the following example. Choose from one of these three profiles.

a	Age	22
	Job title	Machine tool development fitter
	Education	Name of college, full Technician's Certificate by day release over 4 years
	Duties	Works on automatic machines in the machine tool development department, one of 17 millers, turners and grinders. Responsible for maintenance of machines, making jigs and fixtures for specialist jobs and for building and commissioning new machines.
	Responsible to	Foreman
	Like/dislikes	Likes the job because of the variety of work
b	Age	42
	Job title	General foreman/woman
	Education	Name of college, full Technician's Certificate by day release over 4 years
	Duties	Is in charge of 26 people—machine tool operators, tool setters, etc. Based on the shop floor. Controls everyday production jobs.
	Responsible to	Superintendent
	Like/dislikes	Doesn't like having to sack people
c	Age	24
	Job title	Applications engineer
	Education	Technical college. City & Guilds Certificate by day release over 4 years

	Duties	Works in applications department—around 30 people. Responsible for liaison between the company and the customer. Tries to ensure that the customer's requirements can be met by the company's products. Carries out tests on the products and sends results to the customer on how the product performs.
	Responsible to	Department manager
	Like/dislikes	Gets a lot of job satisfaction because he/she gets to see an end result. Finds the systems in the factory a bit cumbersome. They can hold up the work of the section.

Your partner is one of these workers:

Methods engineer

Systems analyst

Tool maker

Student B: Your profile may be organized into the following examples. Choose from one of these three profiles.

a	Age	22
	Job title	Methods engineer
	Education	Name of college, full Technician's Certificate by day release over 4 years
	Duties	Part of a team which plans new components and how they are to be manufactured. Also responsible for specifying or recommending new equipment and new machines. If there is a problem with production or materials used for a project, the team has to sort it out.
	Responsible to	Production engineering manager
	Like/dislikes	Enjoys working as part of a team and solving problems
b	Age	24
	Job title	Systems analyst
	Education	Polytechnic, BA in Business Studies
	Duties	Assistant analyst in management services department. Part of a team composed of analyst and programmer. When a department has a problem, he/she has to analyze it and come up with a solution. If it is a solution which can be solved by a computer, the team design, write and test a computer program for the problem. If it goes well, the program is put into use. This may involve training a "user" in the new system.
	Responsible to	Section leader
	Like/dislikes	Enjoys working with so many departments. Don't like it when a user changes his/her mind about something after hours have been spent on designing a system.

c	Age	23
	Job title	Toolmaker
	Education	Name of college, full Technician's Certificate by day release over 4 years
	Duties	One of 50 who work in the tool room—fitters, turners, millers grinder, jig borers, spends time on each kind of machine—surface grinders, lathe, mill.
	Responsible to	Foreman/woman
	Like/dislikes	Likes working with his/her hands. Enjoys getting experience with different kinds of machines.

Your partner is one of these workers:

Machine tool development fitter

Foreman/woman

Applications engineer

Case 6. Describing Different Types of Mechanical Measuring Tools and Gauges

Work in pairs, A and B. Each student has to name the measuring tools in Figure 3. 3, then describe their functions and demonstrate how to use them to your partner. Combine your answers with your partner's and see how many tools you can recognize.

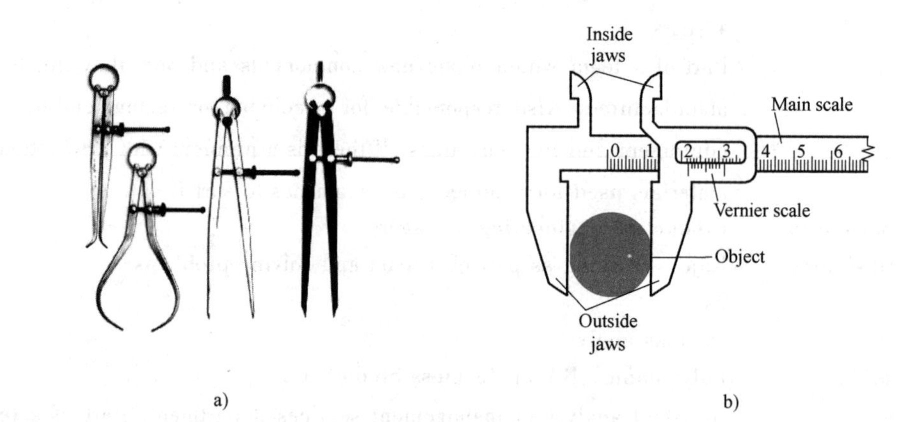

a) b)

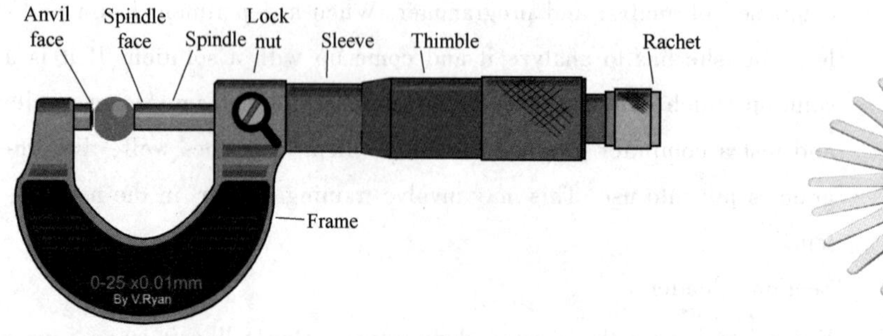

c)

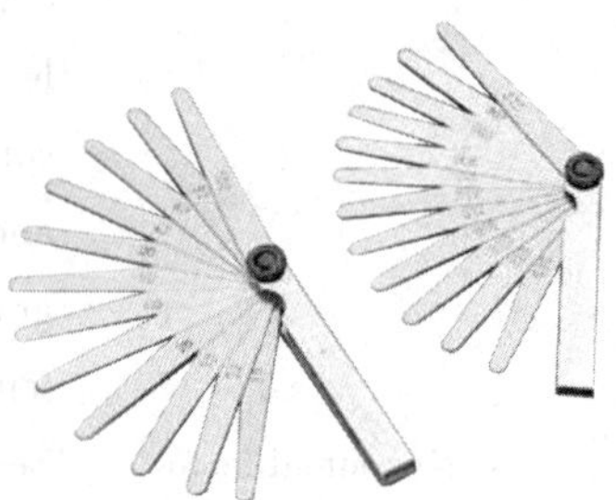

d)

Figure 3. 3 Measuring tools

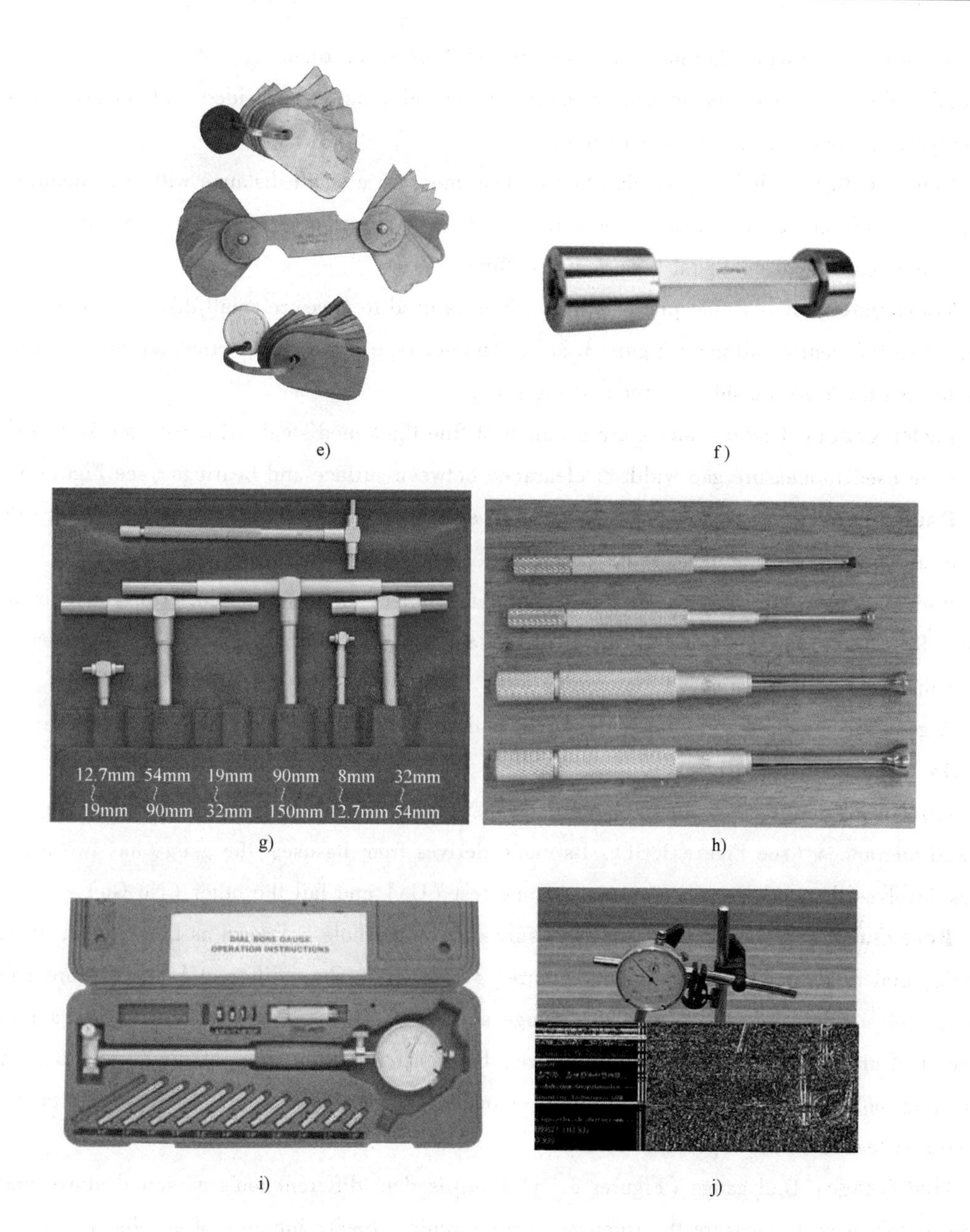

Figure 3. 3 Measuring tools (continued)

Hints: The following text will help you with this practice.

Measuring instruments and gauges are used to measure various parameters such as clearance, diameter, depth, ovality, trueness etc. These are important engineering parameters which describes the condition of the working machinery.

Popular mechanical gauges and tools used are:

Ruler and scales: They are used to measure lengths and other geometrical parameters. They can be single steel plate or flexible tape type tool.

Callipers: They are normally of two types-inside and outside callipers (see Figure 3. 3a).

They are used to measure internal and external size (for e. g. diameter) of an object. It requires external scale to compare the measured value. Some callipers are provided with measuring scale. Other types are odd leg and divider calliper.

Venire calliper: It is a precision tool used to measure a small distance with high accuracy. It has got two different jaws to measure outside and inside dimension of an object (see Figure 3.3b). It can be a scale, dial or digital type venire calliper.

Micrometer: It is a fine precision tool which is used to measure small distances and more accurate than the venire calliper (Figure 3.3c). Another type is a large micrometer calliper which is used to measure large outside diameter or distance.

Feeler gauge: Feelers gauges are a bunch of fine thickened steel strips with marked thickness which are used to measure gap width or clearance between surface and bearings (see Figure 3.3d).

Radius gauge: A radius gauge, also known as a fillet gauge, is a tool used to measure the radius of an object (see Figure 3.3e).

Radius gauges require a bright light behind the object to be measured. The gauge is placed against the edge to be checked and any light leakage between the blade and edge indicates a mismatch that requires correction.

A good set of gauges will offer both convex and concave sections, and allow for their application in awkward locations.

Go/No go gauge: Go/No go gauge is an inspection tool used to check a workpiece against its allowed tolerances (see Figure 3.3f). Its name derives from its use: the gauge has two tests; the check involves the workpiece having to pass one test (Go) and fail the other (No Go).

Bore Gauge: A tool to accurately measure size of any hole is known as bore gauge. It can be a scale, dial or digital type instrument. Figures 3.3g, h, i show different kinds of bore gauges, they are telescopic gauge set, small hole gauge set and dial bore gauge set. These are a range of gauges that are used to measure a bore's size, by transferring the internal dimension to a remote measuring tool. They are a direct equivalent of inside callipers and require the operator to develop the correct feel to obtain repeatable results.

Dial Gauge: Dial gauge (Figures 3.3j) is utilised in different tools as stated above and can be separately used to measure the trueness of the circular object, jumping of an object, etc.

Case 7. Describing Different Types of Tools Used often in Engineering

Task 1

Look at the pictures (Figure 3.4) with the tools that belong to the basic equipment of a toolbox. Use the words from the box to label the tools.

open-ended spanner | screw-driver | Allen keys | pliers | socket wrench | electric drill | metal saw | torque wrench | grip vice pliers | hammer | vernier calipers | electric screw-driver

Task 2

Choose one of the tools and describe it in terms of its material, appearance and what it is used

Figure 3. 4 Different types of tools used often in engineering

for. Your text should be between 60 and 80 words long.

Task 3

Have a look at the sentences, match the correct verb from the box with its definition and then find an appropriate tool from task1 for each task.

drill I saw I screw I loosen I tighten I measure

Verb	Definition	Tool
	to become or make something loose	
	to fasten something or make it tight with the help of screws	
	to find out the dimensions of a work piece	

(continued)

Verb	Definition	Tool
	the opposite of "to loosen"	
	to make a hole in a piece of metal or other materials	
	to separate a piece of material from a whole block	

Case 8. Speaking Practice—Health and Safety at Work

Background Material

Have you ever worked? Was it in a dangerous environment? Did you have to follow any special regulations? Have you ever passed any kind of exam on health and safety at work? How did it look like— just theory or also a practical part? What have you learned?

After this unit students are expected to be able to discuss the rules and regulations about safety at work, build up vocabulary and know more about verb patterns.

Safety and health is an area concerned with protecting the safety, health and welfare of people engaged in work or employment. The goal of all occupational safety and health programs is to foster a safe work environment. As a secondary effect, it may also protect co-workers, family members, employers, customers, suppliers, nearby communities, and other members of the public who are impacted by the workplace environment. The average person finds it difficult to assess risks and that is why work practices need to be regulated.

Safety in the workplace is critical to the success of your business, no matter what size it is. As a business owner you have responsibilities regarding health and safety in your workplace. Even if you don't have any employees, you must ensure that your business doesn't create health and safety problems for your customers and the general public. Knowing and understanding the Occupational Health and Safety laws can help you avoid the unnecessary costs and damage to your business caused by workplace injury and illness.

There are many examples of dangerous activities at your workplace, such as welding without goggles, working at a construction site without the protection of a hard hat, working in noisy environments without ear plugs or mufflers, working in production with different possibly hazardous materials without protective gloves and/or clothes, smoking near inflammable substances etc. There are different risky or hazardous situations, such as combustion, contamination, dust, the possibility of explosion, poisonous fumes, gas leakages, toxic vapors, the danger of electrical shock, and so on and so forth, which can all have effects on us and can cause lethal or very serious damage to our body (for example: vomiting, dizziness, burns, birth defects, cancer, genetic damage).

Task 1

1. Name the safety protection facilities in Figure 3. 5. Can you add more?

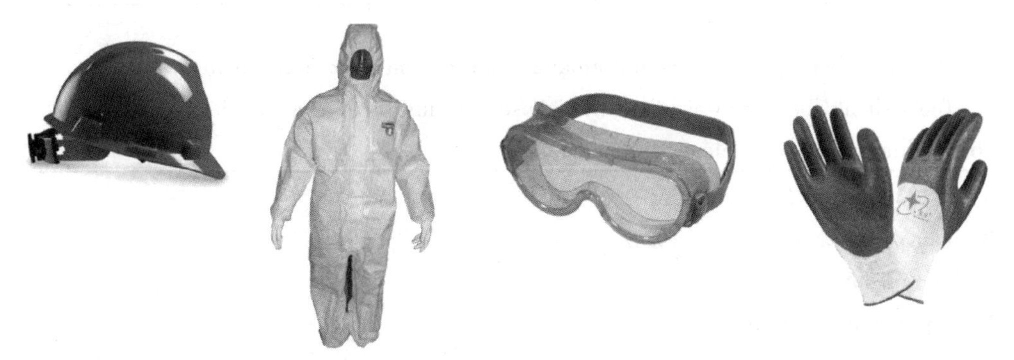

Figure 3. 5 Safety protection facilities

2. Describe when and where workers should wear safety protection facilities shown in Figure 3. 5, and then work in pairs, A and B. Each of you has to describe your explanation to your partner.

Task 2

All around risky environments or materials there are warning signs that people have to take seriously. Figure 3. 6 shows several examples of warning signs.

Figure 3. 6 Warning signs

1. Name the warning signs above, find some other warning signs in public places and explain them.

2. Role play

Work in pairs, A and B. Imagine A is a security guard, B is a visitor, and explain the above signs to each other.

Case 9. Understanding the Safety Instructions at Work

Task 1

Study the safety instructions from a workshop below, and then answer the following questions.

1. Who are the instructions for?
2. Who wrote them?
3. What was the writer's purpose?

> 1. Wear protective clothing at all times.
>
> 2. Always wear eye protection when operating lathes, cutters and grinders and ensure the guard is in place.

3. Keep your workplace tidy.
4. The areas between benches and around machines must be kept clear.
5. Tools should be put away when not in use and any breakages and losses reported.
6. Machines should be cleaned after use.

Task 2: Making Safety Rules

We can make safety rules in these ways:

1. Use an imperative.

Wear protective clothing.

Do not wear loose-fitting clothing.

2. Always/never are used to emphasize that the rule holds in all cases.

Always wear protective clothing.

Never wear loose-fitting clothing.

3. We can use a modal verb for emphasis.

Protective clothing must be worn.

Protective clothing should be worn.

What are the differences in meaning, if any, between these statements?

1. Wear protective clothing.
2. Always wear protective clothing.
3. Protective clothing must be worn.

Study this list of unsafe environmental conditions (hazards). Write safety rules to limit these hazards using the methods given above. For example:

inadequate lighting

Lighting must be adequate. or

Lighting should be adequate.

1. uneven floors.
2. unguarded machinery.
3. untidy workbenches.
4. untidy workplaces.
5. badly maintained machinery.
6. carelessly stored dangerous materials.
7. inadequate ventilation.
8. damaged tools and equipments.
9. machinery in poor condition.
10. equipment used improperly.
11. equipment operated by untrained personnel.
12. apprentices working without supervision.

Case 10. Engineering Drawings

Task 1

Work in pairs and answer the following questions.

1. What is meant by scale on a drawing?

2. Explain how a scale ruler shown in Figure 3.7 is used.

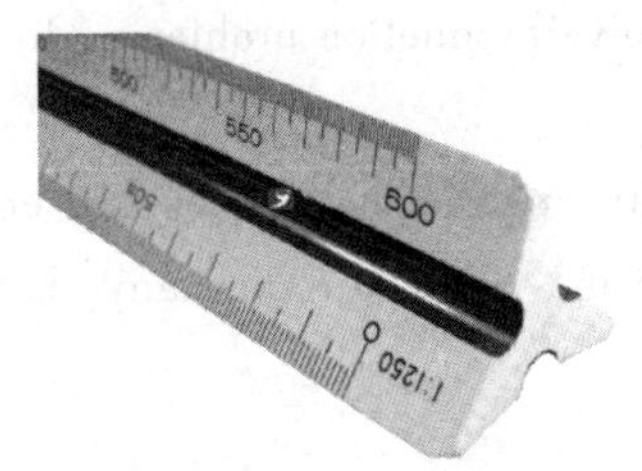

Figure 3.7 A scale ruler

3. Discuss the different types of 2D drawings which may be used to describe a complicated product.

Task 2

Complete the following definitions using the types of drawings below.

cross-section; elevation; exploded view; note; plan; schematic; specification

1. A _________ gives a view of the whole deck from above.
2. An _________ gives a view of all the panels from the front.
3. An _________ gives a deconstructed view of how the panels are fixed together.
4. A _________ gives a cutaway/sectional view of the joint between two panels.
5. A _________ gives a simplified representation of a network of air ducts.
6. A _________ gives a brief description or a reference to another related drawing.
7. A _________ gives detailed written technical descriptions of the panels.

Task 3

1. Which two types of drawing in Task 2 are examples of general arrangement drawings, and which two are examples of detail drawings?

2. Suppose that you are going to provide design information to enablea production team to manufacture a product you know well. Make a list of some drawings that will be needed.

Case 11. Discussing Dimensions and Precision

Task 1

1. Work in pairs and discuss what is meant by precision and accuracy.

2. Work in pairs and discuss what is meant by tolerance in the context of dimensions.

3. In some situations, engineers describe tolerances with plus or minus, for example ±0.01 mm, and in other situations as within, for example within 0.2mm. Work in pairs, discuss the difference in meaning between these two descriptions and give examples of situations where each description might be used.

Task 2

Complete the following definitions using the types of drawings below.

design brief; preliminary drawing; sketch; working drawing

1. A _________ is a rough drawing of initial ideas, also used when production problems require engineers to amend design details and issue them to the workforce immediately.

2. A ________ is a written summary intended to specify design objectives.

3. A ________ is an approved drawing used for manufacture or installation. There is often a need to revise these drawings to resolve production problems. In this case, amended versions are issued to supersede previous ones.

4. A ________ is a detailed drawing that colleagues and consultants are in vited to approve if they accept them, or comment on if they wish to requestany changes.

Case 12. Solving Problems

Problems

Suppose that you work for IPS, a producer of industrial package machinery. As a member of the global service team, your role is to travel abroad and deal with serious technical problems at your clients' plants. Read the following email from a plant in Helsinki and summarise the problems.

Following our phone conversation this morning I confirm that a forklift truck has hit our IPS15 unit. The impact has made a large hole in the main panel on the side of the machine. Our technician who is trained to carry out routine adjustments on the machine has made an external visual inspection. He has advised me that the mechanisms for adjusting the precise alignment of the cutting blades have been damaged. Liquid lubricant is also leaking out from under the machine and a crackling sound can be heard inside the unit when it is switched on—presumably due to earthing/short-circuiting resulting from electrical damage.

Here I confirm my request for intervention by your service team.

Task

Work in pairs and describe the sequence of steps you need take to carry out repairs when you arrive at the plant in Helsinki, using the notes to help you.

IPS is Helsinki

- internal damage
- old parts
- electrical supply: on/off
- lubricant: in/out
- external panels
- alignment of cutting blades
- test
- new parts

Case 13. Discussing Quality Issues

Task 1

Work in pairs and answer the following questions.

1. In the following advertisement (see figure 3.8), which hi-tech, high-performance situations are used to promote watches?

2. What messages are they intended to send about the new features of this wrist watch?

3. What quality issues differentiate higher-quality watches from lower-quality ones?

4. What is the difference between describing something as water-resistant and waterproof?

This very "gadgety" looking watch not only does the essentials like telling the time and date, but also serves as the ultimate universal remote. The six buttons sticking out on the bezel control volume, power on/off, channel up/down, as well as play, rewind, and fast-forward for movies and possibly DVRs. Like any other universal remote, you just need to know the three-digit code for the device you wish to assume control of, enter it in, and the power is in your hands (or, rather, on your wrist).

Figure 3. 8 An advertisement for a wrist watch

Task 2

Work in small groups, choose a well-known consumer product or appliance and discuss it from the perspective of quality, like how suitable the materials are used, how good the product is, compared with others sold by competitors.

Case 14. Product Safety

Task 1

Study the following notes:

The best approach to the prevention of product liability is good engineering in both analysis and design, quality control and comprehensive testing procedures. Advertising managers often make glowing promises in the warranties and sales literature for a product. These statements should be reviewed carefully by the engineering staff to eliminate excessive promises and insert adequate warnings and instructions for use.

With greatly increasing of liability lawsuits and the need to conform to regulations issued by governmental agencies these days, it is very important for the designer and the manufacturer to know the reliability of their product.

The statistical measure of the probability that a mechanical element will not fail in use is called the reliability of that element. The reliability R can be expressed by a number within the range: $0 \leqslant R \leqslant 1$. A reliability of $R = 0.90$ means that there is a 90 percent chance that the part will perform its proper function without failure. The failure of 6 parts out of every 1000 manufactured might be a considered and acceptable failure rate for a certain class of products.

Task 2

Work in pairs, discuss and answer the following questions according to the notes above.

1. Discuss what is meant by product liability and reliability of an element.
2. Discuss what is meant by warranties.
3. If the objective reliability is to be 99. 4 percent, what is meant by that?

Case 15. Describing Shapes and Features

Task 1

Study the following notes.

What do you know about the electrical plugs and sockets used in different countries? If you buy a phone charger at the airport in Florida, you are not able to use it when your flight lands in France. If you buy a three-pronged adapter for your computer in Paris, you might not be able to plug it in when your train drops you off in Germany.

There are currently 15 types of electrical outlet plugs in use today, each of which has been assigned a letter by the US Department of Commerce International Trade Administration (ITA), starting with A and moving through the alphabet, as shown in Table 3.2. These letters are completely arbitrary and they don't actually mandate anything.

Table 3.2 Types of electrical outlet plugs

Type A	Type B	Type C	Type D
1. mainly used in the USA, Canada, Mexico and Japan etc. 2. 2pins 3. not grounded 4. 15A 5. almost always 100 ~ 127V 6. socket compatible with plug type A	1. mainly used in the USA, Canada, Mexico and Japan etc 2. 3pins 3. grounded 4. 15A 5. almost always 100 ~ 127V 6. socket compatible with plug types A and B	1. commonly used in Europe, South America and Asia etc 2. 2pins 3. not grounded 4. 2.5A 5. 220 ~ 240V 6. socket compatible with plug type C	1. mainly used in India etc 2. 3pins 3. grounded 4. 5A 5. 220 ~ 240V 6. socket compatible with plug types C and D (unsafe compatibility with E and F)
Type E	**Type F**	**Type G**	**Type H**
1. primarily used in France, Belgium, Poland, Slovakia and the Czech Republic etc 2. 2pins 3. grounded 4. 16A 5. 220 ~ 240V 6. socket compatible with plug types C, E and F	1. almost used everywhere in Europe and Russia, except for the UK and Ireland 2. 2pins 3. grounded 4. 16A 5. 220 ~ 240V 6. socket compatible with plug types C, E and F	1. mainly used in the UK, Ireland, Malta, Malaysia and Singapore etc 2. 3pins 3. grounded 4. 13A 5. 220 ~ 240V 6. socket compatible with plug type G	1. used exclusively in Israel, the West Bank and the Gaza Strip etc 2. 3pins 3. grounded 4. 16A 5. 220 ~ 240V 6. socket compatible with plug types C and H (unsafe compatibility with E and F)

(continued)

Type I	Type J	Type K	Type L
1. mainly used in Australia, New Zealand, China and Argentina etc 2. 2 or 3pins 3. 2pins: not grounded/ 3pins: grounded 4. 10A 5. 220~240V 6. socket compatible with plug type I	1. used almost exclusively in Switzerland, Liechtenstein and Rwanda etc 2. 3pins 3. grounded 4. 10A 5. 220~240V 6. socket compatible with plug types C and J	1. used almost exclusively in Denmark and Greenland 2. 3pins 3. grounded 4. 16A 5. 220~240V 6. socket compatible with plug types C and K (unsafe compatibility with E and F)	1. used almost exclusively in Italy and Chile etc 2. 3pins 3. grounded 4. 10A and 16A 5. 220~240V 6. 10 A socket compatible with plug types C and L (10 A version) / 16 A socket compatible with plug type L (16A version)

Type M	Type N	Type O
 1. mainly used in South Africa etc 2. 3pins 3. grounded 4. 15A 5. 220~240V 6. socket compatible with plug type M	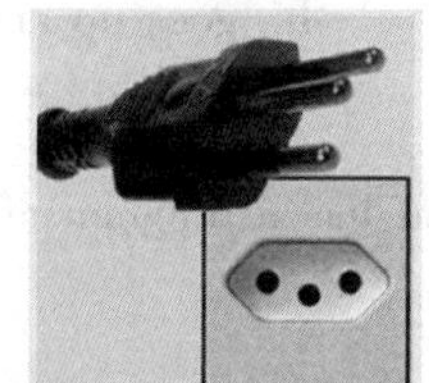 1. used almost exclusively in Brazil etc 2. 3pins 3. grounded 4. 10A and 20A 5. 220~240V 6. socket compatible with plug types C and N	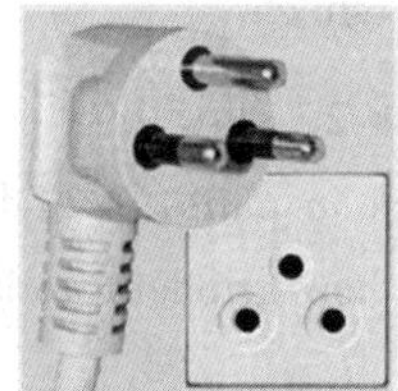 1. used exclusively in Thailand 2. 3pins 3. grounded 4. 16A 5. 220~240V 6. socket compatible with plug types C and O (unsafe compatibility with E and F)

It is important to know how to wire a 3 pin plug correctly. 3 pin plugs are designed so that electricity can be supplied to electrical appliances safely.

A 3 pin plug consists of three pins. Each pin must be correctly connected to the three wires in the electrical cable. Each wire has its own specified color so as it can be easily identified.

The live wire is usually with a letter L. This is connected to a fuse on the live pin. The electric current uses the live wire as its route in. The neutral wire is usually with a letter N. This is the route the electric current takes when it exits an appliance; it is for this reason the neutral wire has a voltage close to zero. The earth wire is usually with a letter E and connected to the earth pin. This is used when the appliance has a metal casing to take any current away if the live wire comes in contact with the casing.

Task 2

Work in pairs and answer the following questions.

1. Why are there so many different electrical plugs and sockets used around world?

2. Describe the different plug and socket formats in the table above.

3. Describe the shape of type A and type H.

4. Why do some plugs and sockets have three pins, while others have two?

5. Imagine you are going to South Africa on business and will bring your laptop with you. Can your adapter work in South Africa? If not, what kind of preparation do you have to do in advance?

Case 16. Describing Motions

Task 1

Match the technical words from the text (1~8) to the more general English from the (a~h).

1. oscillates	a. changes
2. rotates	b. large, thin, flat pieces
3. reciprocates	c. swings backwards and forwards
4. a linear motion	d. goes round and round
5. converts	e. movement
6. motion	f. goes in a line
7. escalator	g. moving stairs
8. sheets	h. goes up and down

Task 2

Work in pairs, A and B. Each of you has a diagram of a robotic gripper (see Figure 3.9). Describe how grippers work.

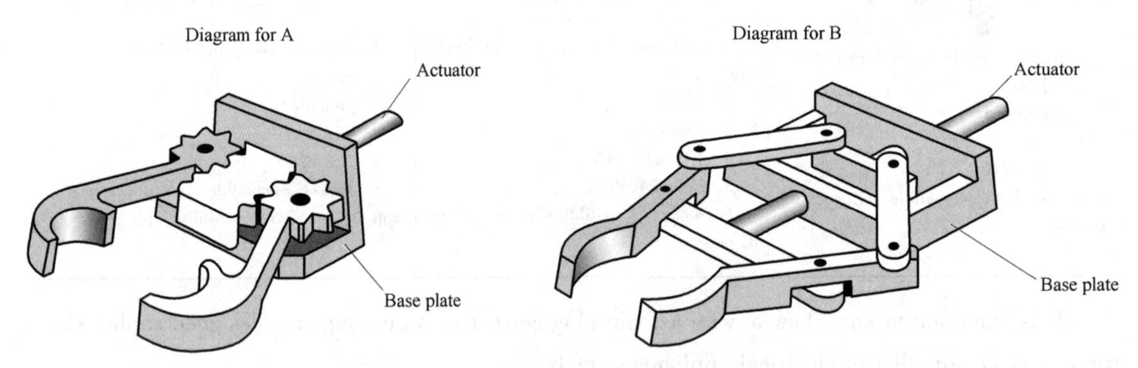

Figure 3. 9 Diagrams of robotic grippers

Hints:

A: The robotic gripper is operated by a rack and pinion mechanism. Rank and opinion open and close the "fingers", permitting them to grasp and release objects. A separate power source (not shown) is required to operate this gripper.

B: The robotic gripper is operated bya reciprocating mechanism. Links open and close the "fingers", permitting them to grasp and release objects. A separate power source (not shown) is required to operate this gripper.

Case 17. Speaking Practice—Describing a Simple Mechanism

Task

Try to explain how this simple mechanism operates using whatever English you know (see Figure 3. 10). Write your explanation down, and then work in pairs, A and B. Each of you has to describe your explanation to your partner.

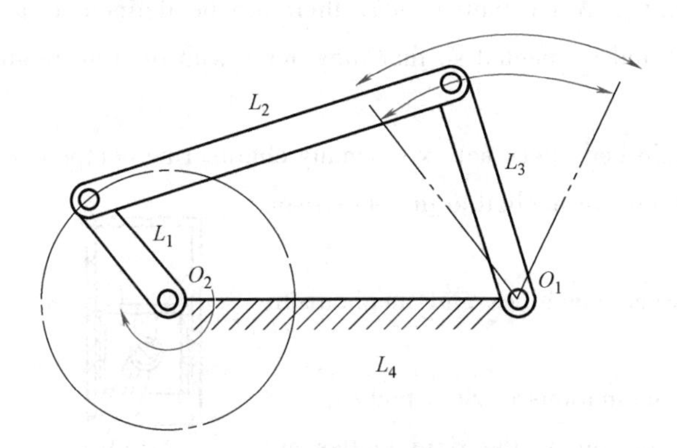

Figure 3. 10 A simple mechanism

Hints: The following texts will help you with the vocabulary you need.

4-bar linkage; crank-rocker linkages;

L_1: a crank

L_2: a coupler or a follower

L_3: a rocker

L_4: fixed link

Case18. Mechanisms and Simple Machines

Background Materials

Mechanism: A mechanism is a device designed to transform input forces and movement into a desired set of output forces and movement. A mechanism generally consists of moving components such as gears and gear trains, belt and chain drives, cam and follower mechanisms, linkages and friction devices such as brakes and clutches, structural components such as the frame, fasteners, bearings, springs, lubricants and seals, as well as a variety of specialized machine elements such as splines, pins and keys.

Machine: A machine is an assemblage of parts that transmit forces, motion and energy in a predetermined manner.

Simple Machine: Any of various elementary mechanisms has the elements of which all machines are composed. Included in this category are the lever, wheel and axle, pulley, inclined plane, wedge and the screw.

The word mechanism has many meanings. In kinematics, a mechanism is a means of transmitting, controlling or constraining relative movement. Movements which are electrically, magnetically, pneumatically operated are excluded from the concept of mechanism. The central theme for mechanisms is rigid bodies connected together by joints.

A machine is a combination of rigid or resistant bodies, formed and connected so that they move with definite relative motions and transmit force from the source of power to the resistance to be overcome. A machine has two functions: transmitting definite relative motion and transmitting force. These functions require strength and rigidity to transmit the forces.

The term mechanism is applied to the combination of geometrical bodies which constitute a machine or part of a machine. A mechanism may therefore be defined as a combination of rigid or resistant bodies, formed and connected so that they move with definite relative motions with respect to one another.

Although a truly rigid body does not exist, many engineering components are rigid because their deformations and distortions are negligible in comparison with their relative movements.

The similarity between machines and mechanisms is that:

1. they are both combinations of rigid bodies.

2. the relative motion among the rigid bodies are definite.

The difference between machine and mechanism is that machines transform energy to do work, while mechanisms do not necessarily perform this function. The term machinery generally means machines and mechanisms. Figure 3. 11a shows cross section of cylinder in a diesel engine. The mechanism of its cylinder-link-crank parts is a slider-crank mechanism, as shown in Figure 3. 11b.

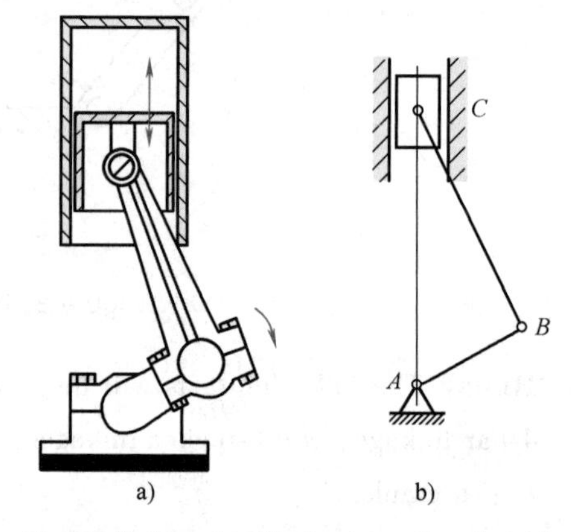

Figure 3. 11 A cylinder in a diesel engine

Task 1

Identify these simple machines (see Figure 3. 12). Work in pairs, A and B. Try to explain the principles on which they operate to each other.

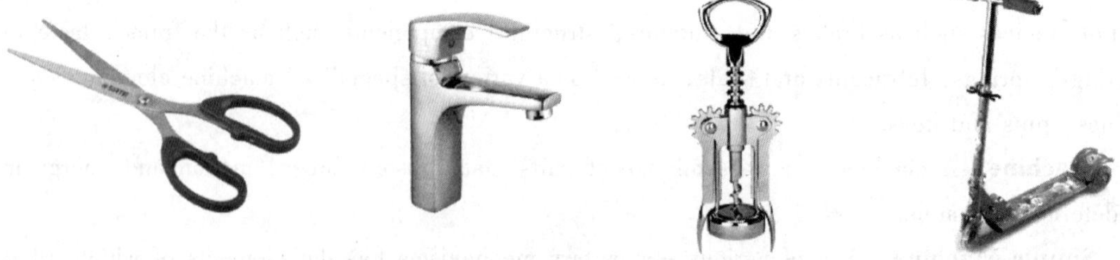
Figure 3. 12 Examples of simple machines

Task 2

Describe the difference between Figure 3. 11a and Figure 3. 11b, why mechanism schematics (Figure 3. 11b) is used in mechanical engineering?

Task 3

Look around you, try to find some simple machines, explain the working principles of them, and then sketch their mechanism-schematics.

Case 19. Shaft (1)

Task 1

Study the following notes.

A shaft is a rotating member, usually of circular across section and used to transmit power or motion (see Figure3. 13a). It provides the axis of rotation, or oscillation, of elements such as gears, pulleys, flywheels, cranks, sprockets, and the like and controls the geometry of their motion. Shafts are subjected to bending moments and twisting moments and sometimes to axial loads. It twists and transmits power.

Axles are rotating or non-rotating members which are subjected to only bending moments due to members supported by it (see Figure 3. 13b). It does not transmit torque. In other words, an axle is not twisted, it only bends.

Spindle is a rotating member that transfers power but does not support any driven member/normal load (see Figure 3. 13c). Term such as lineshaft, headshaft, stub shaft, transmission shaft, countershaft and flexible shaft are names associated with special usage. So we can say that every spindle or axle is a shaft but not every shaft or axle is a spindle.

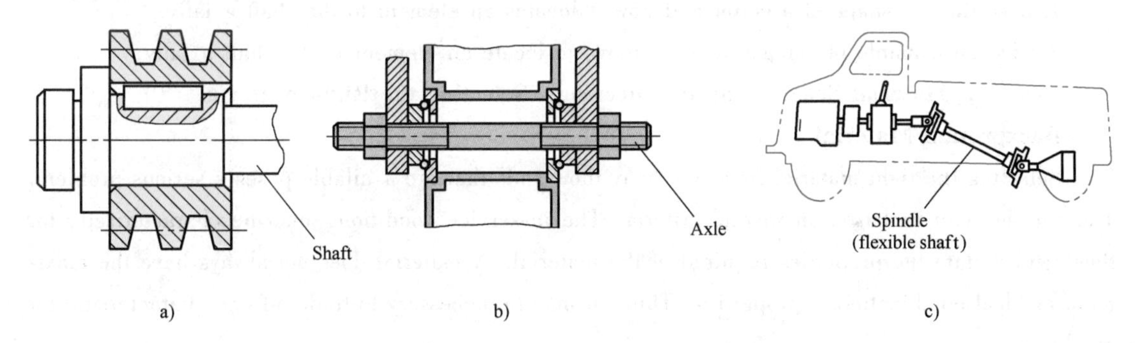

Figure 3. 13 Shafts, axles and spindles

Task 2

Work in pairs, discuss and answer the following questions according to the notes above.

1. Discuss the difference among shaft, axle and spindle.

2. Discuss what is meant by stepped cylinder or stepped shaft and what the function of a shaft shoulder is.

3. Discuss what is meant by pre-load.

Case 20. Shaft (2)

Task 1

Study the following notes.

Many shaft-design situations include the problem of transmitting torque from one element to another on the shaft. Common torque-transfer elements are: keys; splines; setscrews; pins; press or shrink fits; tapered fits.

All the above torque-transfer means solving the problem of securely anchoring the wheel or de-

vice to the shaft, but not all of them solve the problem of accurate axial location of the device. Some of the most-used locational devices include (see Figure 3. 14):

1. Cotter and washer
2. Nut and washer
3. Sleeve
4. Shaft shoulder
5. Ring and groove
6. Setscrew
7. Pins

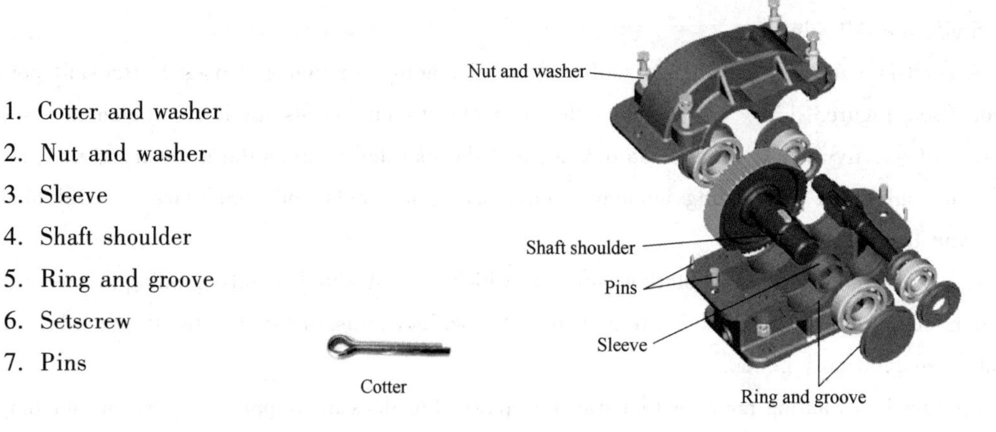

Figure 3. 14 Examples of the most-used locational devices

Task 2

Work in pairs, discuss and answer the following questions according to the notes above.

1. Describe the shape of a cotter and how it locates an element to the shaft axially.
2. Give an example of using ring and groove to locate an element to the shaft axially.

Case 21. Material Science and Engineering—Selecting the Right Material (1)

Background Material

Selecting the right material from the many thousands that are available poses a serious problem. The decision can be based on several criteria. The in-service conditions must be characterized, for these will dictate the properties required of the material. A material does not always have the maximum or ideal combination of properties. Thus, it may be necessary to trade off one characteristic for another.

The classic example includes strength and ductility. Normally, a material with high strength has only a limited ductility. A second selection consideration is any deterioration of material properties that may occur during service operation. For example, significant reductions in mechanical strength may result from exposure to elevated temperatures or corrosive environments. If a compromise concerning desired in-service properties can't be reached, new materials have to be developed.

Probably the most important consideration is that of economics. A material may be found that it has the ideal set of properties but it is extremely expensive. Some compromise is inevitable. The cost of a finished piece also includes any cost occurring during fabrication to produce the desired shape. For example: commodity plastics like polyethylene or polypropylene cost much less than engineering resins or nylon.

Task 1

Write short answers to the following questions.

1. What are necessary steps when considering a material for a certain application?
2. Which trade-offs are unavoidable when choosing a particular material?

Task 2

Work with a partner. Fill the gaps in the text with words from the box in their correct form.

alloy; clay; crystal; housing; manipulate; metal; pottery; structure; wood

Materials used in food, clothing, ________________, transportation and recreation influence virtually every segment of our everyday lives.

Historically, materials have played a major role in the development of societies, whose advancement depended on their access to materials and on their ability to produce and ________ them. In fact, historians named civilizations by the level of their materials development, e. g. the Stone Age (beginning around 2.5 million BC), the Bronze Age (3500 BC), and the Iron Age (1000 BC). The earliest humans had access to only a very limited number of materials, those that occur naturally, e. g. __________, __________, __________ and ________________. With time they discovered techniques for producing materials that had properties superior to those of the natural ones; these new materials included ________________ and various ________________. Furthermore, early humans discovered that the properties of a material could be altered.

Task 3

Skim the following text.

The discipline of material science and engineering includes two main tasks. Material scientists examine the structure-properties relationships of material and develop or synthesize new material. Material engineers design the structure of a material to produce a predetermined set of properties on the basis of structure-property relationships. They create new products or systems using existing materials and/or develop techniques for processing materials.

Most graduates in material programs are trained to be both material scientists and material engineers.

Read the text above. Then decide whether the following statements are true or false. Rewrite the false statements if necessary.

1. Material scientists do research on finished materials.
2. New products are based on new materials only.
3. Material science can be subdivided because different approaches to materials are employed.
4. Material engineers investigate the correlation between structure and property.

Case 22. Material Science and Engineering—Selecting the Right Material (2)

Task 1

List the materials you know which are used in engineering. Combine your list with the others in your group and classify the materials as metals, thermoplastics, etc.

Table 3.3 will help you with the vocabulary you need.

Table 3.3 Materials in engineering

Materials	Properties	Uses
Metals		
Aluminum	Light, soft, ductile, highly conductive, corrosion-resistant	Aircraft, engine components foil, cooking utensils

(continued)

Materials	Properties	Uses
Copper	Very malleable, tough and malleable, highly conductive, corrosion-resistant	Electric wiring, PCBs, tubing
Brass (65% copper, 35% zinc)	Very corrosion-resistant, casts well, easily machined, can be work hardened, good conductor	Valves, taps castings, ship fittings, electrical contacts
Mild steel (iron with 0.15% to 0.3% carbon)	Hardest of the carbon steels but less ductile and malleable, can be hardened and tempered	General purpose
High carbon steel (iron with 0.7% to 1.4% carbon)	Hardest of the carbon steels but less ductile and malleable, can be hardened and tempered	Cutting tools such as drills, saws
Thermoplastics		
ABS	toughness, scratch-resistant, light and durable	Safety helmets, car components, telephones, kitchenware
Acrylic	Stiff, hard, very durable, clear, can be polished and formed easily	Aircraft canopies, baths, double glazing
Nylon	Hard, tough, wear-resistant, self-lubricating	Bearings, gears, casings for power tools
Thermosetting plastics		
Epoxy resin	High strength when reinforced, good chemical and wear resistance	Adhesives, encapsulation of electronic components
Polyester resin	Stiff, hard, brittle, good chemical and heat resistance	Moulding, boat and car bodies
Urea formaldehyde	Stiff, hard, strong, brittle, heat-resistant and a good electrical insulator	Electrical fittings, adhesives

Task 2

Study these facts from the table about aluminum:

1. Aluminum is a light metal.
2. Aluminum is used to make aircraft.

We can link these facts to make a definition of aluminum.

1+2 : Aluminum is a light metal **which** is used to make aircraft.

Use Table 3.4 to make definitions of each of the materials in column A. Choose the correct information in column B and C to describe the materials in column A.

Table 3.4 Definitions of materials

A	B	C
An alloy		allows heat or current to flow easily
A thermoplastic		remains rigid at high temperatures
Mild steel		does not allow heat or current to flow easily
A conductor	a metal	contains iron and 0.7% to 1.4% carbon
An insulator	a material	becomes plastic when heated
High carbon steel	an alloy	contains iron and 0.15% to 0.3% carbon
Brass		formed by mixing metals or elements
Athermosetting plastic		consists of copper and zinc

Task 3

Study the following text about aluminum.

Aluminum is used to make aircraft, engine components and many items for the kitchen.

We can add extra information to the text like this:

Aluminum, **which is light, soft and ductile**, is used to make aircraft, engine components—**for example, cylinder heads**—and many items for the kitchen, **such as pots.**

Note that the extra information is marked with commas or dashes:

, which…,

—for example, …—

such as…,

Now add the extra information (from 1 to 10) to the following text (from A to E) about plastics.

1. Plastics can be moulded into plates, car components and medical aids.
2. Thermoplastics soften when heated again and again.
3. Thermosetting plastics set hard and do not alter if heated again.
4. ABS is used for safety helmets.
5. Nylon is self-lubricating.
6. Nylon is used for motorized drives in cameras.
7. Acrylic is a clear thermoplastic.
8. Acrylic is used for aircraft canopies and double glazing.
9. Polyester resin is used for boat and car bodies.
10. Polyester resin is hard and has good chemical and heat resistance.

A Plastics are synthetic materials. They can be softened and moulded into useful articles. They have many applications in engineering. There are two types of plastics: thermoplastics and thermosetting plastics.

B ABS is a thermoplastic which is tough and durable. Because it has high impact strength, it has applications where sudden loads may occur.

C Nylon is a hard, tough thermoplastic. It is used where silent, low-friction operation is required.

D Acrylic can be formed in several ways. It is hard, durable, and has many uses.

E Polyester resin is a thermosetting plastic used for castings. It has a number of useful properties.

Case 23. Material Science and Engineering—Selecting the Right Material (3)

Task 1

Match the materials (1~8) to the definitions (a~h).

1. stainless steel	a. a metal used to make brass, and in galvanized coatings on steel
2. zinc	b. the predominant metal in steel
3. iron	c. a type of steel not needing a protective coating, as it doesn't rust
4. bronze	d. a dense, poisonous mental
5. lead	e. rocks from which metals can be extracted
6. hardwood	f. an alloy made from copper and tin
7. ore	g. timber from pine trees
8. softwood	h. timber from deciduous tree

Task 2

Complete the following sentences using from, with or of.

1. Bronze contains significant amounts ________ copper.
2. Galvanised steel is steel coated ________ zinc.
3. Steel is an alloy derived ________ iron.
4. Pure metals can usually be recovered ________ alloys.
5. To produce stainless steel, iron is mixed ________ other metals.
6. Stainless steel contains quantities ________ chromium and nickel.
7. Glass tableware contains traces ________ metals, such as lead.
8. When new metal is extracted ________ ore, the costs can be high.

Task 3

Work in pairs, ask and answer questions about different materials listed in task 1 using the following phrases.

1. Can…be recycled?
2. What's…made from?
3. Where does…come from?

Case 24. Material Science and Engineering—Selecting the Right Material (4)

Task 1

Study the following notes.

As motor racing goes green materials, Formula I is aiming to lead automotive research in finding hi-tech efficiency gains. One of the keys to this ecological drive is regenerative braking (also known as kinetic energy recovery), which recovers energy generated during deceleration, and stores it as a source of power for subsequent acceleration.

Regenerative brakes limit the energy loss inherent in traditional braking systems. In most vehicles, conventional brakes comprise pads previously made from asbestos-based composites, but now consisting of compounds of exotic, non-hazardous materials, and discs made off errous metal. The resulting friction generates heat, which is wasted. In performance cars, this phenomenon is taken to extremes, and due to the high temperatures generated, brake discs are often made out of ceramics.

The carbon discs and pads used on Formula I cars generate so much heat that they glow red hot. High temperatures are, in fact, necessary for the effective operation of carbon brakes. But there's still plenty of potential for recovering the kinetic energy, rather than merely dissipating it in the form of heat.

The potential for recovering energy also extends to the heat generated by engines and exhaust systems. This area has also been discussed as a possible area for future exploitation in motor racing. Heat recovery might offer the added benefit of reducing heat soak (thermal absorption by the chassis) as delicate alloy parts and sensitive non-metallic materials, such as polymers, are susceptible to heat damage.

Task 2

Work in pairs and answer the following questions.

1. Why do most braking systems waste energy?
2. What are regenerative braking systems and how do they save energy?
3. What characteristics are required of materials used for the brakes on racing cars?

4. What is meant by "heat soak" and why is it a problem in racing cars?

Task 3

Match the materials from the text (1~7) to the descriptions (a~g).

1. compounds	a. materials that are not metal
2. exotic	b. iron and steel
3. ferrous	c. combinations of materials
4. ceramic	d. mixture of metals
5. alloy	e. plastic materials
6. non-metallic	f. minerals transformed by heat
7. polymer	g. rare or complex

Task 4

Work in pairs and take turns to describe an object using the words from Task 3 and the phrases below. Ask your partner to guess what it is.

Comprise; consist of; made from; made of; made out of

Task 5

Imagine you are presenting a product or appliance you know well to a potential client. Describe the categories of materials used to make different parts.

Hint: take your smart phone as an example or any other products you know well.

Task 6

Think of a product you know well. Work in pairs, discuss materials used in it and what properties make materials suitable. Discuss whether alternative materials could be used.

Case 25. Material Science and Engineering—Corrosion

Task 1

Skim the following text to identify the paragraphs which contain:

1. Conditions in which corrosion occurs.
2. Need to consider corrosion in design.
3. A definition of corrosion.
4. Factors which limit corrosion.
5. Effects of rust.

A major consideration in engineering design is maintenance. One of the commonest causes of failure in the long term is corrosion. This means any deterioration in the component's appearance or physical properties.

Corrosion covers a number of processes whereby a metal changes state as a result of some forms of interaction with its environment. It often occurs where water, either as a liquid or vapour in air of high humidity, is present.

In general, corrosion becomes worse when impurities are present in damp conditions. It never starts inside a material, and there is always surface evidence that indicates corrosion exists, although close examination may be needed.

A common example of corrosion is the rusting of steel where a conversion of metallic iron to a mixture of oxides and other compounds occurs. This not only changes the appearance of the metal but also results in a decrease in its cross-section.

It is imperative that a design takes into account whether a material will be affected in a particular environment, and if corrosion is likely, at what rate.

Many factors can intervene in a way to restrain its progress. An example is aluminum and it salloys which perform satisfactorily in many engineering and domestic applications when exposed to air and water. This is due to the rapid production of a tough adherent film of oxide which protects the metal from further attack so that corrosion halts.

Task 2

Answer these questions with the help of the text above.

1. In corrosion, why do metals change state?
2. Name two factors which encourage corrosion.
3. Where can signs of corrosion always be found?
4. What is rust? Why may rust be dangerous to a structure?
5. What must designers consider regarding corrosion?
6. Why does aluminum perform well when exposed to air and water?

Case 26. Advanced Manufacturing—Rapid Prototyping

Task 1

Study the following notes.

Rapid Prototyping (RP) is the name given to a host of related technologies that are used to fabricate physical objects directly from CAD data sources. These methods are generally similar to each other in that they add and bond materials in layer-wise fashion to form objects. This is directly the opposite of what classical methods such as milling or turning do. Objects are formed in those processes by mechanically removing material.

Although several RP techniques exist, all employ the same basic five-step process. The steps are:

1. CAD model creation. First, the object is modelled using a CAD software package.

2. Conversion to STL format. The various CAD packages use a number of different algorithms to represent solid objects. To establish consistency, the STL (Standard Triangulation Language-Stereolithography, the first RP technique) format has been adopted as the standard of the rapid prototyping industry. The second step, therefore, is to convert the CAD file into STL format. This format represents a three-dimensional surface as an assembly of planar triangle, like the facets of a cut jewel. The file contains the coordinates of the vertices and the direction of the outward normal of each triangle. Since the STL format is universal, this process is identical for all of the RP build techniques.

3. Slice the STL file. In the third step, a pre-processing program prepares the STL file to be built. Several programs are available, and most allow the user to adjust the size, location and orientation of the model. The pre-processing software slices the STL model into a number of layers from 0. 01 mm to 0. 7mm thick, depending on the building technique. The program may also generate an auxiliary structure to support the model during the build. Each RP machine manufacturer supplies their own proprietary pre-processing software.

4. Layer by layer construction. The fourth step is the actual construction of the part. Using one of several techniques RP machines build one layer at a time from polymers, paper or powdered metal.

5. Clean and finish. The final step is post-processing. This involves removing the prototype from the machine and detaching any supports. Some photosensitive materials need to be fully cured before use. Prototypes may also require minor cleaning and surface treatment. Sanding, sealing, and/or painting the model will improve its appearance and durability.

RP is widely used in the automotive, aerospace, medical and consumer products industries. Although the possible applications are virtually limitless, nearly all fall into one of the following categories: prototyping, rapid tooling, or rapid manufacturing.

Prototyping: As its name suggests, the primary use of RP is to quickly make prototypes for communication and testing purposes. Prototypes dramatically improve communication because most people, including engineers, find three-dimensional objects easier to understand than two-dimensional drawings. Such improved understanding leads to substantial cost and time saving.

Rapid Tooling: A much-anticipated application of RP is rapid tooling, the automatic fabrication of production quality machine tools. Tooling is one of the slowest and most expensive steps in the manufacturing process, because of the extremely high quality required.

Rapid Manufacturing: A natural extension of RP is Rapid Manufacturing (RM), the automated production of saleable products directly from CAD data. Currently only a few final products are produced by RP machines, but the number will increase as metals and other materials become more widely available. RM will never completely replace other manufacturing techniques, especially in large production runs where mass-production is more economical.

Task 2

1. Work in pairs and discuss why RP has been used extensively.

2. Work in pairs and discuss the advantages and disadvantages between RP and classical manufacturing methods.

Task 3

Work with a partner. Fill the gaps in the text with words from the box in their correct forms.

global market; time-to-market; fixture; visualization; mould fabrication; lower cost

By building three-dimensional parts in a layer-by-layer additive manner, the RP techniques allow freeform fabrication of parts of complex geometry directly from their CAD models automatically, without having to use special ________ as in the material removal processes. The RP technology has helped product developers to develop their products more rapidly at ________. In the ever changing and more competitive ________, it was initially used to make physical prototypes of three-dimensional parts as ________ and communication aids in design as well as for examining the fit of various parts in assembly. It has provided substantial reduction of ________________, hence widely called as rapid prototyping in industry. Thanks to intensive research and development in the areas of material, process, software and equipment, applications to rapid tooling have also been developed by directly or indirectly employing RP processes in the tool, die and ________________.

Case 27. Describing Graphs

In engineering, graphs and charts are common ways of giving information. They allow a great deal of data to be presented easily in visual form.

Task 1

Label the following graphs and charts (see Figure 3.15) with the correct term from this list: graph; pie chart; bar chart; bar chart (column chart)

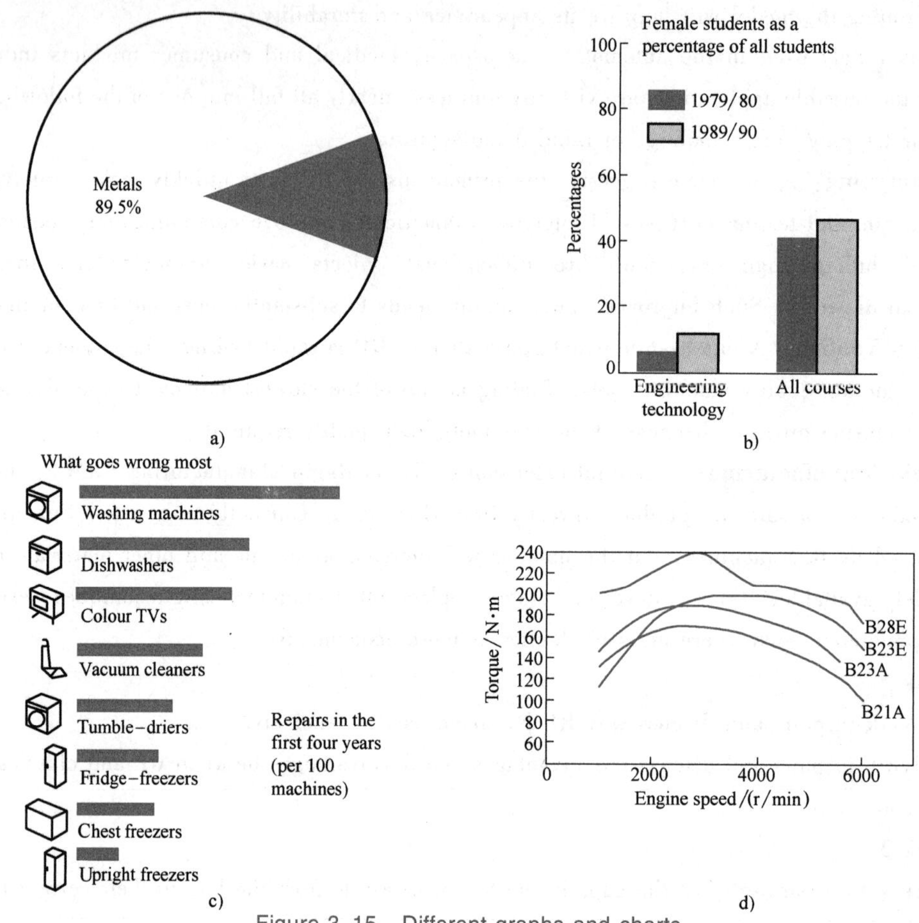

Figure 3.15 Different graphs and charts

Task 2

Study the graph opposite (see Figure 3.16) which shows typical daily load curves for a power station. Answer these questions about the graph for weekdays.

1. When is the peak load?
2. When is there least demand?
3. When is the load 65% of capacity?
4. What is the load at 1 p.m.?

Describe changes in load for these periods:

1. Between 6 a.m. and 10 a.m.
2. Between 7 p.m. and midnight.

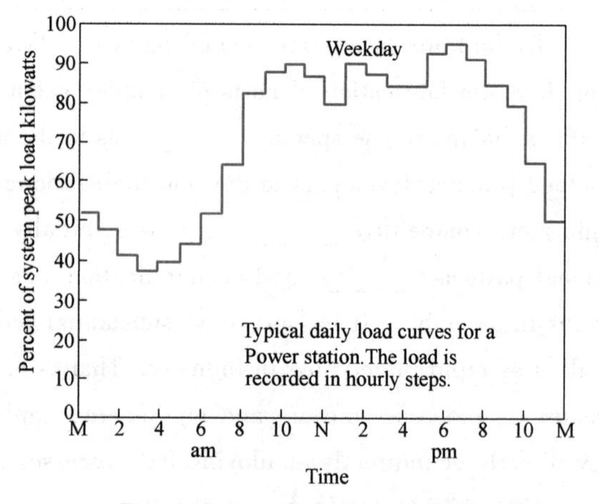

Figure 3.16 Typical daily load curves for a power station

3. Between 3 p. m. and 5 p. m.

Hints: The following texts will help you with finishing this exercise.

Look at the period 6 a. m. to 10 a. m. We can describe the change in load in two ways:

1. The load rises.
2. There is a rise in load.

We can make our description more accurate like this:

1. The load rises sharply.
2. There is a sharp rise in load.

Study this table of verbs and related nouns of change. The past form of irregular verbs is given in brackets.

Direction	Verb	Noun
UP	Climb	
	go up (went up)	
	increase	increase
	rise (rose)	rise
Down	decline	decline
	decrease	decrease
	dip	dip
	drop	drop
	fall (fell)	fall
	go down (went down)	
Level	not change	no change
	remain constant	

These adjectives and adverbs are used to describe the rate of change:

Adjective	Adverb
slight	slightly
gradual	gradually
steady	steadily
steep	steeply
sharp	sharply
sudden	suddenly
fast	fast

Task 3

Study this graph which shows the load at weekends (see Figure 3.17).

Write sentences to describe the load during these periods.

1. Saturday. 8 a. m. to noon.
2. Saturday. 6 p. m. to 10 p. m.
3. Saturday. Noon to 5 p. m.
4. Saturday. Noon to 1 p. m.

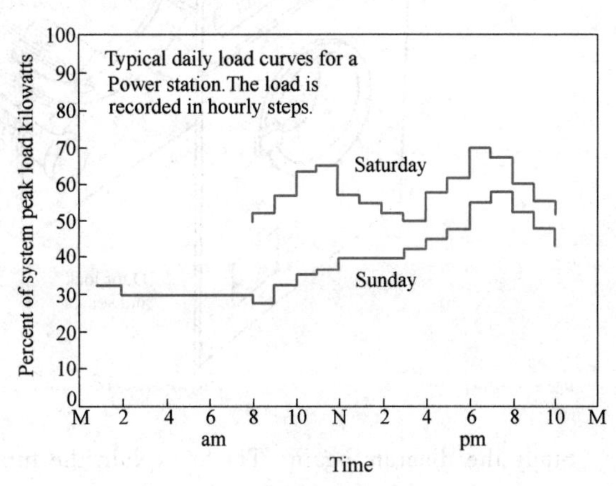

Figure 3.17 Load curves at weekends

5. Sunday. 2 a. m. to 8 a. m.
6. Sunday. 8 a. m. to 9 a. m.
7. Sunday. Noon to 3 p. m.
8. Sunday. 5 p. m. to 10 p. m.

Task 4

Look at Figure 3. 16 and Figure 3. 17. Make comparisons of these periods. For example:

Sunday. 4 a. m. to 8 a. m. /weekdays at the same time.

On Sunday the load remains constant between 4 a. m. and 8 a. m., but on weekdays it rises sharply.

1. Sunday. Noon to 3 p. m. /Saturday at the same time.
2. Weekdays. 10 p. m. to 11 p. m. /Saturday at the same time.
3. Saturday peak load/Sunday peak load.
4. Sunday. Noon to 1 p. m. /the rest of the week at the same time.

Case 28. Describing a Machine—Washing Machine

Task 1

Many items which are found in the home contain control systems. The washing machine is one of the most complex. List some of the factors the control system of a washing machine must handle. The following diagram may help you.

Reading diagrams (see Figure 3. 18): In engineering, diagrams carry a great deal of information. They can also help you to understand the accompanying text. For this reason, it is helpful to try to understand any diagram before reading the text.

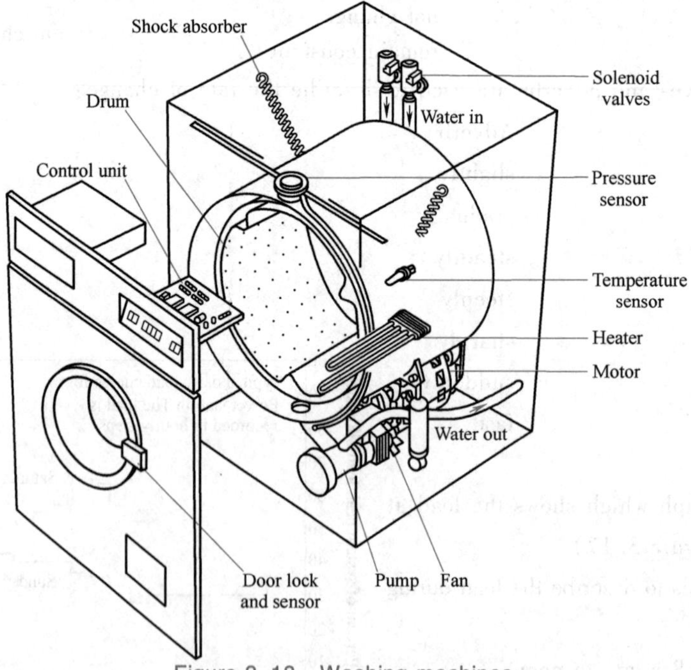

Figure 3. 18 Washing machines

Study the diagram again. Try to explain the function of each of these items.

1. Pump

2. Motor
3. Shock absorber
4. Solenoid valves
5. Heater
6. Pressure sensor
7. Door lock and sensor
8. Temperature sensor
9. Fan

Hints: The following texts will help you with finishing this exercise.

Control systems in the home

Most devices in the home have some sort of control. For example, you can control the volume of a TV by using a remote control. The building blocks of a control system are (see Figure 3.19):

The input can be any movement or any change in the environment. For example, a drop in temperature may cause a heating system to come on.

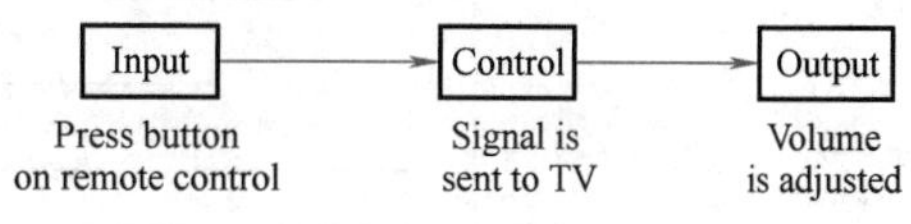

Figure 3.19 The building blocks of a control system

The control may change the size of the output (for example, adjusting the sound of a TV). Often this involves changing one kind of input into a different kind of output. For example, opening a window may set off a burglar alarm.

Outputs can be of many kinds. An alarm system may ring a bell, flash lights, and send a telephone message to the police.

Most control systems are closed loops. That means they incorporate a way of checking whether the output is correct. In other words, they have feedback. The thermostat in a central heating system provides constant feedback to the control unit. (see Figure 3.20)

The control system of a modern washing machine has to take into account several different factors. These are door position, water level, water temperature, wash and spin times, and drum speeds. Most of them are decided when you select which washing program to use.

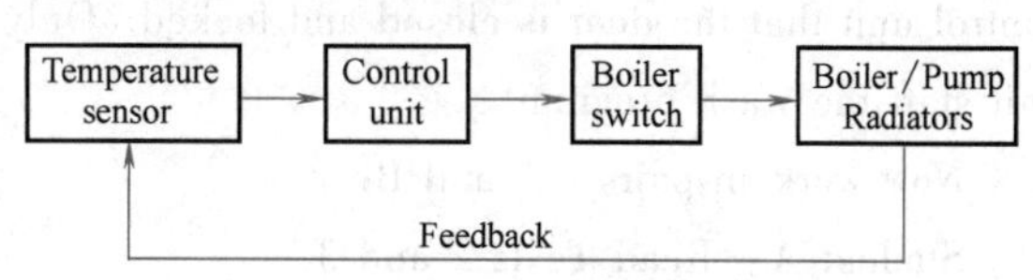

Figure 3.20 Feedback control

Figure 3.21 shows a block diagram of a washing machine control system. You can see that this is quite a complex closed loop system using feedback to keep a check on water level, water temperature and drum speeds.

The control unit is the heart of the system. It receives and sends signals which control all the activities of the machine. It is also capable of diagnosing faults which may occur, stopping the program and informing the service engineer what is wrong. It is a small, dedicated computer which, like other computers, uses the language of logic.

Task 2

Read the following text to find the answers to these questions:

1. What device is used to lock the door?

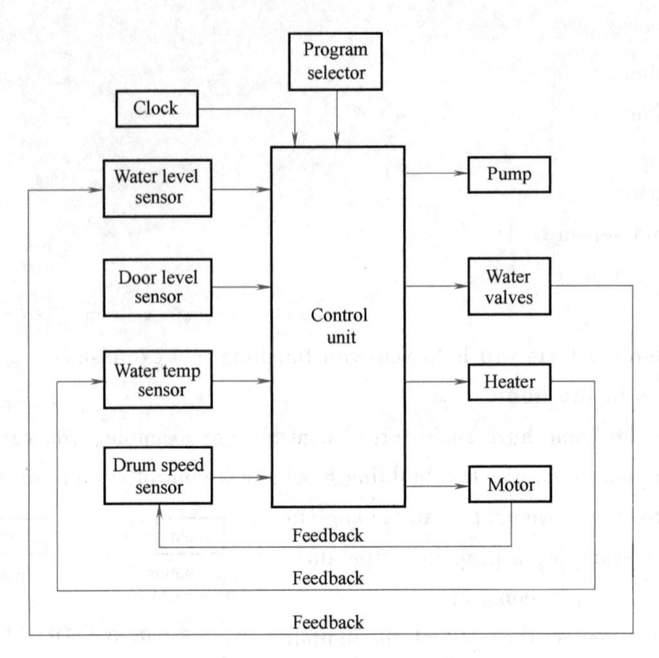

Figure 3. 21 A block diagram of a washing machine control system

2. What provides feedback to the control unit about the door position?

Text 1

Door position

The machine will not start any program unless the door is fully closed and locked. When the door is closed, it completes an electrical circuit which heats up a heat-sensitive pellet. This expands as it gets hot, pushing a mechanical lock into place and closing a switch. The switch signals the control unit that the door is closed and locked. Only when it has received this signal will the control unit start the wash program.

Now work in pairs, A and B.

Student A: Read Texts 2 and 3.

Student B: Read Texts 4 and 5.

Complete your section of the table opposite. Then exchange information with your partner to complete the whole Table 3. 5.

Table 3. 5 Complete the table

Control factor	Operating device	Feedback by
1. Door position	heat-sensitive pellet	switch
2. Water level	________	________
3. Water temperature	________	________
4. Wash and spin times	________	________
5. Drum speeds	________	________

Text 2

Water Level

When a wash program first starts, it has to open the valves which allow the water in. There are usually two of these valves, one for hot water and one for cold. Each must be controlled separately, depending on the water temperature needed for that program. The valves are solenoid operated, i. e. they are opened and closed electrically.

The rising water level is checked by the water level sensor. This is a pressure sensor. The pressure of the air in the plastic tube rises as it is compressed by the rising water. The pressure sensor keeps the control unit informed as to the pressure reached, and the control unit uses the information to decide when to close the water inlet valves.

Text 3

Water temperature

The temperature sensor, a type of thermometer which fits inside the washer drum, measures the water temperature and signals it to the control unit. The control unit compares it with the temperature needed for the program being used. If the water temperature is too low, the control unit will switch on the heater. The temperature sensor continues to check the temperature and keep the control unit informed. Once the correct temperature is reached, the control unit switches off the heater and moves on to the next stage of the program.

Text 4

Clock

The control unit includes a memory which tells it how long each stage of a program should last. The times may be different for each program. The electronic clock built into the control unit keeps the memory of the control unit informed so that each stage of each program is timed correctly.

Text 5

Drum speed

During the washing and spinning cycles of the program, the drum has to spin at various speeds. Most machines use three different speeds: 53 r/min for washing; 83 r/min for distributing the load before spinning; 100 r/min for spinning.

The control unit signals the motor to produce these speeds. The motor starts up slowly, and then gradually increases speed. The speed sensor, a tachogenerator, keeps the control unit informed as to speed that has been reached. The control unit uses the information to control the power to the motor and so controls the speed of the drum at all times.

Case 29. Describing a Machine—Refrigerator

Task 1: Describing the function of components

Study this diagram (see Figure 3.22). It explains how a refrigerator works. In your group try to work out the function of each of the numbered components using the information in the diagram.

Task 2: Dealing with unfamiliar words

You are going to read a text about refrigerator. Your purpose is to find out how they operate. Read the first paragraph of the text below. Underline anywords which are unfamiliar to you.

Refrigeration preserves food by lowering its temperature. It slows down the growth and reproduction of micro-organisms such as bacteria and the action of enzymes which cause food to rot.

You may have underlined words like micro-organisms, bacteria or enzymes. These are words which are uncommon in engineering. Before you look them up in a dictionary or try to find translations in your own language, think the following question: Do you need to know the meaning of these words to understand how refrigerators operate?

You can ignore unfamiliar words which do not help you to achieve your reading purpose.

Now read the text to check your explanation of how a refrigerator works. Ignore any unfamiliar words which will not help you to achieve this purpose.

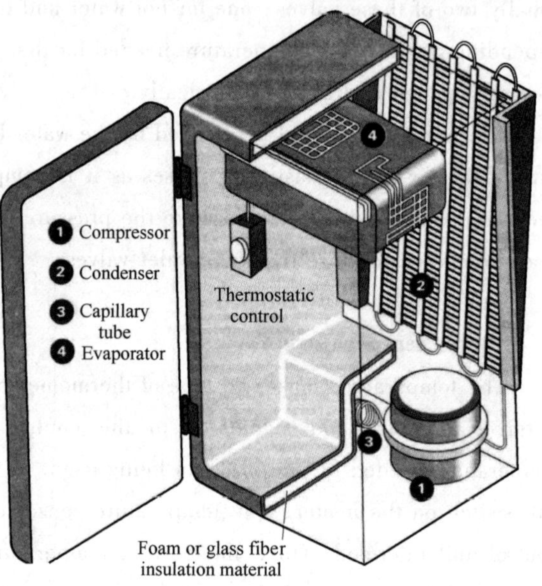

Figure 3. 22 How a refrigerator works

Fridge

Refrigeration preserves food by lowering its temperature. It slows down the growth and reproduction of micro-organisms such as bacteria and the action of enzymes which cause food to rot.

Refrigeration is based on three principles. Firstly, if a liquid is heated, it changes to a gas or vapor. When this gas is cooled, it changes back into a liquid. Secondly, if a gas is allowed to expand, it cools down. If a gas is compressed, it heats up. Thirdly, lowering the pressure around a liquid helps it to boil.

To keep the refrigerator at a constant low temperature, heat must be transferred from the inside of the cabinet to the outside. A refrigerant is used to do this. It is circulated around the fridge, where it undergoes changes in pressure and temperature and changes from a liquid to a gas and back again.

One common refrigerant is a compound of carbon, chlorine, and is fluorine known as R12. This has a very low boiling point: -29℃. At normal room temperature (about 20℃) the liquid quickly turns into gas. However, newer refrigerators which are less harmful to the environment, such as KLEA 134a, are gradually replacing R12.

The refrigeration process begins in the compressor. This compresses the gas so that it heats up. It then pumps the gas into a condenser, a long tube in the shape of a zigzag. As the warm gas passes through the condenser, it heats the surroundings and cools down. By the time it leaves the condenser, it has condensed back into a liquid.

Liquid leaving the condenser has to flow down a very narrow tube (a capillary tube). This prevents liquid from leaving the condenser too quickly, and keeps it at a high pressure.

As the liquid passes from the narrow capillary tube to the larger tubes of the evaporator, the pressure quickly drops. The liquid turns to vapor, which expands and cools. The cold vapor absorbs heat from the fridge. It is then sucked back into the compressor and the process begins again.

The compressor is switched on and off by a thermostat, a device that regulates temperature, so that the food is not over-frozen.

Task 3: Language study—Principles and laws

Study these extracts from the text above. What kind of statements are they?

1. If a liquid is heated, it changes to a gas or vapor.
2. If a gas is allowed to expand, it cools down.
3. If a gas is compressed, it heats up.

Each consists of an action followed by a result. For example:

Action	**Result**
a liquid is heated	it changes to a gas or vapor

These statements are principles. They describe things in science and engineering which are always true. The action is always followed by the same result.

Principles have this form:

If/When (action—present tense), (result—present tense).

Link each action in column A with a result from column B to describe an important engineering principle.

A Action	**B Result**
1. a liquid is heated	a. it heats up
2. a gas is cooled	b. there is an equal and opposite reaction
3. a gas expands	c. it changes to a gas
4. a gas is compressed	d. it extends in proportion to the force
5. a force is applied to a body	e. it is transmitted equally throughout the fluid
6. a current passes through a wire	f. a current is induced in the wire
7. a wire cuts a magnetic field	g. it cools down
8. pressure is applied to the surface of an enclosed fluid	h. it sets up a magnetic field around the wire
9. a force is applied to a spring fixed at one end	i. it changes to a liquid

Task 4: Word study—Verbs and related nouns

Each of the verbs in column A has a related noun ending in—er or—or in column B. Complete the blanks. You have studied these words in this and earlier units. Use a dictionary to check any spellings which you are not certain about.

A Verbs	**B Nouns**
For example: refrigerate	refrigerator
1. condense	________
2. ________	evaporator
3. compress	________

4. resist ____________
5. ____________ charger
6. generate ____________
7. conduct ____________
8. ____________ exchanger
9. radiate ____________
10. control ____________

Task 5: Writing—Describing a process

Study this diagram (See Figure 3.23). It describes the refrigeration process.

When we write about a process, we have to:

1. Sequence the stages.
2. Locate the stages.
3. Describe what happens at each stage.
4. Explain what happens at each stage.

For example:

The refrigeration process begins in the compressor (*sequence and location*). This compresses the gas (*description*) so that it heats up (*explanation*).

In this exercise we will study ways to locate the stages.

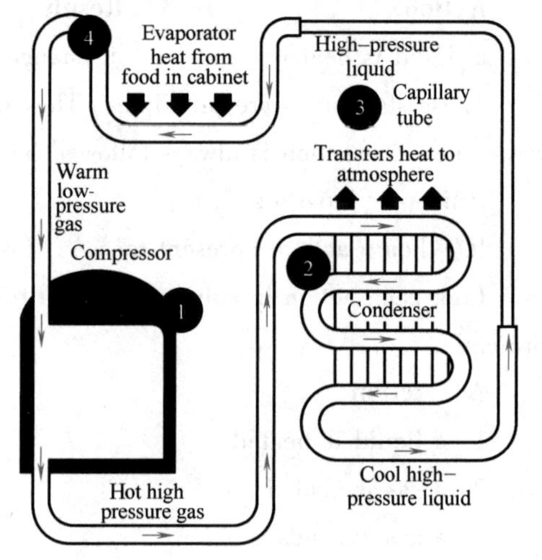

Figure 3.23 Refrigeration process

Put these stages in the refrigeration process in the correct sequence with the help of the diagram above. The first one has been done for you.

1. The liquid enters the evaporator.
2. The gas condenses back into a liquid.
3. The vapor is sucked back into the compressor.
4. The gas is compressed.
5. The liquid turns into a vapour.
6. The gas passes through the condenser.
7. The liquid passes through a capillary tube.
8. The high pressure is maintained.

Hints: There are two ways to locate a stage in a process.

1. Using a preposition+noun phrase. For example:

The liquid turns to vapor **in the evaporator**.

The gas cools down **in the condenser**.

2. Using a where-clause, a relative clause with where rather than which or who, to link a stage, its location and what happens there. For example:

The warm gas passes through the condenser, **where it heats the surroundings and cools down.**

The refrigerant circulates around the fridge, **where it undergoes changes in pressure and temperature.**

Task 6: Describing a process using where-clause

Complete each of these statements.

1. The gas passes through the compressor, where________________.
2. It passes through the condenser, where______________________.
3. The liquid passes through a capillary tube, where______________.
4. The liquid enters the evaporator, where______________________.
5. The cold vapor is sucked back into the compressor, where________.

Case 30. Presentation (1)

Hints: Some phrases for academic presentations

Introduction (after greeting the audience and introducing yourself or being introduced)

1. The subject/topic of my presentation today is…
2. Today I would like to present recent results of our research on …
3. What I want to focus on today is …

Outlining the structure of the presentation

1. I will address the following three aspects of …
2. My presentation will be organized as can be seen from the following slide.
3. I will start with a study of …. Next, important discoveries in the field of … will be introduced.
4. Finally, recent findings of … will be discussed.

Introducing a new point or section

1. Having discussed …, I will now turn to …
2. Let's now address another aspect.

Referring to visual aids

1. As can be seen from the next slide/diagram/table …
2. This graph shows the dependency of … versus …
3. The following table gives typical values of …
4. In this graph we have plotted … with …

Concluding/Summarizing

1. Wrapping up …
2. To summarize/sum up/conclude …

Inviting questions

1. Please don't hesitate to interrupt my talk when questions occur.
2. I'd like to thank you for your attention.
3. I'll be happy/pleased to answer/take questions now.

Dealing with questions

1. I can't answer this question right now, but I'll check and get back to you.
2. Perhaps this question can be answered by again referring to/looking at table …

Task

Work with a partner, study the following notes, then refer to the above phrases for academic presentation and give a short presentation about the subject.

How does a turbofan engine work? The incoming air is captured by the engine inlet. Some of the incoming air passes through the fan and continues on into the core compressor and then the burner, where it is mixed with fuel and combustion occurs. The hot exhaust passes through the core and fan turbines and then out the nozzle, as in a basic turbojet. The rest of the incoming air passes through the fan and bypasses, or goes around the engine, just like the air through a propeller. The air that goes through the fan has a velocity that is slightly increased from free stream. So a turbofan gets some of its thrust from the core and some from the fan. The ratio of the air that goes around the engine to the air that goes through the core is called the bypass ratio.

Because the fuel flow rate for the core is changed only a small amount by the addition of the fan, a turbofan generates more thrust for nearly the same amount of fuel used by the core. This means that a turbofan is very fuel-efficient. In fact, high bypass ratio turbofans are nearly as fuel-efficient as turboprops. Because the fan is enclosed by the inlet and is composed of many blades, it can operate efficiently at higher speed than a simple propeller. That is why turbofans are found on high speed transports and propellers are used on low speed transports. Low bypass ratio turbofans are still more fuel-efficient than basic turbojets. Many modern fighter planes actually use low bypass ratio turbofans equipped with afterburners. They can then cruise efficiently but still have high thrust when dogfighting. Even though fighter planes can fly much faster than the speed of sound and the air going into the engine must travel less than the speed of sound for high efficiency. Therefore, the airplane inlet slows the air down from supersonic speeds.

In short, in the turbofan engine, which is used to power large planes, air is propelled past and into the engine by the turbofan. The air is further compressed by compressor blades, then mixed with fuel and burnt in the combustion chamber/burner. The expanding gases drive the turbine blades, which provide power to the turbofan and the compressor blades, and finally pass out of the rear of the engine, adding to the thrust, as shown in Figure 3. 24.

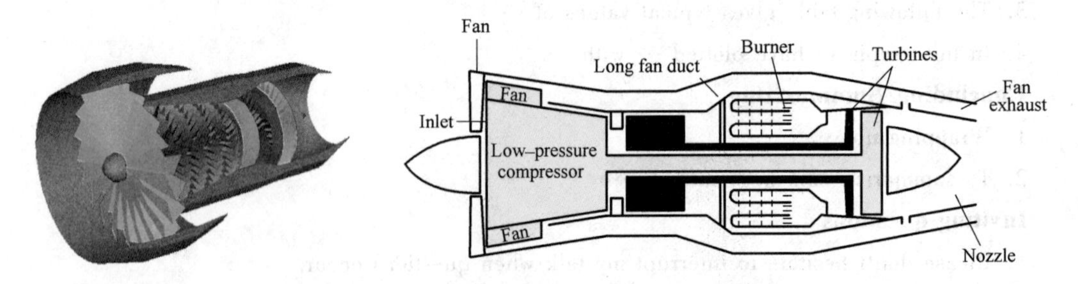

Figure 3. 24 A diagram of the low-bypass turbofan

Case 31. Presentation (2)

Figure 3. 25 shows a rollaway folding guest bed with memory foam mattress produced by a manufacturer.

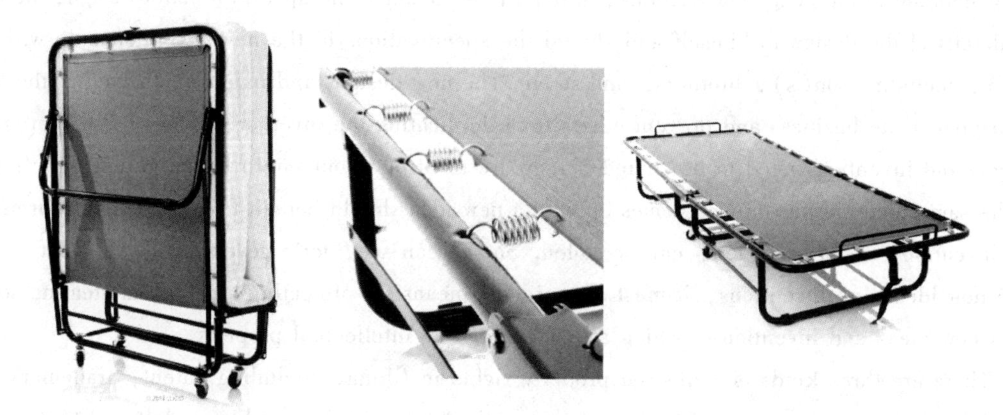

Figure 3. 25 A rollaway folding guest bed with memory foam mattress

Task 1

You can imagine you were the designer of the bed, so now make a PPT-based presentation to introduce it.

Task 2

Work with a partner, then according to the above PPT presentation, deliver an oral report to your partner.

Background Material—About this bed

The following notes provide an introduction to the bed.

Need to accommodate an extra guest? Give them the hospitality of a comfortable bed that's convenient for you to store. This rollaway bed folds up for easy storage, and the four-inch memory foam mattress will give your guest the comfort they need for a good night's rest. The rayon from bamboo cover is ideal for sensitive skin and will help regulate sleeping temperature. Bed frame is 12 inches high with 11 inches of clearance. Supporting the memory foam mattress is a helically suspended poly deck. This deck will absorb impact and the sleeper's weight, successfully avoiding that "bouncy" feeling. The rollaway has two wheels that are able to lock for safety and the structure is specifically engineered for strength. The powder-coated steel has a unique, lightweight design for easy transportation, but is optimized for durability and support. The rollaway is quick and easy to assemble and 25 year warranty is provided.

Hints:

Your presentation should include the following sections.

An introduction—Why do you design this bed?

Main features—Tell the audiences main features of the bed, and how you assure the safety and comfortability of sleepers in the perspective of a mechanical engineer.

Future plan—Do you think the design could be improved further more? If yes, tell your future plan.

Case 32. Patent

Background Material

In the early days, trade information and invention were kept in the family and passed on from one generation to the next. For example, when a plow maker came up with a better design, he kept the details of the design to himself and shared the specifications of the new invention only with his family, including son(s), brothers, and so on. The new designs and inventions stayed in the family to protect the business and prevent others from duplicating the inventor's design. However, new designs and inventions need to be shared if they are to bring about improvement in everyone's life. At the same time the person who comes up with a new idea should benefit from it. Trade information and invention, if not protected, can be stolen. So you can see, for a government, in order to promote new ideas and inventions, it must also provide means for protecting others from stealing someone's new ideas and inventions, which are considered as intellectual property.

There are three kinds of industrial property rights in China, including patent, trademark and copyright. Patent is composed of "patents for invention", "patents for utility model" and "patents for design".The duration of patent for invention is twenty years, the duration of patent for utility model and design is ten years, counted from the application date in China.

According to the China Patent Law, the following items are unpatentable in China:

1. Any invention-creation that is contrary to the laws of the state or social morality or that is detrimental to public interest.

2. Scientific discoveries.

3. Rules and methods for mental activities.

4. Methods for the diagnosis or for the treatment of diseases.

5. Animal and plant varieties.

6. Substances obtained by means of nuclear transformation.

Computer programs as such can't be patented, but may be protected under the "Regulations on Computers Software Protection", formulated in accordance with the Copyright Law. An invention containing a computer program may be patentable if the combination of software and hardware as a whole can really improve prior art, bring about technical results and constitute a complete technical solution.

Task 1

Work with a partner, ask and answer the following questions.

1. How many types of industrial property rights exist in China?

2. What is the duration of patent in China?

3. What kind of invention can't be patented in China?

4. Can computer software be patented in China?

Task 2

Search information related to the Patent Law of the People's Republic of China. Write short answers to the following questions.

1. What is the difference between three kinds of patents?

2. Who decides whether patents are granted or not?

3. What are the conditions of granting patent right?

Hints:

What is the difference between three kinds of patent?

The Rule 1 of Chapter 1 of the Implementing Regulations of the China Patent Law says that invention means any new technical solution relating to a product, a process or improvement.

According to the Rule 2 of Chapter 1 of the Implementing Regulations, utility model in the China Patent Law represents any new technical solution relating to the shape, structure or their combination, of a product, which is fit for practical use.

Design, in the word of the China Patent Law, means any new design of the shape, pattern or their combination, or the combination of the color with shape or pattern, of a product, which creates an aesthetic feeling and is fit for industrial application.

Who decides whether patents are granted or not?

Applications, other than provisional applications, filed with the China Patent Office and accepted as complete applications, are assigned for examination to the respective examining technology centers having charge of the areas of technology related to the invention.

Applications are taken up for examination by the examiner to whom they have been assigned in the order in which they have been filed or in accordance with examining procedures established by the Director.

The examination of the application consists of:

1) A study of the application for compliance with the legal requirements.

2) A search through granted patents, publications of patent applications, foreign patent documents and available literature, to see if the claimed invention is new, useful and non-obvious.

What are the conditions of granting patent right?

Any invention or utility model for which patent right may be granted must possess novelty, inventiveness and practical applicability, and any design for which patent right may be granted must not be identical with and similar to any design which, before the date of filing, has been publicly disclosed in publications in the country or abroad or has been publicly used in the country, and must not be in conflict with any prior right of any other person.

Case 33. Trademark

Task 1: Study the following notes.

The Trademark Law of the People's Republic of China is enacted for the purposes of improving the administration of trademarks, protecting the exclusive right to use trademarks, encouraging producers and operators to guarantee the quality of their goods and services, and maintaining the reputation of their trademarks, with a view to protecting the interests of consumers, producers and operators and promoting the development of the socialist market economy.

The Trademark Office of the administrative authority for industry and commerce under the State Council shall be responsible for the registration and administration of trademarks throughout the country.

The Trademark Review and Adjudication Board, established under the administrative authority for industry and commerce under the State Council, shall be responsible for handling matters of

trademark disputes.

Registered trademarks mean trademarks that have been approved and registered by the Trademark Office, including trademarks, service marks, collective marks and certification marks. The trademark registrants shall enjoy the exclusive right to use the trademarks, and be protected by law.

Collective marks mean signs which are registered in the name of bodies, associations or other organizations to be used by the members thereof in their commercial activities to indicate their membership of the organizations.

Certification marks mean signs which are controlled by organizations capable of supervising some goods or services and used by entities or individual persons outside the organization for their goods or services to certify the origin, material, mode of manufacture, quality or other characteristics of the goods or services.

Regulations for the particular matters of registration and administration of collective and certification marks shall be established by the administrative authority for industry and commerce under the State Council.

Any natural person, legal entity or other organization intending to acquire the exclusive right to use a trademark for the goods produced, manufactured, processed, selected or marketed by it or him, shall file an application for the registration of the trademark with the Trademark Office.

Two or more natural persons, legal entities or other organizations may jointly file an application for the registration for the same trademark with the Trademark Office, and jointly enjoy and exercise the exclusive right to use the trademark.

Any user of a trademark shall be responsible for the quality of the goods in respect of which the trademark is used. The administrative authorities for industry and commerce at different levels shall, through the administration of trademarks, stop any practice that deceives consumers.

Task 2: Work with a partner, ask and answer the following questions.

1. What's the purpose of the Trademark Law of the People's Republic of China?
2. Who is responsible for the registration and administration of trademarks?
3. Who is responsible for handling matters of trademark disputes?
4. How many types of registered trademarks exist in China? What are the differences among them?
5. Who is eligible to apply for and use the trademark?
6. Who shall be responsible for the quality of the goods with an authorized trademark?

Case 34. Copyright

Task 1: Study the following notes.

The Copyright Law of the People's Republic of China is enacted, in accordance with the Constitution, for the purposes of protecting the copyright of authors in their literary, artistic and scientific works and the copyright-related rights and interests, encouraging the creation and dissemination of works which would contribute to the construction of socialist spiritual and material civilization, and promoting the development and prosperity of the socialist culture and science.

Works of Chinese citizens, legal entities or other organizations, whether published or not, shall enjoy copyright in accordance with this law.

Any work of a foreigner or stateless person, which is eligible to enjoy copyright under an agreement concluded between the country to which the foreigner belongs or in which he has habitual residence and China, or under an international treaty to which both countries are parties, shall be protected in accordance with this law.

Works of foreigners or stateless persons first published in the territory of the People's Republic of China shall enjoy copyright in accordance with this law.

Any work of a foreigner who belongs to a country which has not concluded an agreement with China or which is not a party to an international treaty with China, or a stateless person first published in an country which is a party to an international treaty with China, or in such a member state or nonmember state, shall be protected in accordance with this law.

For the purposes of this law, the term "works" includes works of literature, art, natural science, social science, engineering technology and the like which are expressed in the following forms:

1) Written works.

2) Oral works.

3) Musical, dramatic, quyi, choreographic and acrobatic works.

4) Works of fine art and architecture.

5) Photographic works.

6) Cinematographic works and works created by virtue of an analogous method of film production.

7) Drawings of engineering designs and product designs; maps, sketches and other graphic works and model works.

8) Computer software.

9) Other works as provided for in laws and administrative regulations.

Works of publication or distribution which are prohibited by law shall not be protected by this law.

Copyright owners, in exercising their copyright, shall not violate the Constitution or laws, or prejudice the public interests.

The copyright administration department under the State Council shall be responsible for the nationwide administration of copyright. The copyright administration department of the People's Government of each province, autonomous region and municipality directly under the Central Government shall be responsible for the administration of copyright in its administrative region.

Task 2: Work with a partner, ask and answer the following questions.

1. What's the purpose of the Copyright Law of the People's Republic of China?
2. Who is responsible for the registration and administration of copyright?
3. Who is eligible to apply for copyright? Can a foreigner or a stateless person do it?
4. What kind of work is eligible to apply for copyright?
5. Have you ever applied for copyright for your work? If yes, describe the application process.

Case 35. Standards and Codes

Task 1: Study the following notes.

Standards and codes have been developed over the years by various organizations to ensure product safety and reliability in services. The standardization organizations set the authoritative standards for safe food supplies, safe structures, safe water systems, safe and reliable electrical systems, safe and reliable transportation systems, safe and reliable communication systems, and so on. In addition, standards and codes ensure uniformity in the size of parts and components that are made by various manufacturers around the world.

In today's globally driven economy where parts for a product are made in one place and assembled somewhere, standards and codes ensure that parts manufactured in one place can easily be combined with parts made in other places on an assembly line. An automobile is a good example of this concept. It has literally thousands of parts that are manufactured by various companies in different parts of the world, and all of these parts must fit together properly. There are existing international standards that are followed by many manufacturers around the world.

A good example of a product that uses international standards is your credit card or your bankcard. It works in all the ATM machines or store credit card readers in the world. The size of the card and the format of information on the card conform to the International Organization of Standards (ISO), thus allowing the card to be read by ATM machines everywhere.

There are many standardization organizations in the world, among them various engineering organizations. Most national/international engineering organizations create, maintain and distribute codes and standards that deal with uniformity in size of part and correct engineering design practices so that public safety is ensured.

Task 2: Finish the following tasks according to the above notes.

1. Why do we need standards and codes?

2. Write a brief report to explain what is meant by ISO 9000 and ISO 14000 certification.

3. Which organization is responsible for testing drinking water in your city? And which standards and codes the organization follows when testing?

4. When you draft mechanical drawings, what kind of drawing standards do you have to follow? Who create, maintain and distribute this standard? Write a brief note to list several items that designers have to follow while drafting.

5. Investigate the mission of each of the following standards organizations. For each of the organizations listed, write a one-page memo to your instructor about its mission and role.

1) Standardization Administration of the People's Republic of China.

2) China Machinery Industry Federation.

3) China Association of Automobile Manufactures.

4) China National Coal Association.

Case 36. Translation Practice

In the future, when you work in a company, the company may distribute some translation work to you. For example, translate some materials regarding the introduction of the company or a product etc. into English or Chinese.

Task 1: The following introduction is provided by a company in Changzhou city. The writer of this textbook does the translation work for the company. Now take it as your homework, translate it into English.

Notice that translating is a polishing process. During this process, the translator has to consider the differences between cultures, languages, ways of thinking etc., especially Chinese language is a high-context language which carries implicit meanings with more information than the actual spoken parts, while in low-context English language, the messages have a clear meaning, with nothing implied beyond the words used. So it is very important for the translator to be explicit in order to be fully understood. This is also a great challenge for the translator.

Take the following translation as an example, the writer has finished the first version, and then polished it again and again until the last version. Here the writer has provided the Chinese version and two of her translated versions to share with you, you can do a comparison between two translated versions, engage in the polishing process and enjoy it. Obviously, the second English version is better than the first one. What's your take on it?

1. Introduction provided by a Chinese company

企业介绍

常州XX科技有限公司始建于2003年，是一家专业从事铝辊与钢辊制作及金属表面处理工艺的企业。公司占地23,000平方米，建筑面积13,000平方米，固定资产3200万元。公司位于常州市东大门，东邻沪宁高速，南枕京杭大运河，水陆空交通十分便捷。

公司采用先进的技术来指导生产，具有严格的检验流程与品质控制方法，具有较强的机械生产、电镀加工及产品质量检测的能力。公司拥有各类设备180多台套，数控车床、磨床、珩磨机等精密加工机床二十多台，拥有全自动电镀流水线，镀槽18只，并有超长型抛光设备、吊装运输、各类非标专用设备30多台套，生产和检测能力完全满足机械加工和表面处理的供货质量要求。

公司秉承严谨的管理制度，精良的制造工艺，所生产的各类铝辊、钢辊、气缸、油缸、活塞等产品广泛应用于涂装、印刷、造纸、印染、塑料、纺织等行业。精准、稳固、美观、快速为公司产品的四大目标，从产品原料的选用，产品的制作工艺，到不同的表面处理与包覆层，以及精准的动平衡校正，公司将为您提供从原料采购、制作到表面处理一条龙的优质服务。

在发展生产的同时，公司十分注重本地域的生态管理和环境保护。2006年公司通过了ISO 14001环境管理体系认证和ISO 9001质量管理体系认证。2009年公司通过了市环保部门《清洁生产》审核验收。为贯彻落实科学发展观，2009年公司建成了专用废水深度净化回收处理系统，对排放废水处理后再回收利用。

我们热忱地期待您的光临！

2. Translation manuscript as a reference

First version:

Introduction to Our Company

Changzhou XX Technology co., Ltd. was founded in 2003. As a professional manufacturer of aluminum rollers, steel rollers as well as a specialized enterprise of metal surface treatment, our company covers an area of 23,000 square meters, construction area of 13,000 square meters, possessing fixed assets of 32 million RMB. Our company is located in the east of Changzhou. Shanghai-Nanjing expressway is on the east of our company, and the Beijing-Hangzhou Grand Canal is on the south of our company, therefore, the transportation is very convenient.

Our company adopts advanced technology to guide the production, and possesses strong capacity in mechanical processing, electroplating processing as well as rigorous process of quality inspection and control. In addition, our company possesses more than 180 sets of different kinds of equipment, including more than 20 machine tools, such as CNC lathes, grinding machines, honing machines etc., automatic electroplating lines, 18 plating tanks, overlength polishing machines, lifting transportation equipment, and over 30 sets of different special-purpose equipment for machining non-standard parts. Therefore, our production and testing capacity is fully able to meet the quality requirements from our clients on machining and surface treatment.

Our company always adheres to a strict management system and superior manufacturing process. Our products, such as different kinds of aluminum rollers, steel rollers, air cylinders, oil cylinders, pistons etc., are widely used in coating, printing, paper making, printing and dyeing, plastic making, and textile industries. Four goals of our products are accurate, stable, artistic and rapid. All the above goals are ensured through rigorous raw material selection, manufacturing process, surface treatment and coating process, as well as precise dynamic balancing correction. Our company provides a wide range of superior one package service for customers, including raw material purchase, fabrication and surface treatment.

Our company holds regional ecological management and environmental protection paramount concerns while developing our business. In 2006 our company was authorized by ISO 14001 certification (Environmental Management System) and ISO 9001 certification (Quality Management System). In 2009 our company passed evaluation and acceptance of "Cleaner production" which was enacted by municipal environmental protection department. Based on the idea of scientific development, in 2009 our company built up a special wastewater treatment system to purify wastewater, so that wastewater can be recycled and reused.

We are expecting your visit with our sincere service!

Second version:

Introduction to Our Company

Changzhou XX Technology co., Ltd., with an area of 23,000 square meters and fixed assets of 32 million RMB including construction area of 13,000 square meters, was founded in 2003. Our company specializes in manufacturing of aluminum rollers, steel rollers as well as metal surface treatment, locating in the east of Changzhou. Shanghai-Nanjing expressway is on the east of our company, and the Beijing-Hangzhou Grand Canal is on the south of our company, therefore, the transportation is very convenient.

Our company adopts advanced technology to guide the production, and has strong mechanical processing capacity, electroplating processing capacity as well as rigorous process of quality inspection and control. In addition, our company possesses more than 180 sets of different kinds of equipment, including more than 20 machine tools with high precision, such as CNC lathes, grinding machines, honing machines etc., automatic electroplating lines, 18 plating tanks, overlength polishing machines, lifting transportation equipment, and over 30 sets of different special-purpose equipment for machining non-standard parts. Therefore, our production and testing capacity is fully able to meet the quality requirements from our clients on machining and surface treatment.

Our company has always been adhering to rigorous management and superior manufacturing process. Our products, such as different kinds of aluminum rollers, steel rollers, air cylinders, oil cylinders, pistons etc., are widely used in coating, printing, paper making, printing and dyeing, plastic making, and textile industries. Four goals of our products are accurate, stable, artistic and rapid. All the above goals are ensured through rigorous raw material selection, manufacturing process, surface treatment and coating process, as well as precise dynamic balancing correction. Besides, our company provides a wide range of superior one package service for customers, including raw material purchase, fabrication and surface treatment.

Our company holds regional ecological management and environmental protection paramount concerns while developing our business. In 2006 our company was authorized by ISO 14001 certification (Environmental Management System) and ISO 9001 certification (Quality Management System). In 2009 our company passed evaluation and acceptance of "Cleaner production" which was enacted by municipal environmental protection department. In order to carry out the scientific outlook on development, in 2009 our company built up a special wastewater treatment system to purify wastewater, so that wastewater can be recycled and reused.

We are expecting your visit with our sincere service!

Task 2: Translate the following introduction into English. The translation done by the writer is attached as a reference.

学 院 简 介

河海大学是一所拥有百年办学历史、以水利为特色、以工科为主、多学科协调发展的教育部直属全国重点大学，是实施国家“211 工程”重点建设的高校之一。近年来，以科技部批准成立水文水资源与水利工程科学国家重点实验室、国家发改委批准成立水资源高效利用与工程安全国家工程研究中心、教育部先后批准成立研究生院和建设“优势学科创新平台”为标志，学校的建设与发展跨上了一个新的台阶。

河海大学机电工程学院以机械工程学科为引领，以机械设计与制造、材料加工、热能动力、工业设计等为主要特色专业，面向工程应用，培养现代机械设计、制造自动化与信息化、机电系统测试与控制、电厂热能动力工程、空调与制冷、材料加工工程、工业设计、数字媒体等方面的高级工程技术人才。

学院设有机械工程、热能与动力工程、材料科学与工程、工业设计四个系，力学、工程图学两个教研室；设有机械基础江苏省实验教学示范中心、常州校区工程训练中心（江苏省实践教学示范中心）、机电工程学院实验中心三个实践教学平台；设有疏浚技术教育部工程研究中心、机电控制及自动化水利部重点实验室、常州市数字化制造技术重点实验室、常州市光伏系统集成与生产装备技术重点实验室四个学科科研平台；拥有国家工科大学生教学实践基地——"河海大学——宝菱重工教学实践基地"。

学院拥有水利机械二级学科博士点、机械工程一级学科硕士点和材料加工工程二级学科硕士点，具有机械工程领域工程硕士授予权。其中，"机械电子工程"为部级重点学科。学院设有机械工程、能源与动力工程、金属材料工程、工业设计、数字媒体艺术五个本科专业。机械工程本科专业为国家综合改革试点专业、江苏省品牌专业、第一批国家"卓越工程师教育培养计划"试点专业；能源与动力工程本科专业为国家特色专业；机械工程和工业设计本科专业为江苏省"十二五"重点建设专业类。

学院在校本科生 1962 人，在读研究生 445 人，成人教育学生 1028 余人，各类在校生 3445 余人。现有教职工 133 人，其中专任教师 99 人，教授 18 人，副教授 33 人，博士研究生导师 5 人，硕士研究生导师 60 人。教授和副教授占专任教师总人数的 52%。专任教师中江苏省"333 高层次人才培养工程"5 人。

Translation manuscript as a reference

Introduction to the College

Hohai University, which is directly administered by the Ministry of Education, is a state-level key university with 100 years history. With the study and research of water conservancy as its main characteristics, the university mainly focuses on the teaching and research of engineering and also relies on a multi-disciplinary education and development. It is also among the national key universities supported by the state "Project 211". In recent years, a State Key Laboratory of Hydrology-water Resources and Hydraulic Engineering has been established in the university at the approval of the Ministry of Science and Engineering. The establishment of a State Key Engineering Research Center in Efficient Utilization of Water Resources and Engineering Safety has been approved by the National Development and Reform Commission. The Ministry of Education has also approved of the founding of the Graduate School as well as an innovation platform based on preponderant disciplines. All these promote the development of the school to a higher level.

College of Mechanical and Electrical Engineering focuses on mechanical engineering and the programs aim to educate students who can apply solid theoretical knowledge and the expertise to make decision, analyze and solve practical engineering problems in the following fields of modern mechanical design, including automated manufacturing, information technology, mechanical and electrical system test and control, thermal energy and power engineering, air conditioning and refrigeration, materials processing engineering, industrial design, and digital media.

The college consists of 4 departments, 2 teaching and research divisions. The 4 departments are Mechanical Engineering, Thermal Energy and Power Engineering, Materials Science and Engineering, Industrial Design. The 2 divisions are Engineering Graphics and Mechanics. The college has 3 teaching platforms: Jiangsu Experimental Teaching Demonstration Center of Mechanical Foundation Courses, an Engineering Training Center (Jiangsu teaching demonstration center), and an

Experiment Center. The college has also established 4 research platforms: Engineering Research Center of Dredging Technology (Ministry of Education of China), Key Laboratory of Mechatronics (Ministry of Water Resources), Changzhou Key Lab of Digital Technology, Changzhou Key Lab of Photovoltaic System Integration and Equipment Technology. In addition, the college has established a training base for undergraduates under the cooperation of Changzhou Baoling Heavy & Industrial Machinery Co. Ltd.

The college offers Ph. D degrees at first-level discipline in hydraulic mechanical engineering, M. S. degrees at first-level discipline in mechanical engineering, M. S. degrees at second-level discipline in materials processing engineering, and M. S. degrees of engineering in mechanical engineering. Mechatronic engineering is recognized as a key discipline at ministry-level.

The college provides 5 undergraduate degree programs: mechanical engineering, energy and power engineering, metallic materials engineering, industrial design, and digital media. The program of Mechanical Engineering is one of "the National Comprehensive Pilot Reform of Majors", one of the first national pilot majors under the program of "Outstanding Engineers Training Project", and the top-quality program in Jiangsu Province. Energy and power engineering is a state level major with characteristics. Mechanical engineering and industrial design are key majors supported by "the Twelfth Five Year Plan" program of Jiangsu Province.

Currently the college has 3445 students, including 1962 full-time undergraduates, 455 graduate students, and 1028 students for further education. The college has 133 faculty members, including 99 full-time teachers, of whom 18 are professors (8 are supervisors of students studying in a Ph. D program). 33 are associate professors. 60 are supervisors of students studying in a M. S program. 5 are supported by Jiangsu 333 high-level talents training project. Professors and associate professors account for 52% of the total number of full-time teachers.

Case 37. Writing Practice—Resume and Cover letter

Background

Most professional positions require applicants to submit a resume and a cover letter as part of the application process. Your resume is a true "first impression" for a hiring manager. Accordingly, it's important to put time and effort into developing and maintaining an updated, accurate resume. The main goal of any resume is to provide a brief snapshot of your basic skills, talents and ambitions. It is usually a good idea to provide basic contact information, as well as an overview of your educational and work experience, honors and awards, skills and career objective or self-assessment. In most cases, job applicants choose to tailor their resumes to meet the specific criteria of the job for which they are applying. This often includes industry-specific skills and past work experience that may be particularly relevant to the prospective employer.

Task: Imagine you will be graduating from your current college soon. Now write a resume and a cover letter for your future job search. Take the following resume format and a cover letter designed by the writer as references, and you are encouraged to design your own style according to your thought.

<table>
<tr><th colspan="3">Resume</th></tr>
<tr><td>Name</td><td></td><td rowspan="6">Photo</td></tr>
<tr><td>Date of Birth</td><td></td></tr>
<tr><td>Mobile Phone</td><td></td></tr>
<tr><td>Gender</td><td></td></tr>
<tr><td>Email</td><td></td></tr>
<tr><td>Address</td><td></td></tr>
<tr><td>Postal Code</td><td></td><td>QQ</td></tr>
<tr><th colspan="3">EDUCATION AND TRAINING</th></tr>
<tr><td>Oct 2010-Jun 2014</td><td colspan="2">MS, Mechanical Engineering, XX University, 2018</td></tr>
<tr><td>Sep 2014-Apr 2017</td><td colspan="2">BS, Mechanical Engineering, XX University, 2015</td></tr>
<tr><td>…</td><td colspan="2"></td></tr>
<tr><th colspan="3">WORK EXPERIENCE</th></tr>
<tr><td>10 Jul-30 Aug 2011</td><td colspan="2">One month work and study experience in XX</td></tr>
<tr><td>Mar 2016-Present</td><td colspan="2">……</td></tr>
<tr><th colspan="3">COMMUNICATION SKILLS</th></tr>
<tr><td colspan="3">1. Confident, articulate and professional speaking abilities (and experience)
2. Empathic listener and persuasive speaker
3. Speaking in public, to groups, or via electronic media (worked as part time teaching jobs)
4. Excellent presentation and negotiation skills gained by attending many engineering seminars</td></tr>
<tr><th colspan="3">COMPUTER SKILLS</th></tr>
<tr><td colspan="3">1. Good command of OFFICE suit (Word, PowerPoint, Excel)
2. Experienced in using CAD/CAM, AutoCad, C++, Matlab, Adams
…</td></tr>
<tr><th colspan="3">HONOURS AND AWARDS</th></tr>
<tr><td colspan="3">2014-2015: National Undergraduates Scholarship
…</td></tr>
<tr><th colspan="3">CAREER OBJECTIVE</th></tr>
<tr><td colspan="3">To be a part of the challenging team which strives for the better growth of the organization, explores my potential and provides me with the opportunity to enhance my talent with an intention to be an asset to the company</td></tr>
<tr><th colspan="3">SELF-ASSESSMENT</th></tr>
<tr><td colspan="3">Physically fit, mentally sound, academically qualified, technically skilled, highly motivated, decisive and results-oriented dynamic individual with good work ethics and moral characteristics is seeking for a challenging role as a mechanical engineer with the background in design, analysis and modeling and in related areas inside a competitive environment to explore adventurous working opportunities where I am allowed to use my knowledge and skills toward the growth of the organization</td></tr>
<tr><th colspan="3">LANGUAGE SKILLS</th></tr>
<tr><td colspan="3">English (Excellent): Speaking (Fluent), Listening & Reading & Writing (Excellent)
…</td></tr>
</table>

A cover letter as a reference:

Name

Changzhou, China

Mobile Phone: 12345678

xxx@ hotmail. com

04/02/2018

Dear Sir/Madam,

Thanks for the interest shown to my application. I am in search for a suitable position in the field of mechanical engineering. I will complete my undergraduate study from XX University, Changzhou, China in April 2018. I major in mechanical and electrical engineering with a background in design, analysis and modeling. I have knowledge of part and assembly design, analysis and testing. I also have comprehensive knowledge of principles of product design and model analysis, engineering mechanics, thermodynamics, environment engineering, modern system engineering, machine theory and mechanism, electric circuit and electronic technology and economics. Beside oral and written communication skills, good knowledge of computer software and hardware, ability to work in a team, multitasking and interpersonal skills are also my strengths.

At present, I am doing my thesis project on "Kinematics Analysis of a Crawling Robot with Four Feet" My thesis project focuses on constructing a crawling robot with four feet by using Solidworks under the guidance of my supervisor, and then inputting the digital model into Adams to do kinematic simulation, finally checking whether different mechanisms work well according to the results of the analysis.

Additionally, as a socially active, physically fit, mentally sound, academically qualified, technically skilled, highly motivated, decisive and detail-oriented hard-working person with the character of honesty and integrity, I would like to seek a challenging position in mechanical engineering or related areas where I am allowed to use my knowledge and technical skills to get more deeply involved in the growth of the organization.

I do believe my knowledge and experiences make me a suitable candidate for your organization. I have enclosed my resume for your consideration. Please feel free to contact me via phone or email at any time of your convenience for any further information that you might need.

Thanks for your consideration. I am looking forward to your response.

Yours sincerely

Name

Assignments

3.1 How do you interpret the concept of "soft" skills and "hard" skills? List at least four characteristics of them respectively, share your answers with others and take a look at other student's responses.

3.2 List your experiences with design, projects or teamwork. Have you ever been on a team that designed a prototype for a product? If so, describe the situation, especially the characteristics of the team and project that led to success. What were some of the factors?—hard skills? Soft skills? Or any other factors?

3.3 As junior students, you should have had experiences with "course project" in mechanical design. Review problems that teachers gave you, and then write them down in English according to the examples shown in section 3.3.

3.4 Based on your personal experience, which factors could easily cause damage to mobile phones (or to any other products, for example, a camera, a computer, etc.)? List warnings or tips for users. Organize these materials in a clear and concise way.

3.5 Speaking practice: Work in pairs, A and B.

Student A: Play the part of the interviewer. Base your questions on the topics:

What are your career plans after your graduation?

Student B: Play the part of the interviewee.

3.6 Speaking practice: Work in pairs, A and B. Each of you has a diagram of a cam (see Figure 3.26). Describe your diagram to your partner. Your partner should try to reproduce your diagram from the spoken description you provide.

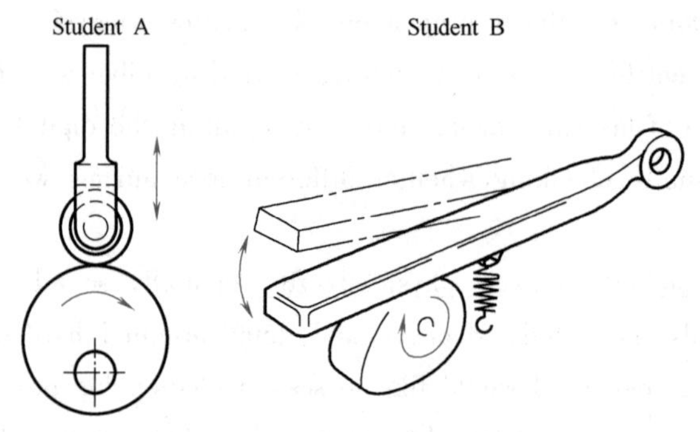

Figure 3.26 A diagram of a cam

Hints:

(1) The following texts will help you with the vocabulary you need:

Cams are shaped pieces of metal or plastic fixed to, or part of, a rotating shaft. A "follower" is held against the cam, either by its own weight or by a spring. As the cam rotates, the follower moves. The way in which it moves and the distance it moves depends on the shape of the cam. Rotary cams are the most common type. They are used to change rotary motion into either reciprocating or oscillating motion.

(2) If you do not understand what your partner says, these questions and phrases may be helpful.

Could you say that again/repeat that, please?

What do you mean by X?

Where exactly is the X?

What shape is the X?

How does the X move?

If your partner does not understand you, try to rephrase what you say.

3.7 Writing practice: Write a short paper in a logic and reasonable way to describe the working principle of washing machine according to the diagram and text given in section 3. 7.

3.8 Imagine you have to give an oral report on a specific topic related to mechanical engineering, the audience is your classmates, and the length of the presentation is about 10 minutes with an additional 5 minutes for questions. Prepare for the presentation, use slides with graphics as an aid. Decide the topic based on your personal interests. The following topics maybe used as a guide.

(1) Present any experiences you have had using 3D CAD software. Have you used solidworks, pro/Engineering, UG. etc.? If so, introduce main features of the software and demonstrate them.

(2) Review text books of machine design which contain gear chains, study planetary gear trains, then give an oral presentation about them, illustrate their types, functions as well as applications. You may search information on the Internet.

(3) Based on your understanding, what is mechatronics? Give an example of mechatronics products or machines, demonstrate how the system works under the control of subsystems.

(4) What is engineering? How does engineering differ from science? What role does design play in engineering?

(5) Imagine you are the human resources manager for a large company. Your company is looking for mechanical engineers, your task is to give a lecture to students. The topic is: What professional skills and personality traits should engineers have in the modern world? Prepare for this topic and try to deliver the lecture to your classmates or dorm mates.

(6) Describe the characteristics of a good presentation in regard to mechanical engineering, give a specific example and show the steps to produce a good presentation.

相关词汇注解

1 problem-solving skill 解决问题的技能
2 cross-cultural communication/intercultural communication 跨文化交流
3 anthropology *n.* 人类学
4 technical communication skill 技术方面进行交流与沟通的技能
5 oral communication skill 口头交流与沟通的技能
6 problem finding 发现（找到）问题
7 written technical report 书面技术报告
8 effective communication 有效的交流与沟通
9 oral report 口头报告（汇报）
10 slide presentation of technical material 技术资料的演示文稿
11 an annotated sketch 包含注解的草图
12 acceleration *n.* 加速度
13 down hill 下坡

14 stand still 静止不动

15 Pugh chart

Pugh chart 是由 Stuart Pugh 发明的一种决策矩阵法，此法可应用于各种领域的不同决策中。它是用于当一个选项中有多个选项存在时对其进行分级的一种定量分析方法，在机械工程中，当要做出设计决定时常用此方法进行定量分析。基本的分析过程是在基于一个评估标准的基础上，将所有潜在的选项进行分解，给每一分项打分并得到每一选项的总分，以此作为对各选项进行分级的评判标准。

16 be quantitative in your exposition to the extent possible 在展示中尽可能用定量分析

17 capstan *n.* 绞盘

18 scenario *n.* 不同的情景（情况）

19 permanent magnet motor 永磁发动机

20 torque *n.* 转矩，扭矩

21 parametric adj. 参数化的

22 layman *n.* 外行

23 academic performance 学习成绩，学业表现

24 a lab report 实验报告

25 experimental error 实验误差

26 raw data sheet 原始资料（数据）清单

27 sample calculation 计算实例（样例）

28 peer and self evaluation 同学互评及自评报告

29 project report 项目报告

30 sponsorship *n.* 赞助

31 industry representative 企业代表

32 peer *n.* 同行

33 visual aid 视觉辅助教具（工具）

34 delivery technique 演讲技巧（技能）

35 question time 提问时间

36 graphical presentation 图形演示

37 abscissa *n.* 横坐标

38 aeronautical *adj.* 航空的

39 heating and ventilating 加热和散热

40 mild steel 低碳钢

41 high carbon steel 高碳钢

42 thermoplastics *n.* 热塑性塑料

43 ABS 丙烯腈-丁二烯-苯乙烯

44 acrylic *adj.* 丙烯酸的

45 nylon *n.* 尼龙

46 thermosetting plastics 热固性塑料

47 epoxy resin ［树脂］环氧树脂

48 polyester resin ［树脂］聚酯树脂
49 urea formaldehyde 脲醛
50 malleable *adj.* 可锻的，可塑的，有延展性的，易适应的
51 wear-resistant *adj.* 抗磨损的
52 insulator *n.*［物］绝缘
53 PCB 电路板
54 valve *n.* 阀门
55 moulding *n.* 铸造
56 reciprocate *v.* 往复运动
57 crankshaft *n.* 曲轴
58 self-lubricating *adj.* 自动润滑的
59 fulcrum *n.* 支点
60 pump *n.* 泵
61 4-bar linkage/crank-rocker linkage 四连杆机构/曲柄摇杆机构
62 crank *n.* 曲柄
63 coupler *n.*［电子］耦合器，随动件
64 follower *n.* 随动件
65 rocker *n.* 摇杆
66 fixed link 固定链
67 workbench *n.* 工作台
68 shock absorber 减振器
69 solenoid valve 电磁阀
70 tachogenerator *n.* 测速发电机
71 regolith *n.* 风化层
72 velocity *n.* 速度
73 equations of equilibrium 平衡方程
74 torque-speed curve 转矩-速度曲线
75 power-speed curve 功率-速度曲线
76 calliper *n.* 卡钳
77 venire calliper 游标卡尺
78 micrometer *n.* 千分尺
79 feeler gauge 塞尺
80 radius gauge 半径量规，半径规
81 go/no go gauge 量规块，塞规
82 bore gauge 孔径规
83 dial gauge 千分表
84 course project 课程设计，课程项目

Part4

How a Car Works?

Objective

A car is a good example of complicated machines. After completing this part, you should be able to:

- build your vocabulary library for cars and related machines;
- understand how a car works;
- understand how an engine, a gearbox and a differential work in a car;
- clearly define and describe parts and machines;
- improve your communication skills in mechanical engineering.

4.1 History of Automobiles

By definition an automobile or a car is a wheeled vehicle that carries its own motor and transports passengers. The automobile as we know was not invented in a single day by a single inventor. All the different parts, such as the engine, the wheels, the gears, and all the fiddly bits like the windscreen wipers, came together very gradually. The history of the automobile reflects an evolution that took place worldwide. It is estimated that over 100,000 patents have created the modern automobile. You can point to the many first inventions that occurred along the way to producing the modern car.

1. Prehistoric Times

It all began with the horse or the camel, or perhaps even the dog. No one really knows which animal prehistoric humans picked on first. People tended to stay in some specific areas, and live more locally than they do now. If they needed to move things about, they had to float them down rivers or drag them by sledge. All that started to change when humans realized the animals around them had raw power they could use and tame. These "beasts of burden" were the first engines.

The next big step was to add wheels and turn sledges into carts. The wheel, which first appeared around 3500 BCE, was one of the last great inventions of prehistoric times. No-one knows exactly how wheels were invented.

Huge and heavy, the first solid wheels were difficult to carve and more square than round. When someone had the bright idea of building lighter, rounder wheels from separate wooden spokes, lumbering carts became swift, sleek chariots. The ancient Egyptians, Greeks and Romans all used chariots to expand their empires. They were a bit like horse-drawn tanks.

Earlier civilizations made small steps by trial and error. The ancient Greeks (the first real scientists) took giant leaps. Greek philosophers (thinkers) realized that a wheel mounted on an axle can magnify a pushing or pulling force. So people now understood the science of wheels for the first time. The Greeks also created gears—pairs of wheels with teeth around the edge lock and turn together to increase power or speed.

Carts and chariots were a big advance, but they were useless for going cross country. That's why ancient Middle Eastern people and the Mediterraneans, who lived in open grassy areas and deserts, developed chariots faster than Europeans and Asians. The Romans were the first to realize that the road should be as good as the carts which travelled on the road. So they linked up their empire with a huge highway network. Roman roads were cutting-edge technology. They had a soft base, underneath the base water was drained away and a harder top made from a patchwork of tight-fitting rocks.

The Greeks gave us gears, the Romans gave us roads, but when it came to engines, the world was still stuck with horsepower. And things stayed that way for hundreds of years through a time known as the Dark Ages, the early part of the Middle Ages, when science and knowledge advanced little in the western world.

2. 17th and 18th Centuries

Early research on the steam engine before 1700 was closely linked to the quest for self-propelled vehicles and ships. Ferdinand Verbiest is suggested to have built what may have been the first steam powered car in about 1672, but very little concrete information on this is known to people.

During the latter part of the 18th century, there were numerous attempts to produce self-propelled steerable vehicles. Many remained in the form of models. Progress was hindered by many problems inherent to road vehicles in general, such as suitable power-plant giving steady rotative motion, suspension, braking, steering, adequate road surfaces, tyres, and vibration-resistant bodywork etc. The extreme complexity of these issues can be said to have hampered progress over more than a hundred years.

3. 19th Century

In 1801, Richard Trevithick constructed an experimental steam-driven vehicle which was equipped with a firebox enclosed within the boiler, with one vertical cylinder, the motion of the single piston being transmitted directly to the driving wheels by means of connecting rods. It was reported as weighing 1520kg fully loaded, with a speed of 14.5km/h (9mph) on the flat. During its first trip it was left unattended and "self-destructed". Trevithick soon built the London Steam Carriage that ran successfully in London in 1803 (see Figure 4.1), but the venture failed to attract interest and soon folded up. In the following years, many people tried to build engine powered vehicles through different ways.

The four-stroke petrol internal combustion engine that constitutes the most prevalent form of modern automotive propulsion is a creation of Nikolaus Otto. The battery electric car owes its beginnings to Ányos Jedlik, one of the inventors of the electric motor, and Gaston Planté, who invented the lead-acid battery in 1859.

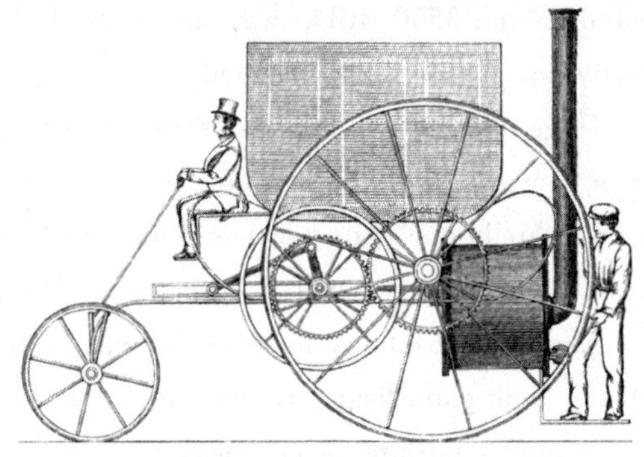

Figure 4.1 Trevithick's London Steam Carriage of 1803

It is generally acknowledged that the first really practical automobiles with petrol/gasoline-powered internal combustion engines were completed almost simultaneously by several German inventors working independently from 1885-1888, such as Karl Benz built his first automobile in 1885 in Mannheim. Benz was granted a patent for his automobile on 29 January 1886, and began the first production of automobiles in 1888. By the end of 19th century, mass production of automobiles had begun in France and the United States.

4. 20th Century

By 1906, steam car development had advanced, and they were among the fastest road vehicles in that period. Throughout this era, development of automotive technology was rapid, due to hundreds of small manufacturers competing to gain the world's attention. Key developments included the electric ignition system, independent suspension and four-wheel brakes. Leaf springs were widely used for suspension, though many other systems were still in use. Transmissions and throttle controls were widely adopted, allowing a variety of cruising speeds, though vehicles generally still had discrete speed settings, rather than the infinitely variable system familiar in cars of later eras. Safety glass also made its debut, patented by John Wood in England in 1905.

Many of today's modern innovations have branched from a man named Preston Tucker, who designed the Tucker 48. Preston Tucker posed his idea of an American-made vehicle in the 1920s and was the man who inspired the idea of a rear-motor and individual torque converters, and went on designing a safety car with innovative features and modern styling. Preston Tucker was the basis of many automotive innovations in the 1920s and had only succeeded in making 50 of these vehicles.

By the 1930s, most of the mechanical technology used in today's automobiles had been invented, although some things were later "re-invented", and credited to someone else. In 1930, the number of auto manufacturers declined sharply as the industry consolidated and matured, thanks in part to the effects of the Great Depression.

Throughout the 1950s, engine power and vehicle speeds rose, designs became more integrated and artful, and automobiles were marketed internationally. Technology developments included the widespread use of independent suspensions, wider application of fuel injection, and an increasing focus on safety in automotive design.

To the end of the 20th century, the United States "Big Three" (GM, Ford, Chrysler) automakers partially lost their leading position, Japan became for a while the world's leader of car production and cars began to be mass manufactured in new Asian, East European, and other countries.

5. Modern era

The modern era is normally defined as 25 years preceding the current year. The modern era has been one of increasing standardisation, platform sharing and computer-aided design, to reduce costs and development time.

Body styles have changed as well in the modern era. The modern era has also seen rapidly rising fuel efficiency and engine output. The automobile emissions concerns have been eased with computerized engine management systems.

Since 2009, China has become the world's largest car manufacturer with sales and production greater than Japan, the United States, and all of Europe. Besides large growth of car production in Asian and other countries, there has been growth in transnational corporate groups, the production of transnational automobiles sharing the same platforms, as well as badge engineering or re-badging to suit different markets and consumer segments.

In recent years, increased concerns over the environmental impact of gasoline cars, higher gasoline prices, improvements in battery technology, and the prospect of peak oil, have brought about interest in all-electric vehicles and plug-in hybrid vehicles, which are perceived to be more environmentally friendly and cheaper to maintain and run.

The auto industry is constantly bringing us new thrilling technologies, whether it is for safety, entertainment, usefulness or simply for pure innovation. Google, Tesla and even Toyota have been trying to develop a self-driving car. In essence, that implies that cars will be connected to the cloud and a specific network that helps people stay safe while getting to and from their destinations automatically.

Some other car makers are trying to let a car "talk to" other cars to avoid a crash. Connectivity gets more exciting when vehicles are connected with other vehicles and even the highways they travel. V2V, or vehicle-to-vehicle communication, allows cars to communicate with each other over a dedicated Wi-Fi band and share information about vehicle speed, direction of travel, traffic flow, and road and weather conditions. The next step in connectivity is V2I, or vehicle-to-infrastructure communication. Urban traffic congestion already costs society billions of dollars in wasted fuel and productivity, and the problem is growing rapidly. When vehicles are connected to smart highways and traffic lights, then linked to highly accurate, real-time traffic updates and navigation systems, we can significantly reduce congestion and urban commute times, in addition to further improving vehicle safety. The rapid advancement of intelligent and connected technologies is also providing the foundation for automated vehicles that make driving safer and easier.

Another pretty promising technology is airbags or braking bags in the vehicle floor or sides for improved safety. This airbag housed within the vehicle floor will deploy when a crash is deemed to be unavoidable, and it uses a friction coating to support the vehicle against the road surface. It'll be a while or even a long time before the braking bag becomes available to the public, but from what some carmakers have done, it can be seen that the airbag is a pretty promising technology indeed.

Task 1

Since 2009, China has become the new world's largest car consuming country with production more than US, Japan or all of Europe. We have witnessed internal and external changes of cars. Now work in pairs to complete the following tasks.

1. According to your perspective, describe changes of cars over the past twenty years to your partner.

2. What kind of car do you like? Why? Tell the reasons to your partner.

3. What are your expectations for future cars?

4.2 Introduction to a Car

A car is a good example of complicated machines. In modern society, cars are so abundant and common that it is hard to imagine a world without cars. In this section you will learn the vocabulary for inside and outside a car using pictures, try to build your vocabulary library for car parts, then explore how a car works.

Task 2

Look at the following pictures (Figure 4.2) to learn vocabulary for parts of a car inside and outside.

Do you have a car? If yes, take a look at your car, find out the following parts in your car, and describe where they are.

Accelerator; ashtray; brake; CD player; clock; clutch; dashboard; door handle; emergency brake; emergency flare; front seat; gasoline level gauge; gearshift; glove compartment; headlight switch; odometer; oil gauge; rear seat; rearview mirror; steering wheel; sun visor; tachometer; temperature gauge; turn signal

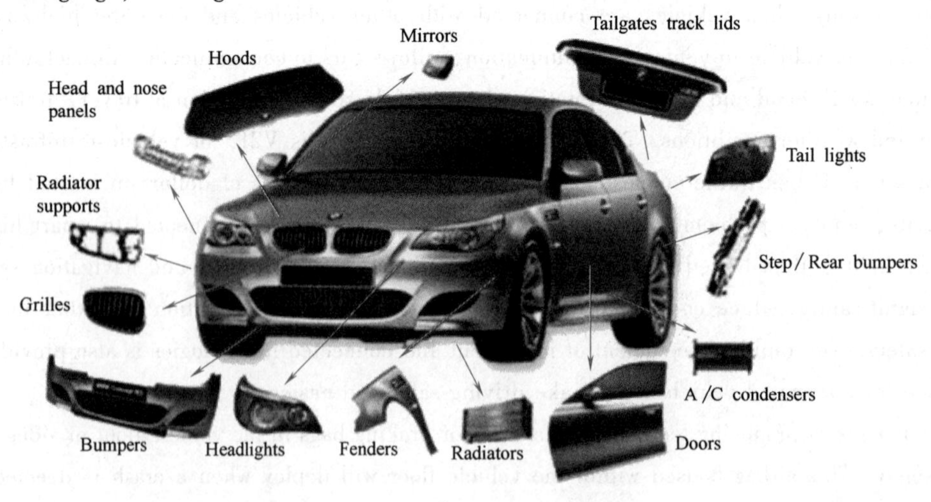

Figure 4.2 Anatomy of a car

a) Outside of a car

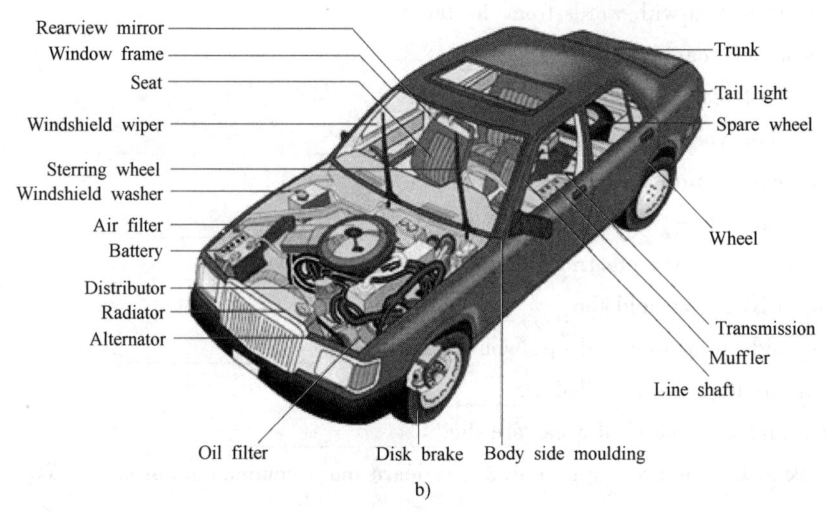

b)

Figure 4. 2 Anatomy of a car (continued)

b) Inside of a car

Task 3: Learn the following car parts vocabulary in alphabetical order (see Table 4. 1).

Table 4. 1 Car parts

letter	
A	accelerator; air bags; air conditioner; air filter; air vent; alarm; antenna; anti-lock brakes; armrest; auto; automatic transmission; automobile
B	baby seat; back-up lights; battery; brake light; brakes; bumper
C	camshaft; car; carburetor; chassis; chrome trim; clutch; cooling system; crankshaft; cruise control
D	dashboard; defroster; diesel engine; differential; dimmer switch; door; door handle; drive shaft
E	emergency brake; emergency lights; emissions; engine; exhaust system
F	fan belt; fender; floor mats; frame; fuel; fuel gauge; fuse; flat tire
G	gas cap; gasket; gasoline engine; gearbox; gearshift; gear stick; glove compartment; grille
H	hand brake; headlight; heater; high-beam headlight; hood; horn; hubcaps; hybrid
I	ignition; interior light; internal combustion engine
J	jack
K	key
L	license plates; lock; low-beam headlight; lugs
M	manifold; manual transmission; mat; mirror; motor; mud flap; muffler
O	odometer; oil filter
P	parking lights; passenger seat; piston; power brakes; power steering
R	radiator; radio; rear-view mirror; rear window defroster; rims; roof; rotary engine
S	seat; seat bags; shock absorber; side mirrors; spare tire; spark plug; speedometer; steering wheel; suspension
T	tachometer; tailgate; thermometer; tire; trailer hitch; trunk; turbocharger; turn signal
U	unleaded gas
V	vents; visor
W	wheel; wheel well; windshield; windshield wiper

Fill the gaps in thc tcxt with words from the table 4. 1.

1. Another word for "car" is________________.
2. A person who fixes a car is a________________.
3. At night, when you drive, you must switch on your________________.
4. When you make a turn, you must switch on your________________.
5. Car fuel is called "gas" in America and in England________________.
6. Your car starts by using electric power from the________________.
7. When you drive, you hold the________________.
8. When you add power to speed up, you change your________________.
9. A car that has no gears is called an________________.
10. The front and back parts of a car are the________________.

Task 4: Look at the picture (Figure 4. 3) to learn main components of a car. Work in pairs and describe how a car works.

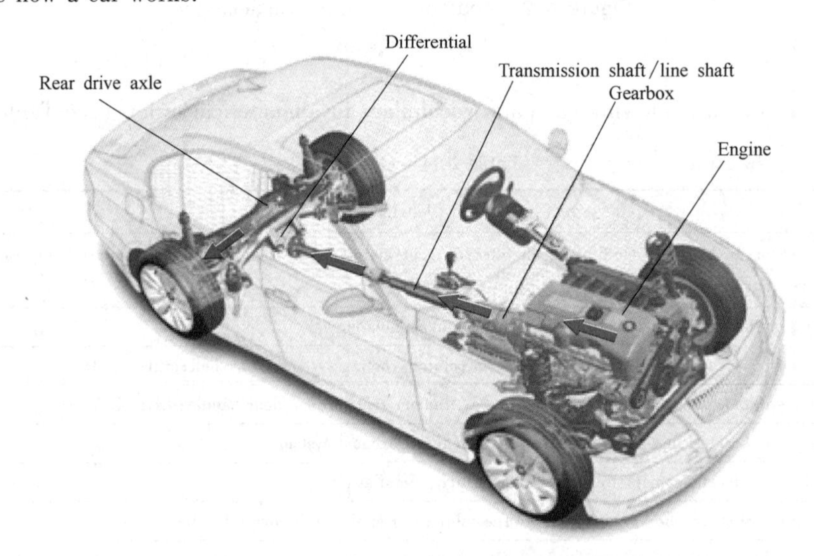

Figure 4. 3 Main components of a car

Hint: How does a car work? When a driver turns a key in the ignition, which ignites the fuel-air mix in the engine. A motor car engine is an Internal Combustion Engine (ICE). Energy is created by burning either diesel or petrol in a combustion chamber. In such an engine, the energy is transferred from the pistons moving up and down at high speeds. This up-down motion is converted to a rotary motion through the crankshaft. The engine and the gearbox are bolted together, with the clutch between them. A gearbox converts the rotational energy of the engine to a rotational speed appropriate for line shaft. In rear-wheel drive cars, the differential converts rotational motion of the transmission shaft which lies parallel to the car's motion to rotational motion of the half-shafts or rear drive axle (on the ends of which are the wheels), which lie perpendicular to the car's motion.

Task 5: Work in pairs and discuss why a differential is used in a car.

Hint: Why a differential is needed in a car? When a car turns a corner, one wheel is on the "inside" of a turning arc, and the other wheel is on the "outside". Consequently, the outside wheel has to turn faster than the inside one in order to cover the greater distance in the same amount of time. Thus, because the two wheels are not driven with the same speed, a differential is necessary. A car differential is placed halfway between the driving wheels, on either the front, rear, or both axes (depending on whether it's a front-, rear-, or 4-wheel-drive car).

4.3 Engines

In this section, you will learn the parts of an engine and its working principle.

4.3.1 Main Parts of a Car Engine

The engine or the motor (see Figure 4.4) is the heart of a car. It is a complex machine built to convert heat from burning gas into the force that turns the road wheels. The chain of reactions which achieve that objective is set in motion by a **spark**, which ignites a mixture of petrol vapour and compressed air inside a momentarily sealed **cylinder** and causes it to burn rapidly. That is why the machine is called an **internal combustion engine**. As the mixture burns, it expands, providing power to drive the car.

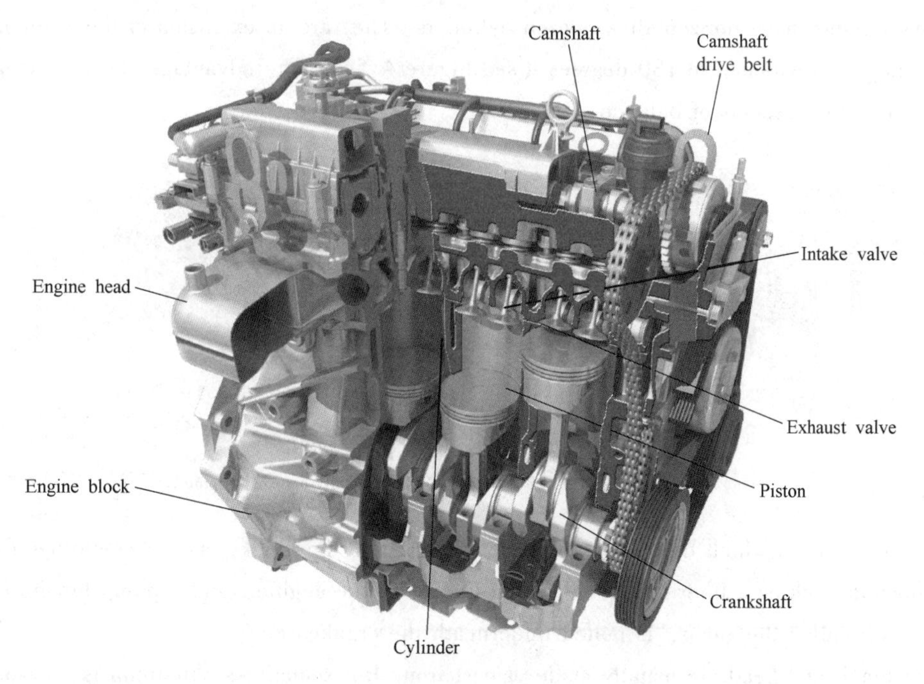

Figure 4.4 Main parts of a car engine

To withstand its heavy workload, the engine must be a robust structure. It consists of two basic parts: the lower, heavier section is the **cylinder block or engine block**, a casing for the engine's main moving parts; the detachable upper cover is the **cylinder head or engine head.**

The cylinder head contains valve-controlled passages, through which the air and fuel mixture enters the cylinders, others and the gases produced by the combustion are expelled.

The block houses the **crankshaft**, which converts the **reciprocating motion** of the **pistons** into **rotary motion** at the crankshaft. Often the block also houses the **camshaft**, which operates mechanisms that open and close the **valves** in the cylinder head. Sometimes the camshaft is in the head or mounted above it.

Task 6: Work in pairs and discuss the following questions.

1. Why is the engine the heart of a car?
2. What are the main parts of a car engine?
3. What is the main function of the crankshaft?

4.3.2 Different Engine Layouts

The simplest and most common type of engine comprises 4 vertical cylinders close together in a row (see Figure 4.5a). This is known as an inline engine. Cars with capacities exceeding 2000cc often have 6 cylinders in line.

The more compact V-engine is fitted in some cars, especially vehicles with 8 or 12 cylinders, and also some with 6 cylinders. The cylinders are arranged opposite each other at an angle of up to 90 degrees (see Figure 4.5b).

Some engines have horizontally-opposed cylinders. They are an extension of the V-engine, the angle having been widened to 180 degrees (see Figure 4.5c). The advantages lie in saving height and also in certain aspects of balance.

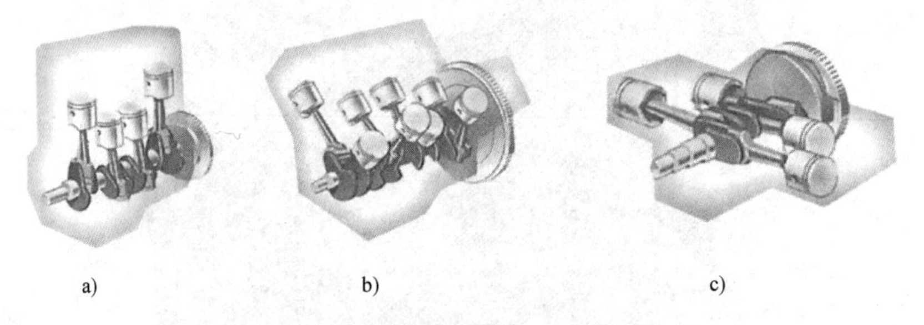

a) b) c)

Figure 4.5 Different engine layouts

a) Inline engine b) V8-engine c) Horizontally-opposed engine

The cylinders in which the pistons operate are cast into the block, as are mountings for ancillary equipment such as a filter for the oil which lubricates the engine, and a pump for the fuel. An oil reservoir, called the sump, is bolted underneath the crankcase.

Both block and head are usually made of cast iron. But sometimes aluminum is chosen for the head, because it is lighter and dissipates heat more efficiently.

Task 7: Work in pairs, complete the following tasks and share your thoughts with your partner.

1. How many engine layouts are there?
2. Discuss and describe the advantages and disadvantages of three engine layouts.
3. Check the engine type in your car, try to search information related to your model on the Internet, see what kind of information you will find and write them down.

4. 3. 3 Types of Car Engines

Car engines vary in design, but certain elements are common to all engines and are used for engine classification. Engines can be classified in several ways such as the number of cylinders, fuel type and cylinder arrangement or type of ignition system used. The two major engine types in use are spark ignition (gasoline engine) and compression ignition (diesel engine) which use different types of fuel. The following are ways engines are classified.

Fuel Type

There are two main fuel types used in car engines: gasoline and diesel. These also classify the type of ignition in the engine as well. More recent alternatives to gasoline and diesel include electricity, ethanol, methanol, hydrogen, propane and natural gas.

Cylinder Arrangement

The cylinders of a car engine are arranged in two different ways, inline or "V". An inline engine indicates that the cylinders are arranged in a row. The "V" arrangement is two rows of cylinders side-by-side that form a "V" shape.

Number of Cylinders

The number of cylinders in a car engine ranges from 3 to 12. The number of cylinders also determines the amount of power produced by the engine. For example, a 6-cylinder engine is more powerful than a 4-cylinder engine.

Ignition

The ignition type of an engine relates to how the fuel is ignited inside the engine. The fuel is either spark-ignited with a spark plug or compression-ignited by compressing air until the temperature reaches ignition. Gasoline engines use spark ignition while diesel engines are compression-ignited.

Strokes per Cycle

Strokes per cycle are the number of times the pistons travel up and down during one cycle. Modern engines have four strokes per cycle: intake, compression, power and exhaust. Two-stroke engines are not used due to their poor power output at low rpm, motor oil mixed with the fuel, less fuel efficient, generate an unacceptable amount of pollution, and require more maintenance.

Cooling

Without a cooling system, car engines will quickly destroy themselves due to extreme temperatures. Car engines are either liquid-or air-cooled. Liquid-cooled engines are the modern industry standard and are cooled by a water pump that circulates water throughout the engine.

Task 8: Work in pairs, discuss and answer the following questions.

1. When a car is referred to as V8, V10 and V12, what does that mean?

2. Except the two main fuel types used in car engines to drive a car—gasoline and diesel, are there any other ways to drive a car?

4. 3. 4 Gasoline Engine

We have used thermal engines widely since J. Watt invented a reciprocating steam engine with a condenser in 1769, and they are used in our life. There are different kinds of engines which use thermal energy. They are: ①Reciprocating steam engine, ②Stirling engine, ③Ericsson engine, ④Gasoline engine, ⑤Steam turbine, ⑥Diesel engine, ⑦Gas turbine, ⑧Rocket engine, ⑨Fuel cell.

Gasoline Engine

Nowadays, the gasoline engine (spark ignition engine) is used widely as the power source of automobiles. A gasoline engine is a type of internal combustion engine that burns gasoline for fuel. A gasoline engine is also referred to as a petrol engine as petrol is another, primarily British term for gasoline.

So how does this engine work? When the piston moves down the cylinder on the intake stroke, it draws air from the cylinder and intake manifold. A vacuum is created that draws air from the carburetor, as shown in Figure 4.6. The airflow through the carburetor causes fuel to be drawn from the carburetor through the intake manifold past the intake valves and into the cylinder. The amount of fuel mixed into the air to obtain the required air to fuel ratio is controlled by the venturi or choke. A mixture gas of the fuel and the air is compressed in the cylinder at the first. And the gas explodes by use of an ignition plug, and generates the output power. So the gasoline engine is also called spark ignition engine. Almost all vehicles on the road today run on a gasoline engine. As good characteristics of the engine, it can be realized a smaller and light-weight engine, and has a possibility of the high engine speed and high power. Besides, the maintenance of the engine is very simple.

All gasoline engines like diesel engines work based upon what is known as the 4-stroke cycle approach, involving the intake of fuel and air, the compression of air and fuel, the generation of power through combustion, and finally, the expulsion of burned gas through an exhaust pipe. In 4-stroke engines the power is produced when piston performs expansion stroke. During 4 strokes of the engine two revolutions of the engine's crankshaft are produced. Other types of gasoline engines include 2-stroke engines and rotary engines.

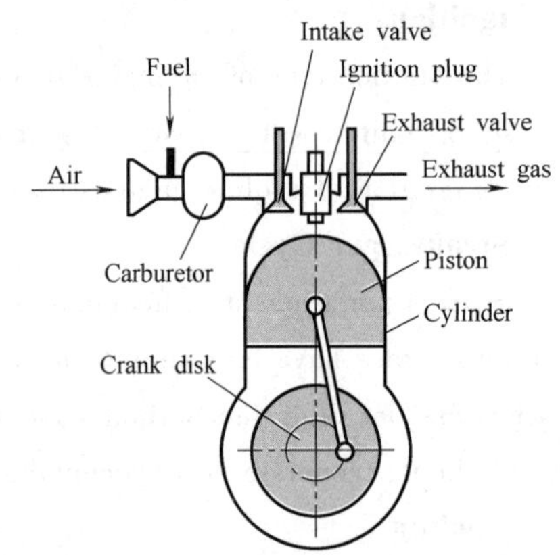

Figure 4.6 Working principle of a gasoline engine

Spark ignition engine can work either on 2-stroke or 4-stroke cycle. Both the cycles are described below.

1. 4-stroke Engine

A 4-stroke engine (also known as 4-cycle) is an internal combustion engine in which the piston completes 4 separate strokes which comprise a single thermodynamic cycle. A stroke refers to the full travel of the piston along the cylinder, in either direction. The 4 strokes of the 4-stroke engine are: suction of fuel or intake stroke, compression of fuel or compression stroke, expansion or power stroke, and exhaust stroke (see Figure 4.7).

Intake: This stroke of the piston begins at top dead center. The piston descends from the top of the cylinder to the bottom of the cylinder, increasing the volume of the cylinder. A mixture of

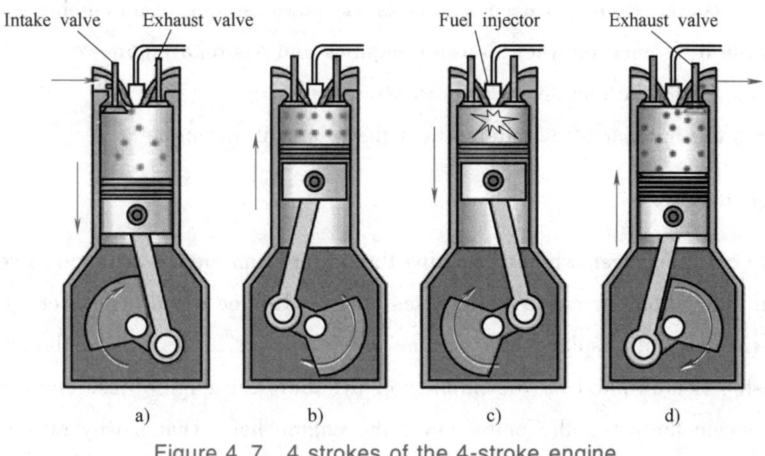

Figure 4. 7 4 strokes of the 4-stroke engine

a) Intake b) Compression c) Power d) Exhaust

fuel and air is forced by atmospheric (or greater) pressure into the cylinder through the intake valve.

Compression: With both intake and exhaust valves closed, the piston returns to the top of the cylinder, compressing the air or fuel-air mixture into the cylinder head.

Power: This is the start of the second revolution of the cycle. While the piston is close to top dead centre (TDC), the compressed air-fuel mixture in a gasoline engine is ignited, by a spark plug or an ignition plug in a gasoline engine. The resulting pressure from the combustion of the compressed fuel-air mixture forces the piston back down toward bottom dead center (BDC).

Exhaust: During the exhaust stroke, the piston once again returns to top dead center while the exhaust valve is open. This action expels the spent fuel-air mixture through the exhaust valve(s).

At the end of the exhaust stroke, the exhaust valve is closed and the engine begins another intake stroke. The whole cycle then repeats itself to generate power by endlessly repeating a series of 4 strokes.

2. 2-stroke Engine

In case of the 2-stroke, the intake/suction and compression strokes occur at the same time. Similarly, the expansion/power and exhaust strokes occur at the same time. Power is produced during the expansion stroke. When 2 strokes of the piston are completed, one revolution of the engine's crankshaft is produced.

In 4-stroke engines the engine burns fuel once for two rotations of the wheel, while in 2-stroke engine the fuel is burnt once for one rotation of the wheel. Hence the efficiency of 4-stroke engines is greater than the 2-stroke engines. However, the power produced by the 2-stroke engines is more than the 4-stroke engines.

Task 9: Work in pairs, discuss and answer the following questions according to the notes above.

1. Describe the working principle of gasoline engine (spark ignition engine).

2. Why the gasoline engine is used widely as the power source of automobiles?
3. What's the difference between 2-stroke engines and 4-stroke engines?
4. How does a 4-stroke engine make power?
5. How does the number of cylinders affect the power of an engine?

4.3.5 Engine Size

What does cc or liter mean when describing the power of an engine or engine size? Engine sizes are measured in either liters or cubic centimetres (cc)—1000cc equals one liter. This number basically is approximately the displacement volume of the engine, i. e., the volume covered by the stroke of the piston is multiplied by the number of cylinders the engine has. As a rough-and-ready rule, the more space there is, the more power the engine has. That's why city cars such as the Skoda Citigo use a tiny 1.0-liter engine to move them around, while the high performance Ferrari F12 uses a 6.3-liter unit to achieve its 217mph top speed. It's worth bearing in mind that older cars usually have bigger engines, but it isn't quite fair to compare them to new cars. Technology moves at such a rapid rate that the latest vehicles can easily make the same power from smaller engines.

But all things aren't always equal. Sometimes a large displacement engine will produce a lot less power than a smaller displacement engine and this can be done with changing compression ratios, larger or smaller fuel and air injection, different exhaust ports, and turbo chargers. Large displacement engines are naturally heavier, so high performance cars will try and get more power from a smaller engine with more costly features. So as a very basic rule, you can expect that a large engine will be more powerful than a small one. Advances in technology mean that there are small turbo-charged engines now available that are both more efficient and more powerful than older, larger engines. However, generally you can still expect that out of two engines, particularly if they're of similar age, the larger one will be more powerful than the smaller one.

If a car is a 2.0-liter model, it has an engine capable of displacing roughly 2000cc or two liters of volume with each stroke, i. e., a 2.0-liter engine has two liters of cylinder space altogether. So if it has 4 cylinders, each cylinder will be 500ml in volume.

Task 10: Work in pairs, discuss and answer the following questions according to the notes above.

1. What does engine size mean?
2. What does 1.6-liter, or 1.6, and V6 mean?
3. How does engine size affect performance of a car?
4. How does engine size affect fuel economy and how does it affect your choice of a car?
5. Discuss factors that affect fuel economy except engine size.

Hints:

How does engine size affect performance of a car?

A car with a large engine is likely to accelerate quicker, have a higher top speed and superior towing ability comparing with a car with a smaller engine. However, this won't be true absolutely in every comparison, particularly if you compare cars from different eras, because advances in technology have meant that smaller engines have become more powerful, and also changing compression ratios, larger or smaller fuel and air injection, different exhaust ports, and turbo chargers will make an engine more efficient and more powerful.

How does engine size affect fuel economy?

A large engine often has more cylinders and a greater capacity to burn fuel than a small one, so it uses more fuel as a result. This is an important thing to consider when buying a new car. Most buyers want to strike a balance between a car that has sufficient power for their needs and one that doesn't cost too much in fuel bills.

Small engines tend to be suited to cars that are used predominantly around the town. They provide enough performance for short journeys, like trips to the supermarket, school and office, where high speed and rapid acceleration aren't really necessary. As the engine isn't regularly needed to produce lots of power, it makes sense to keep it small and take advantage of the gains in economy.

If you buy a car with a small engine but try to run it in a way less suited to its strengths, like on long motorway journeys, it will have to work much harder than a larger engine. Not only will it be louder and less refined than an engine suited to motorway cruising, but also less efficient and likely to suffer from more wear and tear because it's under more strain than a larger engine at higher speed.

Beyond the size of the car's engine, driving style can also contribute to how much fuel the engine uses. Keeping the revs low by changing up to the highest possible gear will help to save fuel as with accelerating and braking gently; keeping tyres correctly inflated could also help to save a number of fuel each year. Car's engine size and power will also have an effect on users' insurance premium. Cars in low insurance groups (i. e. they are cheap to insure) tend to have smaller, less powerful engines.

A V6 Engine

A V6 engine is a V engine with 6 cylinders mounted on the crankshaft in two rows of 3 cylinders, usually set at either a 60 or 90 degree angle to each other. The V6 engine is one of the most compact engine configurations, usually ranging from 2. 0 L to 4. 3 L displacement, shorter than the inline 4 and more compact than the V8 engine. Because of its short length, the V6 fits well in the widely used transverse engine front-wheel drive layout. It is becoming more common as the space allowed for engines in modern cars is reduced at the same time as power requirements increase, and it has largely replaced the inline 6, which is nearly twice as long—too long to fit in many modern engine compartments, and the V8, which is larger, more expensive, and has poorer fuel economy. The V6 engine has become widely adopted for medium-sized cars, often as an optional engine where an inline 4 is standard, or as a standard engine where a V8 is a higher-cost performance option.

Most frequently the cylinders are arranged in a row known as "inline" or "straight", but it's not uncommon especially on larger engines, where alternate cylinders are arranged in a "V" formation, hence car badges state "V6" or "V8", among others. Less common are horizontally-opposed engines, where the cylinders "V" angle has been widened out to 180 degrees, so they're also called "flat" motors. Rarer engines still are "W" format engines, which is similar to a "V" but with cylinders pointing

in three directions rather than two.

4.4 Manual Transmissions (MT)

Do you know the main difference between automatic cars and manual cars? When you drive a stick-shift car, you may have several questions floating in your head. How does the "H" pattern that I am moving this shift knob through have any relation to the gears inside the transmission? What is moving inside the transmission when I move the shifter? When I mess up and hear that horrible grinding sound, what is actually grinding? What would happen if I were to accidentally shift into reverse while I am speeding down the freeway? Would the entire transmission explode?

In this section, we will explore the interior of a manual transmission in a car. A transmission is a gadget, which provides controlled application of the power. Often the term transmission refers simply to the gearbox that uses gears and gear trains to provide speed and torque conversions from a rotating power source to another device. A manual transmission is also known as a manual gearbox, stick shift, n-speed manual (where n is its number of forward gear ratios), or MT (Manual Transmission). It uses a driver-operated clutch engaged and disengaged by a foot pedal (automobile) or hand lever (motorcycle) for regulating torque transfer from the engine to the transmission, and a gear selector operated by hand (automobile) or by foot (motorcycle). A conventional 5-speed manual transmission is often the standard equipment in a base-model car.

4.4.1 Role of a Gearbox

In a front-engined rear-wheel-drive car, power is transmitted from the engine through the clutch and the gearbox to the rear axle by means of a tubular propeller shaft, as shown in Figure 4.8. A transmission/gearbox is the part of an engine assembly that connects the engine to the wheels.

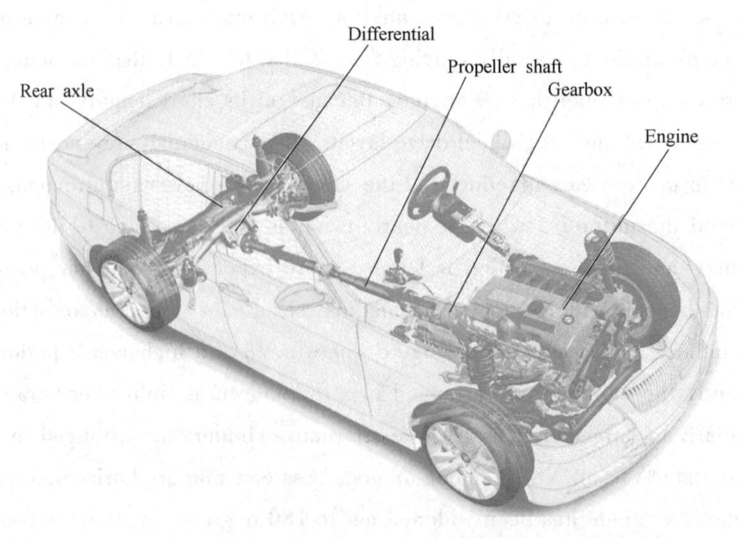

Figure 4.8 Gearbox connecting the engine to the wheels

Why is a gearbox needed in a car? The gearbox adapts the output of the engine to the drive wheels. In a car the engine operates at a relatively high rotational speed, which is inappropriate for starting, stopping and slow travel. The gearbox reduces the higher engine speed to the slower wheel speed, meanwhile increases torque in the process. In a car where multiple speeds are needed, a gearbox with multiple gears can be used to increase torque while slowing down the output speed. For a car with manual gearbox, drivers will start in first gear, gain some speed, then shift to second gear and increase speed more etc. With an automatic transmission, the same process is applied with the assistance of a "torque converter", which automates this acceleration and shift process. Therefore, the gearbox plays a vital role for a car.

In an automobile, there are five types of transmission used commonly: manual (MT), automatic manual (AMT), automatic (AT), dual clutch (DCT) and continuously variable (CVT). A manual transmission vehicle provides the best example of a simple gearbox. In both the automatic and continuously variable transmissions (AT and CVT), the gearboxes are closed systems, requiring very little human interaction.

Task 11: Work in pairs, discuss and answer the following questions according to the notes above.

1. Describe the function of a gearbox in a car.
2. Do you have a car? If yes, what kind of transmission does your car have?
3. What kind of transmission in a car do you like? Why?

4.4.2 Basic Ideas Behind a Manual Transmission

The gearbox provides a gear selector mechanism that disengages one gear and selects another for different driving conditions: standing start, climbing a hill, or running at a high speed. The lower the gear, the slower the wheels turn in relation to the engine speed. The gears are selected by a system of rods and levers operated by the gear lever. The gearbox is connected to the engine through the clutch. The input shaft of the gearbox therefore turns at the same rpm as the engine. Most modern types of this gearbox have 5 or 6 forward (and one reverse) gears.

In order to understand the basic idea behind a standard gearbox, a very simple two-speed gearbox is demonstrated in Figure 4.9.

The input shaft comes from the engine through the clutch. The input shaft and input gear are connected as a single unit. The clutch is a device that connects and disconnects the engine and the transmission. When the driver pushes in the clutch pedal, the engine and the transmission are disconnected so the engine can run even if the car is standing still. When the driver releases the clutch pedal, the engine and the input shaft are directly connected to one another. The input shaft and gear turn at the same rpm as the engine.

The layshaft/counter shaft and gears are also connected as a single piece, so all of the gears on the layshaft and the layshaft itself spin as one unit. The input shaft and the layshaft are directly connected through their meshed gears, so that if the input shaft is spinning, so is the layshaft. In this way, the layshaft receives its power directly from the engine whenever the clutch is engaged.

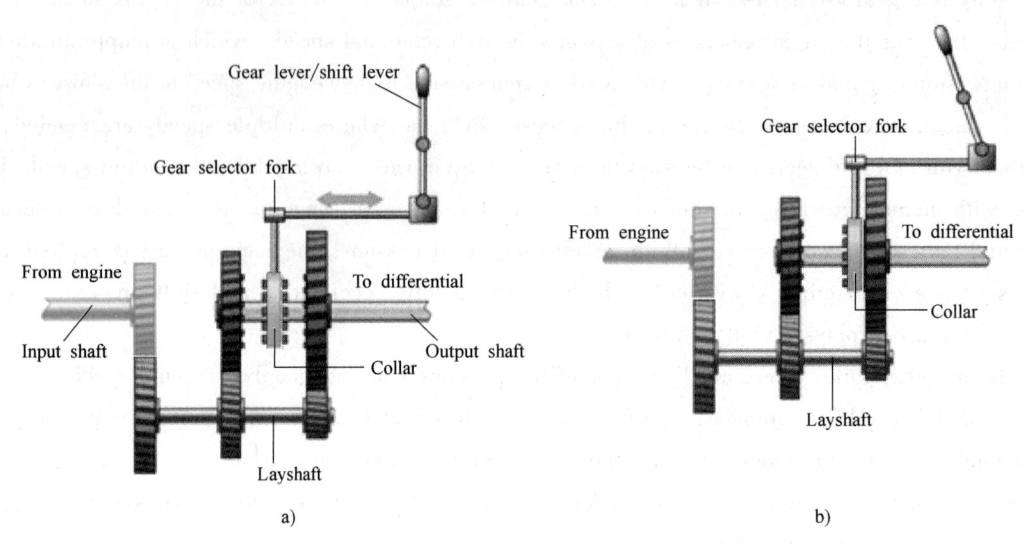

Figure 4.9 Diagram of a simple two-speed gearbox

a) Neutral position b) First gear

The output shaft is a splined shaft that connects directly to the drive shaft through the differential to the drive wheels of the car. If the wheels are spinning, the output shaft is also spinning.

The two gears on the output shaft ride on bearings, so they spin on the output shaft. If the engine is off but the car is coasting, the output shaft can turn inside the gears while the gears on the output shaft and the layshaft are motionless.

The purpose of the collar is to connect one of the two gears on the output shaft to the drive shaft/output shaft. The collar is connected through the splines, directly to the output shaft and spins with the output shaft. However, the collar can slide left or right along the output shaft to engage either of the two gears. Teeth on the collar, called dog teeth, fit into holes on the sides of the two gears on the output shaft to engage them. The gear shift knob moves a rod connected to the fork. The fork slides the collar on the output shaft to engage one of the two gears.

When the collar is between the two gears (see Figure 4.9a), the transmission is in neutral. Both of the two gears freewheel on the output shaft at different rates controlled by their ratios to the layshaft at this moment and no power is transmitted to the output shaft.

Now, let's see what happens when the driver shifts gear. When the driver shifts into first gear, the collar engages the gear on the right (see Figure 4.9b). The input shaft from the engine turns the layshaft, which turns the gear on the right side of the output shaft. This gear transmits its energy through the collar to drive the output shaft. Meanwhile, the gear on the left side of the output shaft is turning, but it is freewheeling on its bearing, so it has no effect on the output shaft. When the driver shifts into second gear, the collar engages the gear on the left and the gear on the right has no effect on the output shaft. The transmission is in second gear.

Task 12: Work in pairs, discuss and answer the following questions.

1. When you make a mistake and hear a horrible grinding sound while shifting, where does the

sound come from?

2. Why the clutch is needed between the gearbox and the engine?

Hint: When you make a mistake and hear a horrible grinding sound while shifting, you are not hearing the sound of gear teeth mis-meshing. As you can see in Figure 4. 9, all gear teeth are fully meshed at all times. The grinding is the sound of the dog teeth trying unsuccessfully to engage the holes in the side of a gear on the output shaft.

4. 4. 3 Five-Speed Manual Transmission

In this section, we'll take a look at a real transmission. The five-speed manual transmission is fairly standard on cars today. Its internal structure is demonstrated in Figure 4. 10. There are three forks controlled by three rods that are engaged by the shift lever, the working principle behind this gearbox is similar to the two-speed gearbox discussed in section 4. 4. 2.

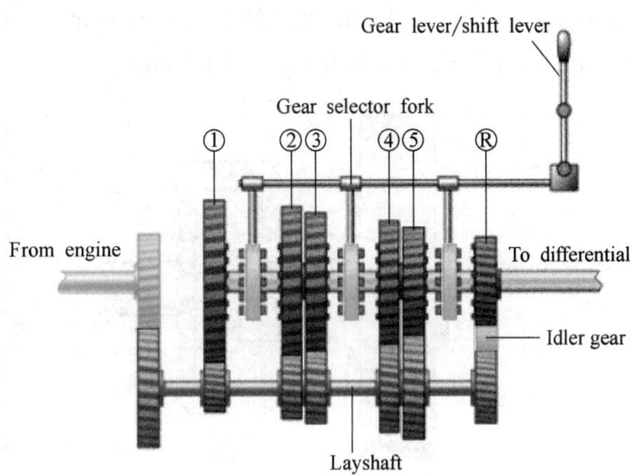

Figure 4. 10 A five-speed manual transmission

It can be seen that as the driver moves the shift lever left and right, different forks (and therefore different collars) are engaged. Moving the knob/lever forward and backward moves the collar to engage one of the gears.

Reverse gear is handled by a small idler gear (see Figure 4. 11). At all times, the reverse gear is turning in a direction opposite to all of the other gears (from ① to ⑤). Therefore, it would be impossible to throw the transmission into reverse while the car is moving forward, because the teeth would never engage. However, they will make a lot of noise.

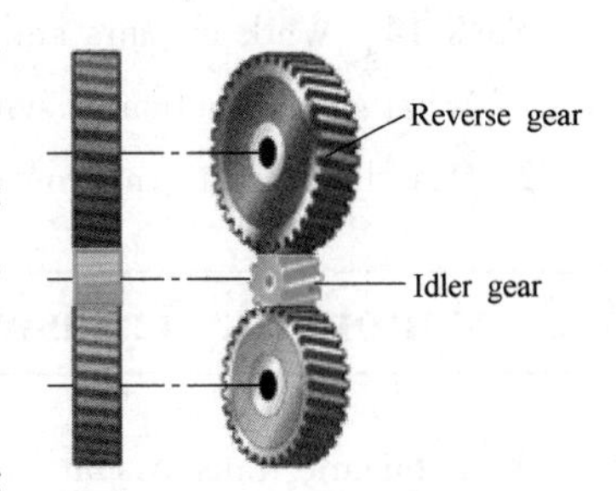

Figure 4. 11 Reverse gear

Task 13: Work in pairs, discuss and answer the following questions.

1. What's the function of the idler gear in Figure 4. 11?

2. When a driver moves the shift lever left or right, then forward or backward, what's going on in the gearbox?

4. 4. 4 Synchronizers

In some old MT cars, double-clutch was common in order to disengage a gear, allow the collar and next gear to reach the same speed, and then to engage the new gear. For double-clutch shift, the driver pushed the clutch pedal to free the engine from the transmission, and then the collar moved into neutral. Then he released the clutch and revved the engine to get it to the right rpm value for the next gear, so the collar and the next gear spun at the same rate to allow the dog teeth to engage the gear. When the engine hit the right speed, the driver depressed the clutch again in order to lock the collar into place on the next gear. So it's a double-clutching process.

Modern cars use synchronizers in order to avoid the need for double-clutch. A synchronizer, or "synchro", lets the collar and gear synchronize their speeds while they're already in contact but before the dog teeth engage. Each manufacturer's synchro is slightly different from the others, but the basic idea is the same. For instance, a cone on one gear will fit into a cone-shaped hole on the collar, as shown in Figure 4. 12a. The gear and collar synchronize their speeds thanks to the friction between the cone and the collar. The outer portion of the collar then slides so that the dog teeth can engage the gear (see Figure 4. 12b). Every manufacturer implements transmissions and synchros in different ways, but this is the general idea.

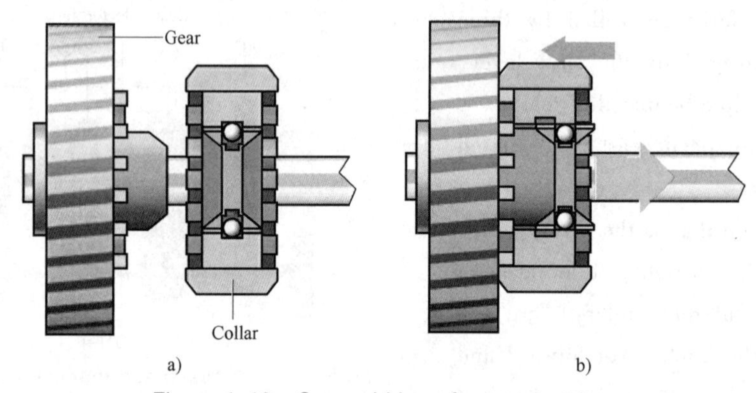

Figure 4. 12 General idea of a synchronizer

Task 14: Work in pairs and discuss the following questions.

1. What's the function of synchronizers?
2. Describe how a synchronizer works according to the diagram shown in Figure 4. 12.

4. 5 Automatic Transmissions

An automatic transmission, or AT, is a type of motor vehicle transmission that can automatically change gear ratios as the vehicle moves, freeing the driver from having to shift gears manually. Like other transmission systems on vehicles, it allows an internal combustion engine, best suited to run at a relatively high rotational speed, to provide a range of speed and torque outputs necessary for vehicular travel.

Currently, there are several types of automated transmissions used in cars, such as semi-automatic transmissions—also known as automated manual transmission (AMT), automatic transmission (AT), continuously variable transmission (CVT) and direct shift gearbox (DSG).

4. 5. 1 Automated Manual Transmission (AMT)

An automated manual transmission (AMT) is also known as a semi-automatic transmission.

The simplest way to describe this type is to call it a hybrid between a fully automatic and manual transmission, i. e. it is automatic with the ability to switch to a manual mode. Similar to a manual transmission, gears are changed via a simple shifter or paddles located behind the steering wheel. This kind of transmission unit's mechanism involves the use of two key parts—a hydraulic actuator system and an electronic control unit that engages and disengages the clutch while executing gear shifts once the driver has signaled the change. Basically it is just a kit that can be added to any regular manual transmission, which makes it a low-cost solution for carmakers.

The design of AMT systems varies, but all semi-automatic transmissions rely on microprocessors to control the change of mechanical gear ratios with the help of electrically operated actuators and servos. These transmissions were limited to high-end supercars at first due to their high cost, but an increasing number of manufacturers are fitting them to mid-range cars.

The operation of semi-automatic transmissions has evolved as vehicle manufacturers experimented with different systems. In an early mass-production example, Ferrari offered their Mondial model with a clutchless manual, which Ferrari called the Valeo transmission. In this system, the gear stick of a traditional manual transmission was retained, moving the shifter automatically engaged the electro-hydraulic clutch. Saab's Sensonic transmission worked in a similar fashion.

In standard mass-production automobiles today, the gear lever appears similar to manual shifts, except that the gear stick only moves forward and backward to shift into higher and lower gears, instead of the traditional H-pattern. The Bugatti Veyron uses this approach for its seven-speed transmission. In Formula One, the system is adapted to fit onto the steering wheel in the form of two paddles, depressing the right paddle shifts into a higher gear, while depressing the left paddle shifts into a lower one. Numerous road cars have inherited the same mechanism.

Hall effect sensors sense the direction of requested shift, and this input, together with a sensor in the gearbox which senses the current speed and gear selected, feeds into a central processing unit. This unit then determines the optimal timing and torque required for a smooth clutch engagement, based on input from these two sensors as well as other factors, such as engine rotation, the electronic stability control, air conditioner and dashboard instruments.

The central processing unit powers a hydro-mechanical unit to either engage or disengage the clutch, which is kept in close synchronization with the gear-shifting action the driver has started. In some cases, the hydro-mechanical unit contains a servomotor coupled to a gear arrangement for a linear actuator, which uses brake fluid from the braking system to impel a hydraulic cylinder to move the main clutch actuator. In other cases, the clutch actuator may be completely electric.

The power of the system lies in the fact that electronic equipment can react much faster and more precisely than a human, and takes advantage of the precision of electronic signals to allow a complete clutch operation without the intervention of the driver.

The clutch is really only needed to get the car in motion. For a quicker upshift, the engine power can be cut and the collar can be disengaged, until the engine drops to the correct speed for the next gear. For the teeth of the collar to slide into the teeth of the rings, both the speed and position must match. This needs sensors to measure not only the speed, but also the positions of the teeth, and the throttle may need to be opened softer or harder.

Task 15: Work in pairs and discuss the following question. What are the advantages of AMT?

4.5.2 Automatic Transmissions (AT)

In an automatic transmission, the hydraulically operated control systems are managed electronically by the vehicle's computer instead of the clutch and gear stick. There is no clutch pedal or gear shift in AT cars. All the driver has to do is to shift the selector from park (P) or neutral (N), into drive (D), and the gear shifting will take place automatically and smoothly, without any additional input from the driver under normal driving conditions. For instance, once the driver put the transmission into drive (D), everything else is automatic.

The automatic transmission is one of the most complex assemblies of the modern automobile. The transmission is bolted directly behind the engine on rear-wheel-drive vehicles and is typically mounted on the driver's side of the vehicle on front-wheel-drive vehicles. The engine's power is transmitted from the crankshaft, through the flywheel, and into the transmission assembly via the torque converter as shown in Figure 4.13. The torque converter is like an "automatic" clutching device. It takes the place of a clutch on a conventional manual transmission. In order to understand how an AT works, we have to understand how a torque converter work and how to change gear ratios without gear shift.

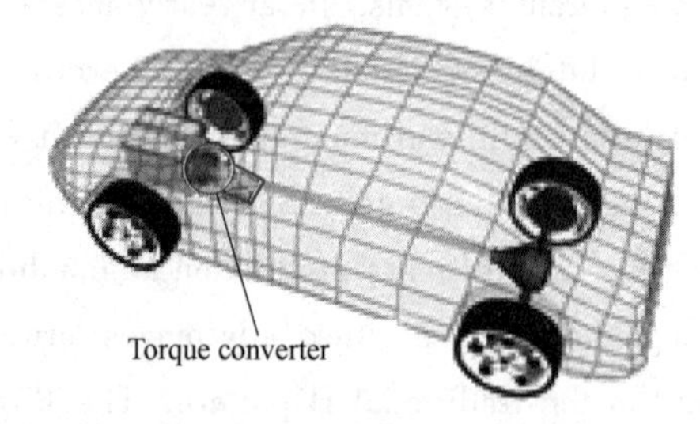

Figure 4.13 Torque converter situated between the engine and the transmission

1. How does a torque converter work?

By far the most common form of torque converters in automobile transmissions is the hydraulic torque converter. A hydraulic torque converter can be broken down into three main components: impeller (or pump), turbine and stator, as demonstrated in Figure 4.14.

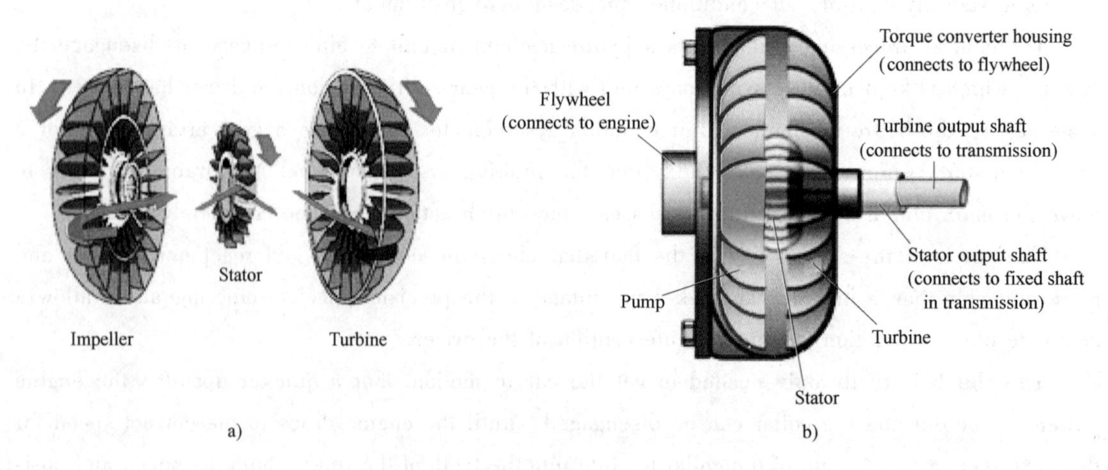

Figure 4.14 Three main components of a hydraulic torque converter

The impeller, or pump, is a ring of metal blades driven by the engine shaft that transfers kinetic energy by flinging fluid around. Its spinning motion also draws more fluid into the center by creating a vacuum. The fluid from the impeller eventually hits the turbine—a set of three-ring blades connected to the engine shaft. The pressure from the fluid turns the turbine, which causes the transmission to spin, thus powers cars; this is the basic principle behind converter. However, the actual structure of a hydraulic torque converter is a more complicated gadget, as shown in Figure 4. 15.

a) b)

Figure 4. 15 Structure of a hydraulic torque converter

a) Three main components b) An exploded model of a torque converter

(**left to right: turbine, stator, pump**)

The stator resides in the very center of the torque converter. Its job is to redirect the fluid returning from the turbine before it hits the pump again. This dramatically increases the efficiency of the torque converter.

The stator has a very ingenious blade design that almost completely reverses the direction of the fluid. A one-way clutch (inside the stator) connects the stator to a fixed shaft in the transmission (the direction that the clutch allows the stator to spin is noted in Figure 4. 14b. Because of this arrangement, the stator can not spin with the fluid, it can spin only in the opposite direction, forcing the fluid to change direction as it hits the stator blades.

The housing of the torque converter is bolted to the flywheel of the engine, so it turns at whatever speed the engine is running at. The blades that make up the pump of the torque converter are attached to the housing, so they also turn at the same speed as the engine.

2. Planetary gearset in AT

Notice that AT came after the manual transmission, and you have known how a MT works, so you have got the basic idea of how an AT works. A MT requires the driver to manually change gear ratios, an AT does this on its own through the use of fluid pressure. Transmission fluid enters the torque converter which then activates clutches and bands in a planetary gearset. In turn, those determine what gear ratio should be engaged. The planetary gearset is what's responsible for changing gear ratios, and it consists of a sun gear surrounded by smaller planet gears carried by a planet carrier enclosed in a ring gear, as shown in Figure 4. 16. The planetary gearset can then be configured to the right gear combination. All the parts in the planetary gearset can be locked or unlocked to determine the gear ratio. The basic idea behind the planetary gearset is as follows:

(1) If the sun gear is locked and the planets are driven by the planet carrier, the output is taken from the ring gear, achieving a speed increase.

(2) If the ring gear is locked and the sun gear is driven, the planet gears transmit drive through the planet carrier and the speed is reduced.

(3) With power input going to the sun gear and the planet carrier locked, the ring gear is driven, but transmits drive in reverse.

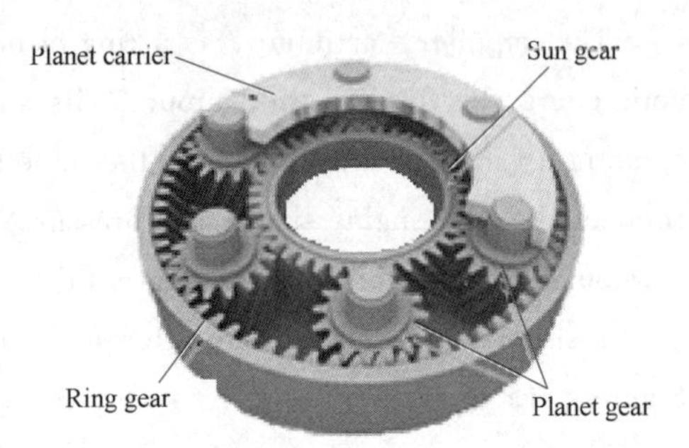

Figure 4. 16 A planetary gearset in AT

(4) To achieve direct drive without change of speed or direction of rotation, the sun gear is locked to the ring gear and the whole unit turns as one.

The illustration in Figure 4. 17 shows how the simple system described above would look in an actual transmission. The input shaft is connected to the ring gear, the output shaft is connected to the planet carrier which is also connected to a "multi-disk" clutch pack. The sun gear is connected to a drum which is also connected to the other half of the clutch pack. Surrounding the outside of the drum is a band that can be tightened around the drum when required to prevent the drum with the attached sun gear from turning.

In this instance, the clutch pack is used to lock the planet carrier with the sun gear forcing both to turn at the same speed. If both the clutch pack and the band were released, the system would be in neutral. Turning the input shaft would turn the planet gears against the sun gear. But since nothing is holding the sun gear, it will just spin free and have no effect on the output shaft. To place the unit in first gear, the band is applied to hold the sun gear from moving. To shift from first to high gear, the band is released and the clutch is applied causing the output shaft to turn at the same speed as the input shaft.

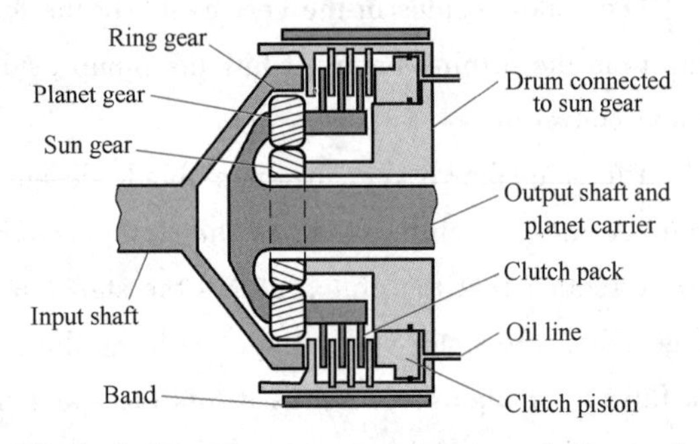

Figure 4. 17 Planetary gear system (side view)

A clutch pack consists of alternating disks that fit inside a clutch drum (see Figure 4. 18). Half of the disks are steel and have splines that fit into groves on the inside of the drum. The other half have a friction material bonded to their surface and have splines on the inside edge that fit groves on the outer surface of the adjoining hub. There is a piston inside the drum that is activated by oil pressure at the appropriate time to squeeze the clutch pack together so that the two components become locked and turn as one.

Many more combinations are possible using two or more planetary sets connected in various ways to provide different forward speeds and reverse that are found in modern automatic transmissions, as the real model demonstrated in Figure 4. 19.

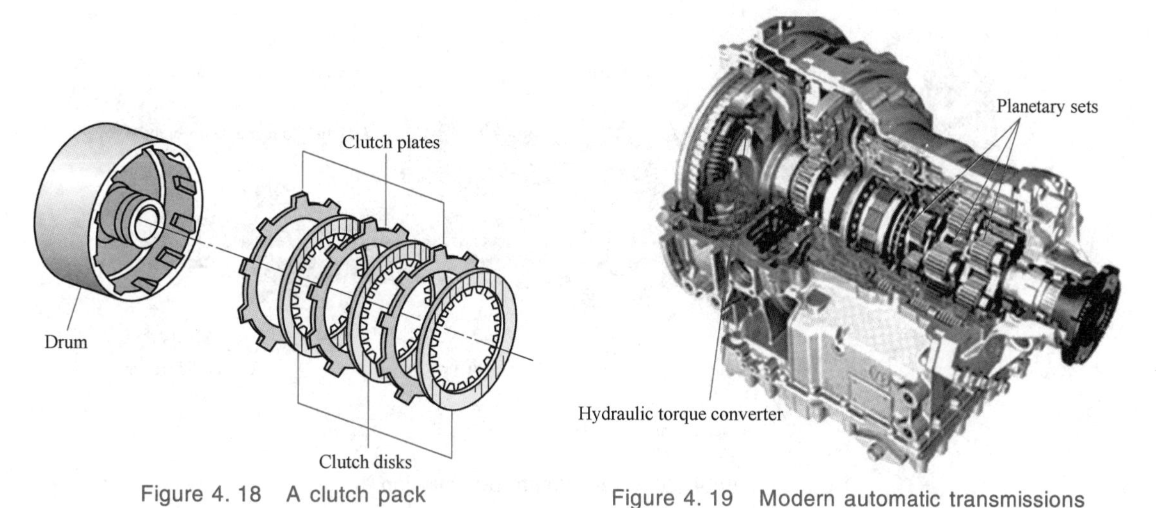

Figure 4. 18 A clutch pack

Figure 4. 19 Modern automatic transmissions

Task 16: Work in pairs, discuss and answer the following questions.

1. What's the function of the torque converter?
2. How does gear ratios change in AT?
3. Do you like an AT car or a MT car? Why?

4. 5. 3 Dual Clutch Transmissions (DCT)

Dual clutch transmissions (DCT), also known as Direct Shift Gearbox (DSG) or twin-clutch transmission, is an automated transmission that can change gears faster than any other geared transmission. Dual clutch transmissions deliver more power and better control than a traditional automatic transmission and faster performance than a manual transmission. Originally marketed by Volkswagen as the DSG and Audi as the S-Tronic, now dual clutch transmissions are offered by several automakers, including Ford, Mitsubishi, Smart, Hyundai and Porsche.

Dual clutch transmissions employ two clutches, as shown in Figure 4. 20. Clutch 1 controls gearshifts in the odd numbered gears and reverse gear, clutch 2 controls gearshifts in the even numbered gears. This arrangement does not interrupt the power flow from the engine. The driver still has to initiate a gearshift via a shifter or paddles located behind the steering wheel, but there is no need for the driver to operate a clutch. Figure 4. 21 shows another example of a dual clutch transmission. Notice that two types of DCT demonstrated in Figure 4. 20 and Figure 4. 21 share similar working principles with different layouts, the latter is more compact. In Figure 4. 21, the 3rd gear is preselected. If clutch 1 is disengaged, power transmitted from the engine to the 3rd gear is interrupted as indicated by the dotted arrow; meanwhile clutch 2 is engaged, and the 2nd gear is in active as indicated by the solid arrow. The output speed to the differential is reduced from the 3rd gear to the 2nd gear. Using a pair of clutches means that as one clutch disengages a gear, the other clutch engages the next at the same time, effectively eliminating the time spent in neutral between changing gears.

Therefore, while one gear is engaged, one of clutches prepares to switch to the gear on the

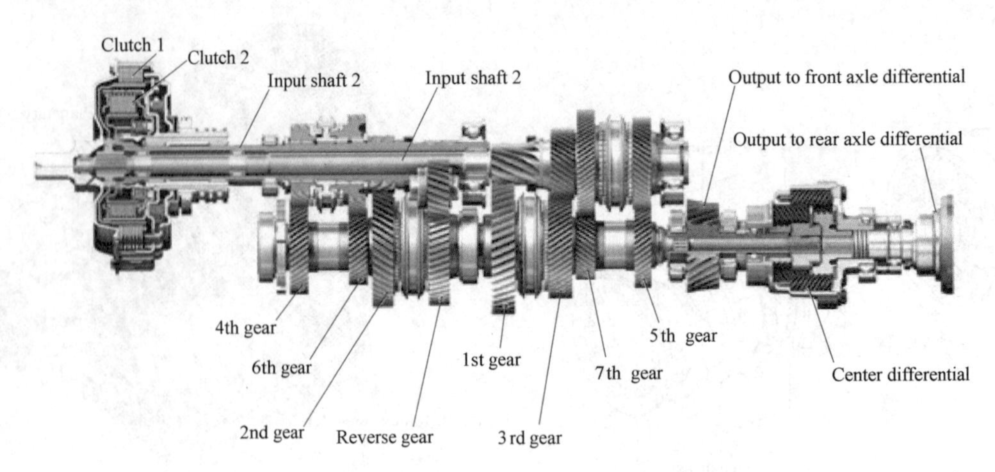

Figure 4. 20 Dual clutch transmission

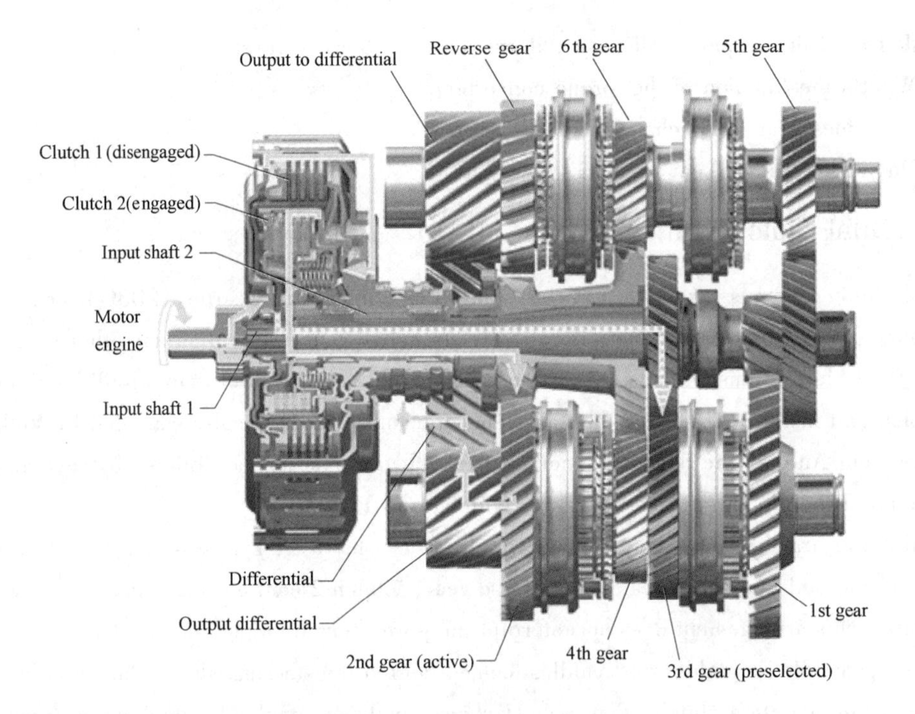

Figure 4. 21 Dual clutch transmission with compact layout

other clutch. Since one clutch holds the odd numbered gears and the other the even numbered gears, the clutches constantly switch back and forth and the transition from one to another happens so quickly that the driver never feels the car has changed gears.

Advantages of the dual clutch transmission include more efficient performance and faster shifting speeds. Since the next gear is already ready to switch into, the change occurs almost instantly and the shift is smooth. No engine power is lost, as it is in automatic transmissions using torque converters to engage and disengage the engine. The torque converter is similar in principle to the electro-hydraulic fluid, but loses some of the power from the engine when engaging and disengaging.

Disadvantages of the dual clutch transmission include its high cost and the complexity of its parts. Cars utilizing this type of transmission cost more and repairs of the car also take more effort, due to the extra parts and depending on how they are aligned with one another. More parts there are, more problems will arise.

Figure 4. 22 shows a six-speed dual clutch gearbox (Ford PowerShift) produced jointly by the Ford, Getrag and Luk.

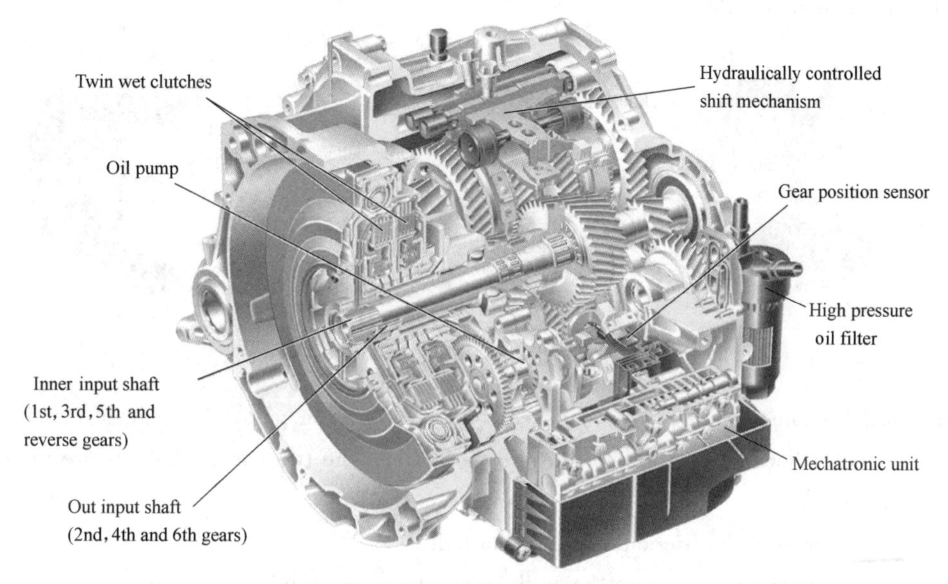

Figure 4. 22 Ford PowerShift gearbox

Task 17: Work in pairs, discuss and answer the following questions.

1. What are the advantages of DCT over AT and MT?
2. Describe the basic ideas behind DCT.

4. 5. 4 Continuously Variable Transmission (CVT)

This transmission doesn't use gears as its means of producing various vehicle speeds at different engine speeds. Instead of gears, the system relies on a rubber or metal belt running over pulleys that can vary their effective diameters. To keep the belt at its optimum tension, one pulley will increase its effective diameter, while the other decreases its effective diameter by exactly the same amount. This action is exactly analogous to the effect produced when gears of different diameters are engaged.

Since one pulley (drive pulley) is driven by the engine and the other is connected to the drive shaft (driven pulley), an infinite number of ratios can be produced. This enables the car to always run at the most efficient speed, regardless of the load placed on it. Microprocessor-controlled sensors quantify load variations and the optimum operating speed for the engine can be maintained without any input from the driver by adjusting both pulleys.

Most CVTs have three basic components: ①a high-power metal or rubber belt, ②a variable-

input "driving" pulley, ③an output "driven" pulley, as shown in Figure 4. 23. One of the pulleys, known as the drive pulley (or drive pulley), is connected to the crankshaft of the engine. The drive pulley is also called the input pulley because it's where the energy from the engine enters the transmission. The other pulley is called the driven pulley because the drive pulley is turning it. As an output pulley, the driven pulley transfers energy to the driveshaft. CVTs also have various microprocessors and sensors, but the three basic components described above are the key elements that enable the technology to work.

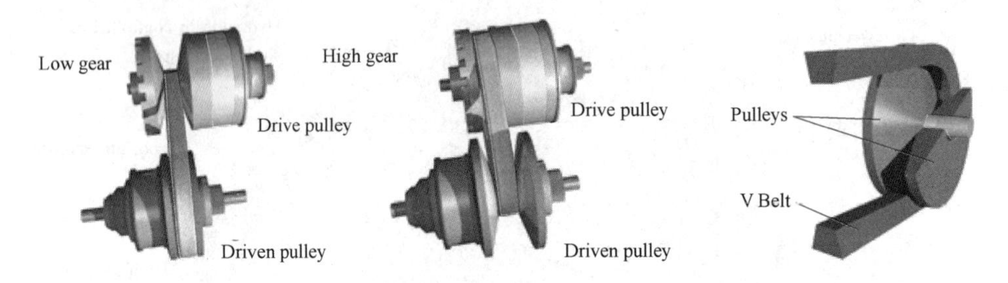

Figure 4. 23 Basic components in CVTs

The variable-diameter pulleys are the heart of a CVT. Each pulley is made of two 20-degree cones facing each other. A belt rides in the groove between the two cones. V-belts are preferred if the belt is made of rubber. V-belts get their name from the fact that the belts bear a V-shaped cross section, which increases the frictional grip of the belt.

The distance from the center of the pulleys to where the belt makes contact in the groove is known as pitch radius. When the pulleys are far apart, the belt rides lower and the pitch radius decreases. When the pulleys are close together, the belt rides higher and the pitch radius increases. The ratio of the pitch radius on the driving pulley to the pitch radius on the driven pulley determines the gears.

When one pulley increases its pitch radius, the other has to decrease its pitch radius to keep the belt tight. As the two pulleys change their radii relative to one another, they create an infinite number of gear ratios—from low to high and every value in between. For example, when the pitch radius is small on the drive pulley and large on the driven pulley, then the rotational speed of the driven pulley decreases, resulting in a lower "gear". When the pitch radius is large on the drive pulley and small on the driven pulley, then the rotational speed of the driven pulley increases, resulting in a higher "gear", as

Figure 4. 24 Inner structure of a CVT

shown in Figure 4. 23. Thus, in theory, a CVT has an infinite number of "gears" so that it can run through at any time, at any engine or vehicle speed. Figure 4. 24 shows the inner structure of a CVT. It's very complicated.

The simplicity and stepless nature of CVTs make them an ideal transmission for a variety of machines and devices, not just cars. CVTs have been used for years in power tools and drill presses. They've also been used in a variety of vehicles, including tractors, snowmobiles and motor scooters. In all of these applications, the transmissions have relied on high-density rubber belts, which can slip and stretch, thereby reducing their efficiency.

The introduction of new materials makes CVTs even more reliable and efficient. One of the most important advances is the design and development of metal belts to connect the pulleys. These flexible belts are composed of several (typically 9 or 12) thin bands of steel that hold together through high-strength, bow-tie-shaped pieces of metal, as shown in Figure 4. 25.

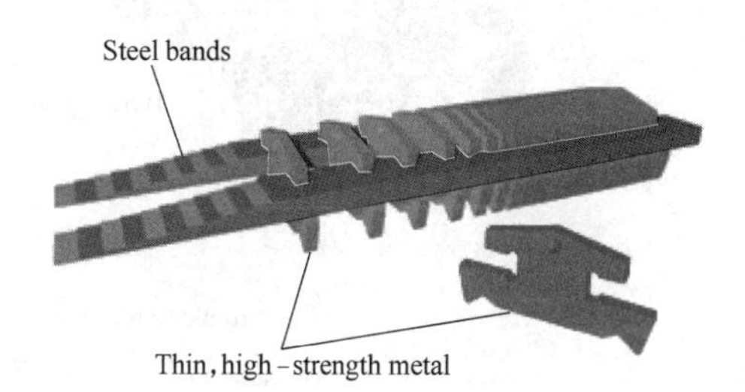

Figure 4. 25 A metal belt

Metal belts don't slip and are highly durable, enabling CVTs to handle more engine torque. They are also quieter than rubber-belt-driven CVTs.

Task 18: Work in pairs, discuss and answer the following questions.

1. What's the main difference between AT and CVT?
2. In your opinion, what could be the main factors to affect the performance of CVT?
3. How does CVT create an infinite number of gear ratios?
4. Which one do you think is the future of the transmission in cars? MT, AMT, AT, DCT or CVT? Why?

4. 6 Differentials

The differential is a part of the power transmission device. In automobiles and other wheeled vehicles, the differential allows the outer drive wheel to rotate faster than the inner drive wheel during a turn, as shown in Figure 4. 26. It is necessary making the wheel that is traveling around the outside of the turning curve roll farther and faster than the other wheel when the vehicle turns. The average of the rotational speed of the two drive wheels equals the input rotational speed of the drive shaft. An increase in the speed of one wheel is balanced by a decrease in the speed of the other.

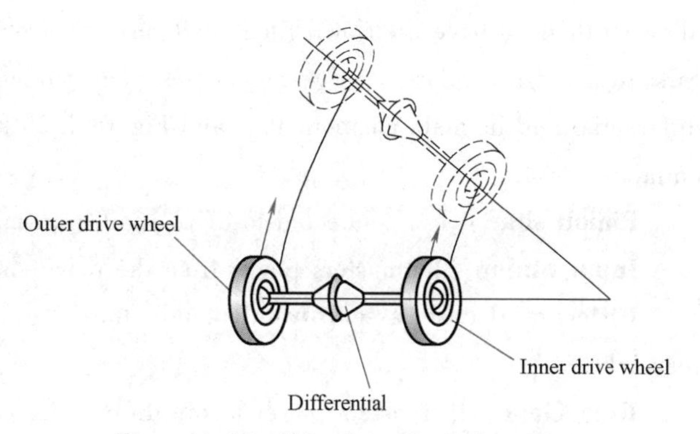

Figure 4. 26 Wheels' path during a turn

The differential is found on all modern cars and trucks, and also in many all-wheel-drive (four-wheel-drive) vehicles, as shown in Figure 4. 27. These all-wheel-drive vehicles need a differential between each set of drive wheels, and they need one between the front and the back wheels as well, because the front wheels travel a different distance through a turn than the rear wheels.

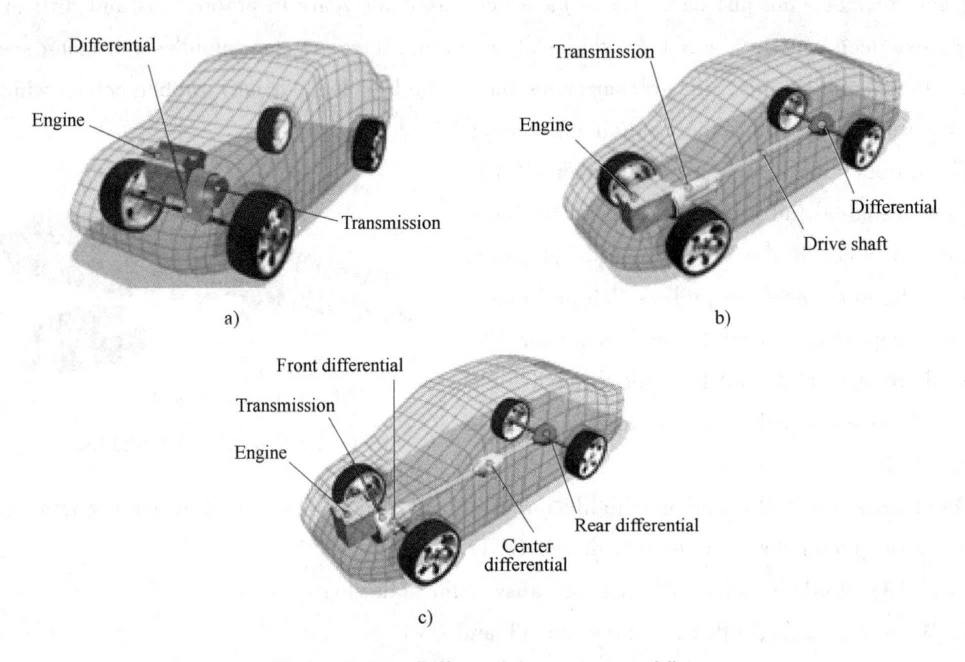

Figure 4. 27 Differentials on automobiles

a) Front-wheel drive b) Rear-wheel drive c) All-wheel drive

How do differentials work? There are different types of differentials. We will start with the simplest type of differential, called an open differential.

4. 6. 1 Open Differentials (Standard Differentials)

1. Some terminologies

The open differential often referred to as just a differential. The ability to rotate both driven wheels at different speeds is the primary objective of the differential; it separates the both wheels by allowing them to have their own final shaft instead of one continuous shaft between the both wheels. First we need to explore some terminologies. The model in Figure 4. 28 shows the inside of an open differential and its main components, and Figure 4. 29 shows a schematic diagram of an open differential.

Pinion shaft: It's connected to drive shaft/transmission shaft through universal joint.

Input pinion: It transfers power from the drive shaft to the ring gear.

Differential case assembly: It holds the ring gear and other components that drive the rear axle.

Ring Gear: It transfers power to the differential case assembly.

Side/Spider gears: They help both wheels to turn independently when turning. Two side

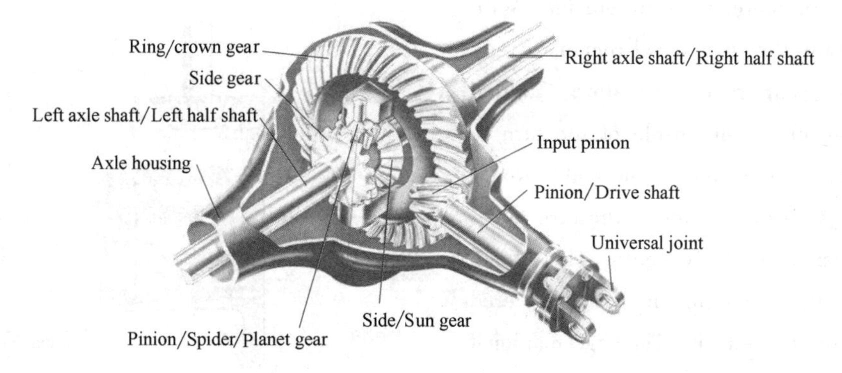

Figure 4. 28 Inside and main components of an open differential

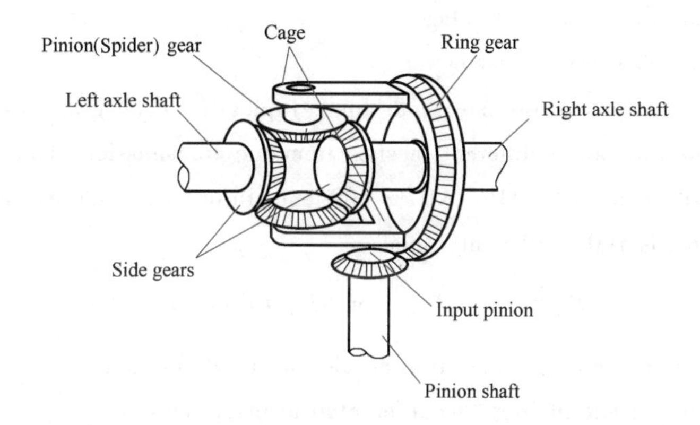

Figure 4. 29 Schematic diagram of an open differential

gears are equal-sized bevel gears, one for each half axle shaft, with a space between them. Two spider gears are also equal-sized bevel gears held by the cage and they spin with ring gear together, meanwhile they can turn around their own shafts.

Axle shafts: They transfer torque from the differential assembly to the drive wheels.

Rear axle bearings: They are ball or roller bearings that fit between the axles and the inside of the axle housing.

Axle housing: It is a metal body that encloses and supports parts of the rear axle assembly.

2. How does a differential work?

How does a differential work? Assuming the wheels don't slip and spin out of control, the following two examples of car motion describe how the differential works when the car is going forward and turning around a corner.

(1) When a car is driving straight down the road, both drive wheels are spinning at the same speed. The input pinion is turning the ring gear and holder/cage. Thus, the spider gears do not spin at all. Instead, as the transmission shaft turns the input opinion, the rotary motion is transmitted directly to the axle shafts, and both wheels spin with the angular velocity of the input opinion (they have the same speed), as shown in Figure 4. 30.

(2) When the car is turning at a corner, the wheels must move at different speeds. The inside

wheel must turn slower than the outside. With constant throttle the cage speed remains fixed, but the spider gears rotate very slowly, which allows the wheel on the inside of the turn to rotate a little slower than it normally would and the wheel on the outside of the turn to rotate faster than it normally would. This allows the speed of the input pinion to be delivered unevenly to the two wheels. The cage rotational speed is always the average of the two sides.

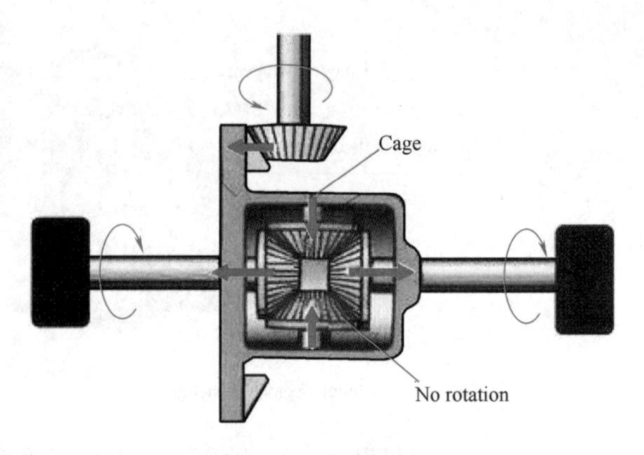

Figure 4. 30 Working principle of a differential

3. Disadvantages of an open differential

In order to explain the main disadvantages of an open differential, let's discuss how power is distributed from input shaft to two axle shafts first. Typically, each gear mesh will have 1%-2% loss in efficiency. So with three different meshes from the transmission shaft to each of the half shafts, the system will actually be 94% to 97% efficient. In order to simplify the discussion, let us assume that the system is 100% efficient, then

$$P_{in} = P_{out1} + P_{out2} \quad \text{or} \quad P_{in} = T_1\omega_1 + T_2\omega_2$$

where P_{in}——the power input from the transmission to the differential;

P_{out}——the power output from the differential to the wheels;

T——the torque supplied to each axle shafts, respectively;

ω——the angular velocity of one gear, respectively.

As stated before, when a car is driving straight down the road, both wheels spin with the angular velocity of the input opinion and the same amount of torque is supplied to each of axle shafts respectively. It means the input power is divided into two equal shares and goes to each of axle shafts. The problem with the open differential happens when one tire is on a low traction surface such as ice, snow or mud, the wheel on the slippery surface will spin fast while the other wheel provides no forward motion. In order to prevent power from going to a wheel with little or no traction, a locking mechanism is needed to shift that power over to the tire with more traction so that the vehicle can continue moving forward. This is why the limited slip differential (LSD) has been invented.

The first differential that automobiles started to use was locking differential, and limited slip differential, electronic or electronically controlled limited-slip differential have come along the way as technical advancements have been made in the field. The most advanced one which is result of much late developments is the torque-vectoring differential. It's being used mostly in sport cars. Each differential has its own set of benefits and limitations.

Task 19: Work in pairs, discuss and answer the following questions.

1. What's the function of differentials in a car?

2. For a rear-wheel drive car, describe the location of the differential.
3. Describe how an open differential works.
4. What are main disadvantages of an open differential?

4.6.2 Locking Differentials

The idea behind locking differentials is very simple. The simplest way is to drill a hole there all the way through one of side gears, and through the other side gear on the other side. Stick a steel pin through the holes. Now both sides of the differential are locked together. This is a simplified "locking differential". When the housing/cage turns, the spider gear pushes on both drive/side gears, and no matter what the traction situation is, they are forced to turn at the same speed. Real locking differentials use a pneumatically or hydraulically activated sleeve that slides across between the drive gears out near their edge to lock the drive gears or spider gears in place, but the effect is the same. This is great when there is no traction on one wheel. Because no matter what happens, both sides turn at the same speed; If one side offers lots of resistance and the other none, then effectively all the "usable" or "useful" torque goes to the side where there is resistance. It's getting 100% of the available torque. The side with no traction doesn't need torque to spin helplessly, so it's not really getting 50% of the torque as it is thought, but close to zero.

The above locking mechanism is usually activated manually by switch, and when activated, both wheels will spin at the same speed. If one wheel ends up off the ground, the other wheel won't know. Both wheels would continue to spin at the same speed as if nothing had changed. The locking differential is useful for serious off-road vehicles.

Task 20: Work in pairs, discuss and answer the following questions.

1. Describe the basic principle behind locking differentials.
2. What are advantages and disadvantages of locking differentials?

4.6.3 Clutch-Type Limited Slip Differentials

The clutch-type LSD is probably the most common version of the limited slip differentials. This type of LSD has all of the same components as an open differential, but it adds a spring pack and two sets of clutch packs, as the model shown in Figure 4.31.

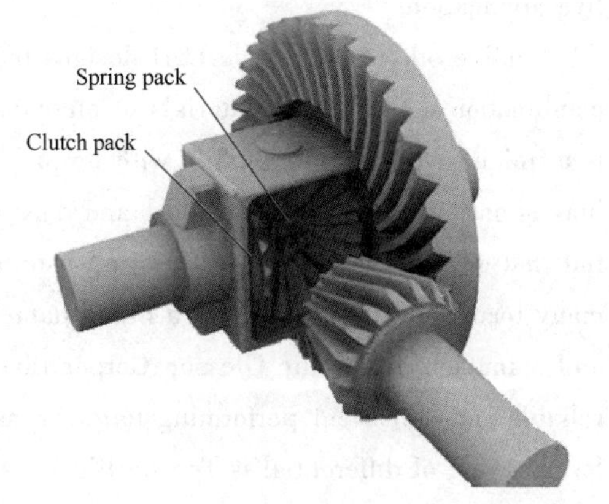

Figure 4.31 Model of clutch-type LSD

The clutch-type LSD has a stack of thin clutch-disks, as shown in Figure 4.32, half of which are coupled to one of the drive shafts, the other half of which are coupled to the spider gear carrier. The clutch stacks may be present on both drive shafts, or on only one. If on only one, the remaining drive shaft is linked to the clutched drive shaft through the spider gears. Nothing happens in normal conditions with equal traction, as both side gears turn at the same speed. When one side loses traction and the differential tries to spin that side, the spring-loaded clutch simply imparts a fixed amount of torque back to the other side.

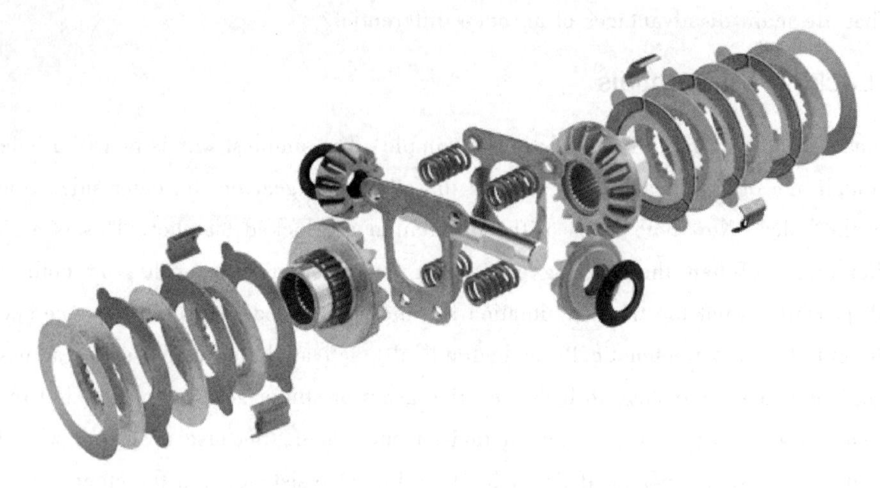

Figure 4. 32 Parts of the clutch-type LSD

The principle of this type of locking differential is that the locking effect is realized by the friction of plates (or clutches). By the amount of clutches and adjusting the pressure we can set how much locking effect the differential is providing. This is expressed in percentage, like 25% for most standard sports cars used on public roads.

Task 21: Work in pairs and complete the following tasks.

1. Describe the basic principle behind clutch-type LSD.

2. Search some information on Internet about clutch-type LSD, find out other types of clutch-type LSD except the one mentioned above, write them down and share your work with your partner.

4. 6. 4 Geared Limited Slip Differentials

Geared, torque-sensing mechanical limited slip differentials use worm wheels and spur gears to distribute and differentiate input power between two drive wheels or front and back axles. This is a completely separate design from the most common beveled spider gear designs seen in most automotive applications.

Unlike other clutch-type LSD designs that combine a common spider gear "open" differential in combination with friction materials or clutches that inhibit differentiation, the torque sensing design is a unique type of differential, with torque bias inherent to its design, not as an add-on. Torque bias is only applied when needed, and does not inhibit differentiation. The result is a true differential that does not bind up like LSD and locking types, but still gives increased power delivery under many road conditions. Torsen, a brand name of differential invented by American Vernon Gleasman and manufactured by the Gleason Corporation, is a good example of this geared LSD. It is a highly reliable and consistent performing torque sensing limited slip differential design. The correct name for this type of differential is Torque Biasing Differential (TBD) and it is more commonly known as Torsen differential. The name Torsen comes from how the differential works in simple terms, a torque sensing differential (Tor/Sen).

The internal components of a Torsen differential are quite different from that of conventional differentials. Figure 4. 33 shows the Torsen differential used in Audi A4 Quattro all-wheel drive system. The key components are gear pairs. It consists of two worms which are connected to the left and right drive shafts (rear drive) or front axle drive and rear axle drive (for 4-wheel drive) respectively, a number of worm wheels/worm gears and spur gears. All worm wheels are fitted with the case/housing, so the engine power received by the case is transferred to the worm wheels and each end of the worm wheels is fitted with a spur gear.

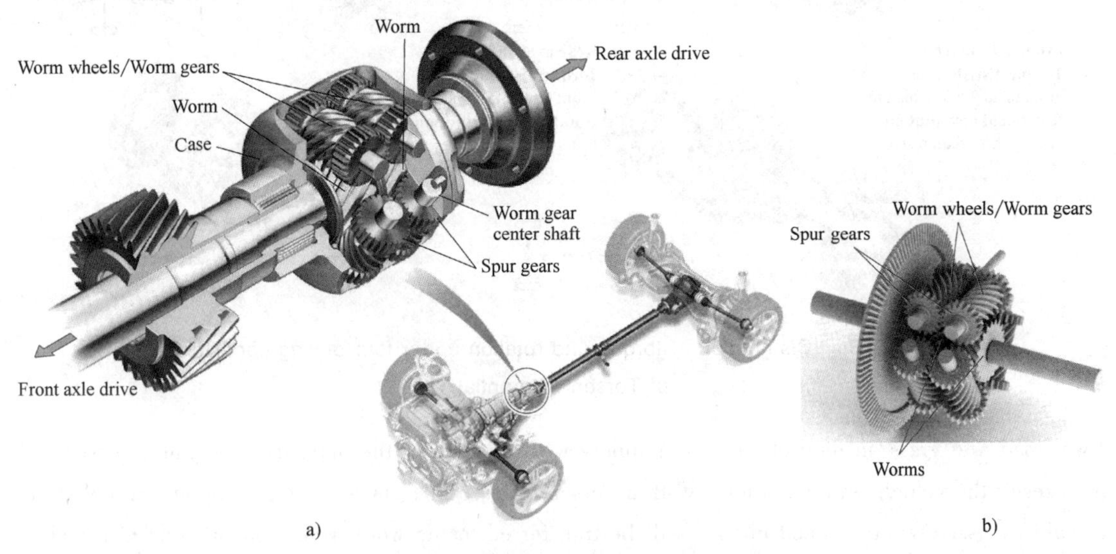

Figure 4. 33 Torsen differential used in Audi A4 Quattro all-wheel drive system

Figure 4. 34 demonstrates the distribution of torque and rotation under four driving conditions of Torsen differential used in Audi A4 Quattro.

Torsen differentials can be used in rear/front drive or four-wheel drive vehicles. The basic idea behind Torsen differential is simple. It works on the principle of worm and worm gear. A worm can rotate a worm gear, but the worm gear can't rotate the worm. Let's consider different driving scenarios for a rear drive vehicle with a Torsen differential. ①When the vehicle moves straight, worm wheels will push and turn the worms, so both the drive wheels will rotate at the same speed. Note that under this circumstance, worm wheels do not spin on its own axles, so the whole mechanism moves as a single solid unit as an open differential works under normal driving. ②When the vehicle makes a left turn, the right wheel needs to rotate at a higher speed than the left wheel. The worm of the fast right axle will make its corresponding worm wheels spinning on its own axis. On the other side, relative to the case the slower rear axle is turning in the opposite direction, thus the worm wheel will spin in opposite direction. The meshing spur gears at the ends of worm wheels will make sure that the worm wheels are spinning at the same speed. That guarantees a perfect differential action. ③If one wheel starts losing traction/spinning faster than the other on a slippery/loose surface

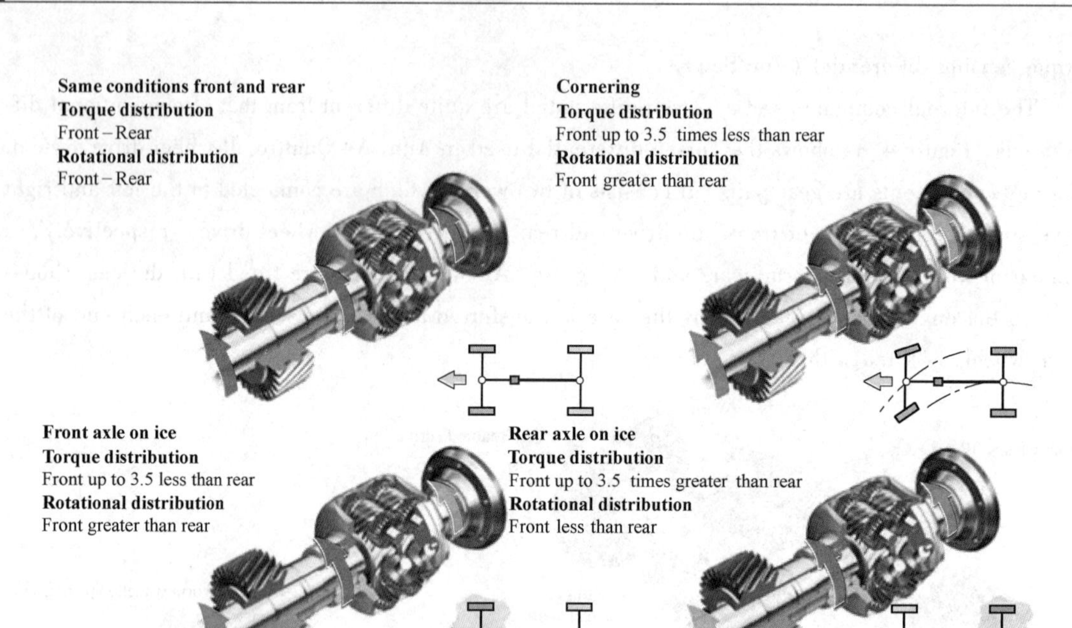

Figure 4. 34 Distribution of torque and rotation under four driving conditions of Torsen differential

(wet road/wet grass/in mud etc.), the slippery wheel will draw the majority of engine's power and as a result the vehicle will get stuck. With a Torsen differential, as soon as the slippery wheel starts to spin excessively, the speed change will be transferred to the worm wheel on the side of the slippery wheel. Then the worm wheel will change the speed to the other side worm wheel which is not able to turn the corresponding worm, because a worm wheel can't drive a worm. As a result, the rear wheels turn together, this allows a large amount of torque to be transferred to the higher traction wheel. In other words, the differential can sense the torque lost to the spinning wheel (torque imbalance). When this happens, gears bind up to limit that loss from the low traction wheel (one spinning most) and send more drive to the wheel with the greater traction, and the vehicle can overcome the traction difference problem. In order to carry more loads, more worm wheel pairs are added. As shown in Figure 4. 34, three worm wheel pairs are added. Comparing with other differentials, the greatest advantage of Torsen differential is that the locking action is instantaneous, that means as soon as the vehicle encounters a traction difference track, the wheels will get locked.

Task 22: Work in pairs, discuss and answer the following questions.

1. What is the purpose of inventing limited slip differentials?
2. How does a Torsen differential work? You can refer to Figure 4. 33.
3. What's the function of spur gears in a Torsen differential?

4. 6. 5 Electronic Controlled Limited Slip Differentials (eLSD)

Electronic controlled limited slip differential is also referred to as eLSD. An eLSD system makes sure each wheel is receiving sufficient torque by use of an electronic control unit. The system electronically monitors input from various wheel sensors and, in the event of slippage, transfers extra torque to the wheel or wheels with the most traction. Some models even allow drivers to choose particular settings for the system. For instance, Mitsubishi's Active Center Differential allows drivers to choose specific presets for driving on the road, gravel and snow. An eLSD also allows for better handling during high-speed curves and lane changes, performing all the tasks of standard differentials, only with computerized speed and precision.

Electronic differential is an advancement in electric vehicles technology along with more traction control. Electronic differential provides the required torque for each driving wheel and allows different wheel speeds electronically. It is used in place of the mechanical differential in multi-drive systems. When cornering the inner and outer wheels rotate at different speeds, because the inner wheels travel a smaller turning radius. The electronic differential uses the steering wheel command signals, throttle position signals and the motor speed signals to control the power to each wheel so that all wheels are supplied with the torque they need. Electronic differential has the advantages of replacing loosely, heavy and inefficient mechanical transmission and mechanical differential with a more efficient, light and small electric motors directly coupled to the wheels using a single gear reduction or an in-wheel motor.

Task 23: Work in pairs and describe the advantages of eLSD to your partner.

4.6.6 Torque-Vectoring Differential

Torque vectoring is a new technology employed in automobile differentials. A differential transfers engine torque to the wheels. Torque vectoring technology provides the differential with the ability to vary the power to each wheel. This method of power transfer has recently become popular in all-wheel drive vehicles. Some newer front-wheel drive vehicles also have a basic torque-vectoring differential. As technology in the automotive industry improves, more vehicles are equipped with torque-vectoring differentials. This allows for the wheels to grip the road for better launch and handling.

The torque vectoring idea builds on the basic principles of a standard differential. A torque-vectoring differential performs basic differential tasks while also transmitting torque independently between wheels. This torque-transfer ability improves handling and traction in almost any situation. Torque-vectoring differentials were originally used in racing. Mitsubishi rally cars were some of the earliest to use the technology. The technology has slowly developed and is now being implemented in a small variety of production vehicles. The most common use of torque vectoring in automobiles today is in all-wheel drive vehicles.

The idea and implementation of torque vectoring are both complex. The main goal of torque vectoring is to independently vary torque to each wheel. Differentials generally consist of only mechanical components. A torque vectoring differential requires an electronic monitoring system in addition to standard mechanical components. This electronic system tells the differential when and how to vary the torque. Due to the number of wheels that receive power, a front or rear wheel drive differential is less complex than an all-wheel drive differential.

In recent times, these systems have taken a fairly radical step forward. Automakers have reinvented front and rear differentials to the point where an engine's torque can be passed around—or vectored—to each corner of the car. In other words, the torque can go from front to back like a traditional all-wheel-drive setup and distribute from left to right on a given axle very quickly. It's like having a computer-controlled, super-speed limited slip differential in each axle.

Task 24: Work in pairs and describe the basic idea behind torque-vectoring differentials to your partner.

4.7 Electric Vehicles (EVs)

An electric vehicle (EV) refers to the broad category of vehicles that use electricity as some or all energy source to drive . Below is a description of EVs that are available for purchase now or in the near future.

1. Battery Electric Vehicles (BEV)

BEV is a type of electric vehicles that must be plugged in to an electrical source to obtain energy to drive. Since they don't run on gasoline or diesel and are powered entirely by electricity, BEVs, also called "pure electric" vehicles or "all-electric" vehicles, are vehicles powered by electric motors fed by batteries. BEVs typically operate with a mileage range of 100-200 miles.

BEVs require a large direct current (DC) electric motor and a large battery pack. The basic design of a BEV is relatively simple and looks very much like that of a front wheel drive internal combustion engine (ICE) vehicle. BEVs contain a fixed gear box, transmission, battery packs and electric motor.

When driven, BEVs don't produce tailpipe pollution—they don't even have a tailpipe. However, the electricity they use may produce heat-trapping gases and other pollution at the source of its generation or in the extraction of fossil fuels. The amount of pollution produced depends on how the electricity is made. BEVs powered by renewable energy sources like wind or solar are virtually emission-free.

2. Plug-in Hybrid Electric Vehicles (PHEV)

PHEVs are hybrid vehicles that use rechargeable batteries, or another energy storage device, that can be restored to full charge by connecting a plug to an external electric power source. A PHEV is called "hybrid" because it contains both an internal combustion engine and an electric motor that plugs into a connection to the electrical grid. PHEVs are sometimes referred to as "extended range electric vehicles" or EREVs. Most PHEVs have a mileage range expectation of 30-40 miles on electric power for use with shorter trips and gasoline powered ICEs for longer trips. Additional characteristics for PHEVs include: ①They contain complex systems of motors and or generators that include a gearing or power split device coupling an internal combustion engine to recharge the battery for electric power. ②Variations on this configuration depend on the number of motors/generators contained in the vehicle and how they are used to power the vehicle.

3. Hybrid Electric Vehicles (HEVs)

HEVs combine an internal combustion engine propulsion system with an electric propulsion system. The presence of the electric power-train is intended to achieve better fuel economy than a conventional gasoline vehicle. HEVs can not be recharged from the electric grid. The presence of the electric motor is to provide increased fuel economy, lower pollutants and an extra power "boost".

A HEV has an electronically controlled power splitter, which directs power from the internal combustion engine to both wheels, and the generator allows the electric motor to run alone or in parallel with the ICE. Maximum speed can be up to 100 mph and some vehicles can accelerate from 0 to 60 mph in 12. 7 seconds. The vehicle switches from the ICE to the electric motor once the vehicle is warmed and it will remain on electric power at low speeds (typically less than 15 mph). The presence of the electric motor can boost the driving range of a HEV to over 600 miles per tank of gasoline in city, driving at an average of 55 mpg.

HEVs come in various different types. In **parallel hybrids**, the engine and the motor both send power to the wheels; in **series hybrids**, only the motor powers the wheels, while the engine simply drives the motor like a generator, recharging the batteries. **Full hybrids** have powerful enough electric motors and batteries to drive the engine independently, while in **mild hybrids**, the motor is too puny to power the car by itself and simply assists the engine (or allows it to switch off when the car is idle in traffic). **Ordinary hybrids** charge up their batteries using power from the engine and energy recovered from the regenerative brakes; **plug-in hybrids** can also be "refueled" from a charging station or domestic power supply, they have much bigger batteries and can be driven by the motor and batteries alone, so they work more like conventional electric cars.

Figure 4. 35a shows the general configuration of a series HEV. The generator includes a charging unit in the form of an alternating-current-to-direct-current (AC/DC) converter. The energy storage medium comprises a large rechargeable battery of the same type used in a BEV. An optional external charger can be used, when the vehicle is not running, to keep the battery "topped off". The electric motor connects mechanically to a transmission that resembles the drive system in a BEV.

A series HEV always runs from the electric motor, never from an internal-combustion engine. For this reason, this type of vehicle has limited speed and acceleration capabilities. Electric motors can't provide the short-burst high performance which an internal-combustion engine can produce. The primary asset of the series design is the fact that the limited combustion ensures low fuel usage and minimal polluting emission.

Figure 4. 35b shows the interconnection of components in a parallel HEV. As in the series design, the generator includes an AC/DC converter to charge the battery. An external charger can be used when the vehicle is parked. The electric motor and primary engine both connect to the transmission. At low speed, or when relatively little power is required to propel the vehicle, the electric motor predominates. At high speed, or under conditions of high power demand (when climbing a steep hill or passing another vehicle, for example), the internal combustion engine delivers.

The outstanding feature of a parallel-design HEV is the fact that both the primary engine and the electric motor contribute directly to propulsion. The two systems operate in tandem (parallel), with the burden automatically shifting from one to the other, depending on driving conditions from moment to moment. The primary engine usually gets its power from conventional gasoline or gasohol, although some primary engines use alternative fuels such as methane or propane.

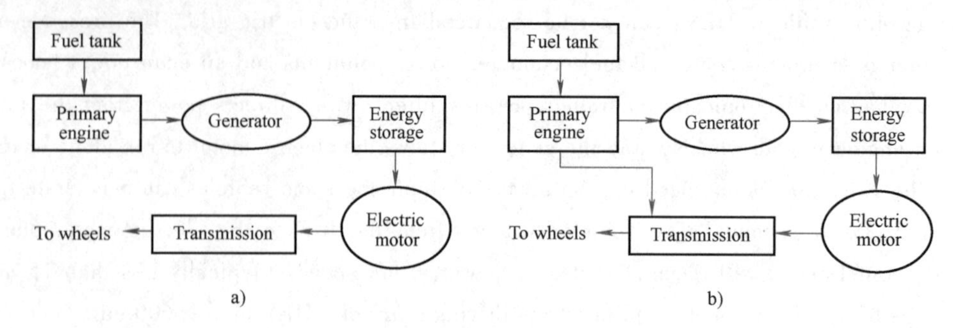

Figure 4.35 Functional block diagram of HEVs

a) series design b) parallel design

Task 25: Work in pairs, discuss and answer the following questions.

1. Describe the types of EVs to your partner.
2. Describe advantages and disadvantages of EVs to your partner.
3. If you will buy a car, what kind of car do you want to buy? Why?
4. In your perspective, what are main obstacles to hinder the mass use of electric cars in China?

4.8 Future Car Technologies

What will the cars of 2050 look like? What will power them? Will they even have a steering wheel? The auto industry is constantly bringing us new technologies, whether it is for safety, entertainment, usefulness or simply for pure innovation. Some of the latest car innovations are truly exciting technologies that could revolutionize not just the automotive industry but human transportation in general. So what will the cars look like in the future? We don't know for sure, but based on what's currently being tested and what's on the road today, we have an idea of some new technologies that will most likely make it into production. Some of them will help keep us safe, some will give us information like never before and some will let us sit back and just enjoy the ride.

1. Cars that communicate with each other and the road

Car manufacturers are seriously looking into and researching two technologies that would enable future cars to communicate with each other and with objects around them. One of developing technologies called Vehicle-to-Vehicle communication, or V2V, is being tested by automotive manufacturers like Ford as a way to help reduce the amount of accidents on the road. Another one called Vehicle-to-Infrastructure communication, or V2I, is being tested as well.

V2V works by using wireless signals to send information back and forth between cars about their location, speed and direction. The information is then communicated to the cars around it in order to provide information on how to keep the vehicles' safe distances from each other. At MIT, engineers are working on V2V algorithms that calculate information from cars to determine what the best evasive measure should be if another car starts coming into its own projected path. A study put out by the National Highway Traffic Safety Administration of USA in 2010 says that V2V has the potential to reduce 79 percent of target vehicle crashes on the road.

V2I would allow vehicles to communicate with things like road signs or traffic signals and provide information to the vehicle about safety issues. V2I could also request traffic information from a traffic management system and access the best possible routes. Reports by the NHTSA of USA say that incorporating V2I into vehicles, along with V2V system, would reduce all target vehicle crashes by 81 percent.

2. Self-driving/driverless cars/autonomous cars

The idea of a self-driving car isn't a new idea. Many TV shows and movies have had the idea and there are already cars on the road that can park themselves. But a truly self-driving car means exactly one that can drive itself.

In California and Nevada, engineers from Google have already tested self-driving cars on more than 200,000 miles (321,869 kilometers) of public highways and roads. Google's cars not only record images of the road, but their computerized maps view road signs, find alternative routes and see traffic lights before they're even visible to a person. By using lasers, radars, cameras, and different types of sensors the cars can analyze and process information about their surroundings faster than a human can.

If self-driving cars do make it to mass production, we might have a little more time on our hands. Americans spend an average of 100 hours sitting in traffic every year. Cars that drive themselves would most likely have the option to engage in platooning, where multiple cars drive very close to each and act as one unit. Some people believe platooning would decrease highway accidents because the cars would be communicating and reacting to each other simultaneously, without the ongoing distractions that drivers face.

In some of Google's tests, the cars learned the details of a road by driving on it several times. When it was time to drive itself, it was able to identify if there were pedestrians crossing and stop to let them pass by. Self-driving cars could make transportation safer for all of us by eliminating the cause of 95 percent of today's accidents: human error.

On November 7 in 2017, Waymo, which is part of Google, announced it would begin regularly testing fully driverless cars without a safety driver on public roads. It was a momentous announcement. A technology that had seemed like science fiction a decade earlier became a reality.

Baidu, China's largest search engine, has joined the driverless car race, announcing in 2015 that its vehicle had completed a fully autonomous test on a route with mixed roads and different weather types. Often referred to as the Google of China, Baidu said the self-driving car, which is a modified BMW 3 series, completed a 30-kilometer test drive route by "executing a comprehensive set of driving actions and accurately responding to the driving environment". With the test, Baidu is joining a host of technology companies looking to establish themselves at the forefront of developments

in automotive technology.

Baidu said its car had "successfully executed driving actions" including turning right and left, U-turns, decelerating while detecting vehicles ahead, changing lanes, passing other cars and merging into traffic from ramps. The autonomous vehicle peaked at a 100 kilometers per hour (62 miles per hour).

3. Augmented reality dashboards

GPS and other in-car displays are great for getting us from point A to point B and some high-end vehicles even have displays on the windshield. But in the near future cars will be able to identify external objects in front of the driver and display information about them on the windshield.

Think of the Terminator, or many other science fiction stories, where a robot looks at a person or an object, automatically brings up information about them and can identify who or what they are. Augmented reality dashboards, AR for short, will function in a similar way for drivers. BMW has already implemented a windshield display in some of their vehicles which displays basic information, but they're also developing augmented reality dashboards that will be able to identify objects in front of the vehicle and tell the driver how far they are away from the object. The AR display will overlay information on top of what a driver is seeing in real life.

So if you're approaching a car too quickly, a red box may appear on the car you're approaching and arrows will appear, showing you how to maneuver into the next lane before you collide with the other car. An augmented reality GPS system could highlight the actual lane you need to be in and show you where you need to turn down the road without ever having to take eyes off the road.

BMW is also researching the use of augmented reality for automotive technicians. They produced a video where a BMW technician uses AR glasses to look at an engine, identify what parts need to be replaced and then shows step-by-step instructions on how to fix it.

AR is also being researched for passengers as well. Toyota has produced working concepts of their AR system that would allow passengers to zoom in on objects outside of the car, select and identify objects, as well as view the distance of an object from the car using a touch-screen window.

4. Airbags that help stop cars

Airbags have been an important safety feature of cars for over twenty years and since that time we have seen them increase throughout the car. Airbags now are not just positioned in front of the driver and passenger seats, they are now to the side and at the knees. In some vehicles, even seat belts have airbags to limit the impact on the passengers in a collision. Maybe all of us don't have them in our cars, but they're on the road.

Accident prevention has evolved so much in modern automobiles that several famous carmakers are now focusing their efforts on diminishing the effects of collisions. The most innovative safety advancement around the world may be the braking bag—an airbag for your car that inflates 80 to 100 milliseconds before collision to dilute the physics of impact.

Mercedes is experimenting with airbags that deploy from underneath the car and will help stop a vehicle before a crash. The airbags are part of the overall active safety system and deploy when sensors determine that air impact is inevitable. The bags have a friction coating that helps slow the car down and can double the stopping power of the vehicle. The bags also lift the vehicle up to 8 centimeters, which counters the car's dipping motion during hard braking, improves bumper-to-bumper contact and helps prevent passengers from sliding under seat belts during a collision. Although Mercedes are thought to have been working on this technology for a number of years, it is not yet at a stage where it is available in any production models.

The idea behind the braking bag is to decelerate the car faster than you can. The braking bag is installed between the front axle carrier and the underbody paneling, according to Mercedes. When a collision is imminent, the bag, coated with rubber and steel for friction, inflates. Another advantage of the braking bag is the vertical movement also makes the vehicle's restraint systems more efficient. The seats move toward the occupants by about an inch, which allows the belt tensioners to grab tighter.

Mercedes-Benz has built an S400 Hybrid called the Experimental Safety Vehicle (ESF in German) as the company's safety technology. One of those tech highlights is the braking bag—the panel under the car as shown in Figure 4.36, which operates both as an air brake and a support in the event of an accident. When the pre-safe system detects that a crash is imminent, it will deploy an airbag located in the vehicle floor above the front axle carrier. This action will thrust a panel to the ground that, in its most simple form, slows the car down fractionally due to aerodynamic and friction drag. But the panel is also thrust down forcefully enough to raise the car up 8 centimeters, reducing dive and possibly helping to prevent the car from submarining under an obstacle in front of it.

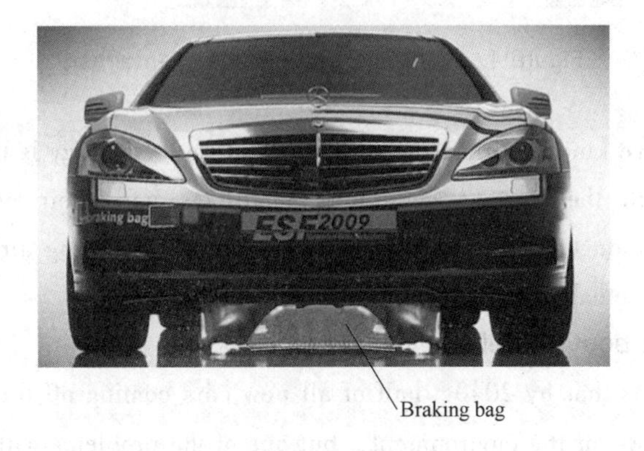

Figure 4.36 The Experimental Safety Vehicle (ESF) of Mercedes-Benz

TRW (Thompson Ramo Wooldridge) Automotive has developed plans for airbags to deploy from the sill beneath the side doors of vehicles, as shown in Figure 4.37. Using radar and camera systems the airbags will deploy a fraction of a second before impact, helping to reduce injuries to the occupants of cars.

Volvo, the Swedish carmaker known for its safety engineering (a Volvo engineer invented the modern 3-point seat belt), has turned their focus to keeping those outside the car safe too. Starting with Volvo V40, they've introduced the first car airbag for pedestrians (see Figure 4.38). First in-

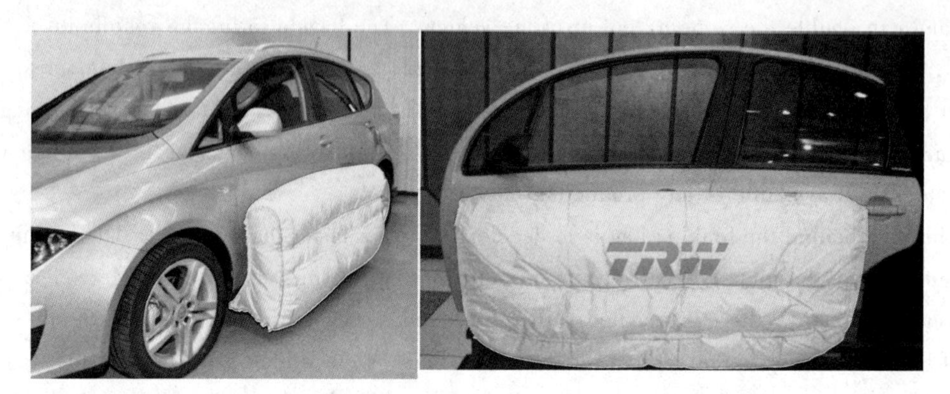

Figure 4. 37 Airbags deploying from the sill beneath the side doors of vehicles

troduced at the Geneva Motor Show in March, the 2013 V40 has now gone on sale in some countries.

Figure 4. 38 Volvo V40's pedestrian airbag

What gives the above kind of airbag potential as a future technology is that it uses existing vehicle safety systems. With the current evolution of airbags and their pervasiveness within the automotive world, it wouldn't be a stretch to imagine future cars using airbags to not only protect passengers, but also to actually stop cars as well.

5. Energy-storing body panels

Exxon Mobil predicts that by 2040, half of all new cars coming off the production line will be hybrids. That's great news for the environment, but one of the problems with hybrids is that the batteries take up a lot of space and are very heavy. Even with advances in lithium-ion batteries, hybrids have a significant amount of weight from their batteries. That's where energy-storing body panels come in.

Imagine a car whose body also serves as a rechargeable battery. A battery that stores braking energy while you drive and also stores energy when you plug in the car overnight to recharge. At the moment this is just a fascinating idea, but tests are currently under way to see if the vision can be transformed into reality.

In Europe, a group of nine auto manufacturers are currently researching and testing body panels that can store energy and charge faster than conventional batteries of today. The body panels being tested are made of polymer resin and carbon fiber that are strong enough to be used in vehicles and pliable enough to be molded into panels. These panels could reduce a car's weight by up to 15 percent.

The panels would capture energy produced by technologies like regenerative braking, or when the car is plugged in overnight and then feed that energy back to the car if needed. Not only would this help reduce the size of hybrid batteries, but the extra savings in weight would eliminate wasted energy used to move the weight from the batteries.

Volvo Car is one out of nine participant in an international material development project. Volvo Car Group has developed a revolutionary concept for lightweight structural energy storage components that could improve the energy usage of future electric vehicles. The material, consisting of carbon fibres, nanostructured batteries and super capacitors, offers lighter energy storage that requires less space in the car, cost effective structure options and it is eco-friendly. The Volvo shows body panels that could replace batteries in electric cars (see Figure 4. 39).

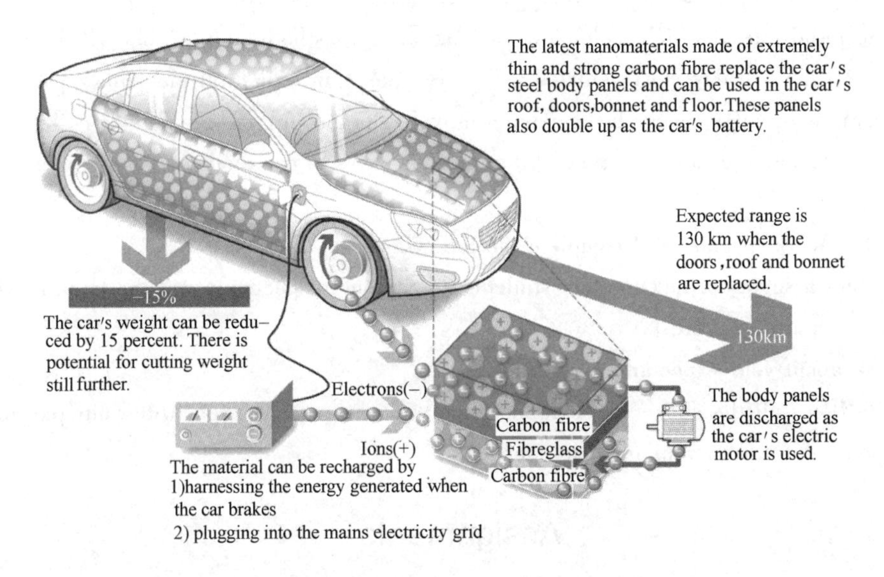

Figure 4. 39 Volvo's body panels that could replace batteries in electric cars

The project, funded as part of a European Union research project, included Imperial College London as the academic lead partner along with eight other major participants. Volvo was the only car manufacturer in the project. The project team identified a feasible solution to the heavy weight, large size and high costs associated with the batteries seen in hybrids and electric cars today, whilst maintaining the efficient capacity of power and performance. The research project took place over 3. 5 years and is now realized in the form of car panels within a Volvo S80 experimental car.

The reinforced carbon fibers sandwich the new battery and are molded and formed to fit around the car's frame, such as the door panels, the boot lid and wheel bowl, substantially saving on space. The carbon fiber laminate is first layered, shaped and then cured in an oven to set and harden. The super capacitors are integrated within the component skin. Then this material can be used around the vehicle, replacing existing components, to store and charge energy.

The material is recharged and energized by the use of brake energy regeneration in the car or by plugging into a mains electrical grid. It then transfers the energy to the electric motor which is discharged as it is used around the car. The breakthrough shows that this material not only charges and stores faster than conventional batteries, but also is strong and pliant.

Toyota is also looking into lightweight energy storing panels, but they're taking it one step further and researching body panels that would actually capture solar energy and store it in a lightweight panel.

The above projects show the new kind of thinking that we need to develop with electric cars, because electricity isn't like liquid fuel, it can be stored in a more distributed fashion. Since researchers are working with carbon fibers, they're probably trying to make some sort of giant hypercapacitor. The main challenge is no doubt to make it safe and inexpensive, because users really don't want all that energy to be unsafely discharged in case of accident, and at current prices, making such a huge mass of hyper capacitors would be prohibitively expensive.

With the rapid rate at which technology is progressing, nobody knows what will be possible in 10-15 years. Maybe one day most vehicles will store energy in structural elements, possibly along with more traditional battery packs for maximum range. Whether future body panels collect energy or just store it, automotive companies are looking into new ways to make our cars more energy-efficient and lightweight.

Task 26: Work in pairs and complete the following tasks.

1. Conduct a survey or talk to other students on campus respectively and the topic is what their expectations for future cars are.
2. Think about your expectations for future cars.
3. Write down all the above expectations for future cars, share them with your partner, then describe visions of a car on your mind to each other.

Assignments

4.1 The invention of the car has changed people's life. Can you imagine how different your life would be if there were no cars in our lives? Discuss it with your partner.

4.2 Fill in the gaps in the text with words from the box in their correct form.

spark plug; engine; moement; piston; internal combustion engine; crankshaft; gasoline

The heart of every car is its______. It produces the power that turns the wheels and the electricity for lights and other systems.

Most automobiles are powered by__________________. Fuel, usually________or petrol, is burned with air to create gases that expand. ______________creates a spark that ignites the gas and makes it burn. This energy moves through cylinders in which________slide up and down. They are attached to rods that move a__________. Normal car engines have 4 to 6 cylinders but there are also models with 8 and 16 cylinders. The turning__________is passed through the drivetrain to the drive wheels.

automatic transmission; clutch; drive shaft; four-wheel drive; drivetrain; torque; joint; transmission; differential; power; lever; manual transmission; driveshaft ; differential; gear

The engine and all parts that carry power to the wheels are called the ________. It includes the ____________, ____________, ______________, the axles and the drive wheels that move the car. While most cars have drive wheels in the front, some have them in the back. Cars that need to drive over all kinds of ground have a ______________.

The __________ controls the speed and __________. When a car travels at a normal speed on a flat road, it does not need so much torque to keep it moving, but when you want to start a car from a hill, the engine must produce more power. ____________ control speed and power of the engine in different driving conditions.

In cars with __________________ you have to change gears by pressing down the ______ with your foot and moving a ________. Cars with ______________ change gears without control by the driver. Lower gears give the car more torque and speed. When the car moves faster, the transmission shifts to higher gears.

The __________ carries the power to the axle which is connected to the wheels. It has several __________ which make the axle and wheels moveable as the car drives on uneven and bumpy roads.

The ____________ is connected to the rear end of the driveshaft. It lets the wheels turn at different speeds because in curves the outer wheels must travel a greater distance than the inner ones.

4.3 When driving a car, drivers must comply with traffic regulations in order to secure the safety of people. Search online to find information about seatbelt regulation, child car seat regulation, and cell phone regulation. Discuss the above regulations with your partner.

4.4 Work in pairs and discuss the following question with your partner: What are revolutions per minute for a 4-stroke cycle engine?

Hints: The 4-stroke cycle repeats itself thousands of times a minute. These repetitions are more commonly known as Revs. A rev counter tells you how many thousand times per minute the cycle is repeated.

4.5 Work in pairs and describe how an engine in a car produces power. You can take diagram in Figure 4.40 as a reference.

4.6 Work in pairs and describe how the gearbox in a car changes driving speed. You can take diagram in Figure 4.41 as a reference.

4.7 Work in pairs and discuss why a cooling system is needed in cars.

Hints: Burning fuel inside a car's engine creates a lot of heat. Most of the heat has to be removed by a cooling system. Liquid cooling systems have a mixture of water and chemicals. A water pump forces this mixture to flow between the cylinders of the engine. The hot water is then pumped through a radiator where the air carries away the heat.

4.8 What is the use of the dashboard in a car?

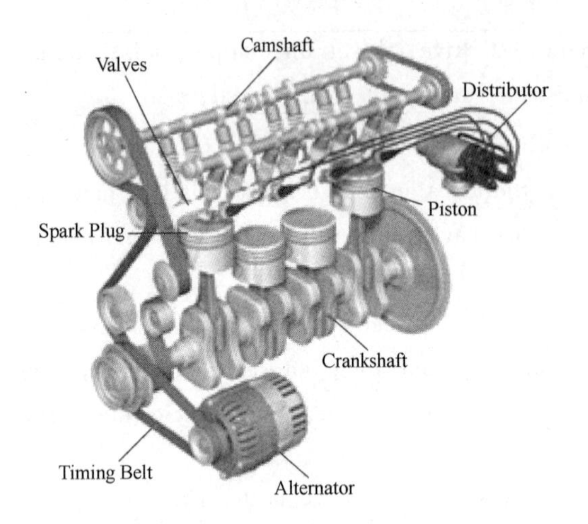

Figure 4. 40　Diagram of an engine

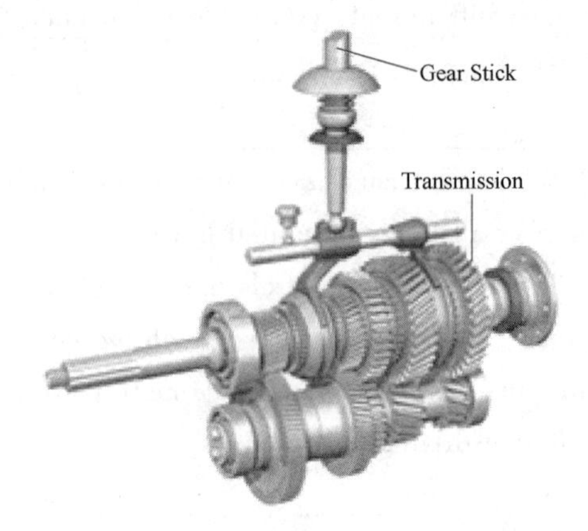

Figure 4. 41　Diagram of a gearbox

Hints: The dashboard has many instruments that show you how fast you are moving, the amount of petrol that is left in the tank, the oil temperature and some other useful information.

4. 9　What is the function of the brake system in a car?

Hints: The brake system slows down or stops the car. Brakes operate on all four wheels. There are two basic types of brakes, drum or disk brakes. In both cases a friction pad is pressed against a drum or disk with the help of a hydraulic system.

All cars have emergency hand brakes which you need to use if the hydraulic system fails. It is also called a parking brake because you use it to stop a vehicle from rolling down a hill. Antilock braking systems (ABS) keep the wheels turning when you step on the brakes. This computer-controlled system prevents skidding if you are on a slippery road.

4. 10　Figure 4. 42 shows a diagram of a typical open differential. Work in pairs, give names of parts (from 1 to 7), and then describe its working principle.

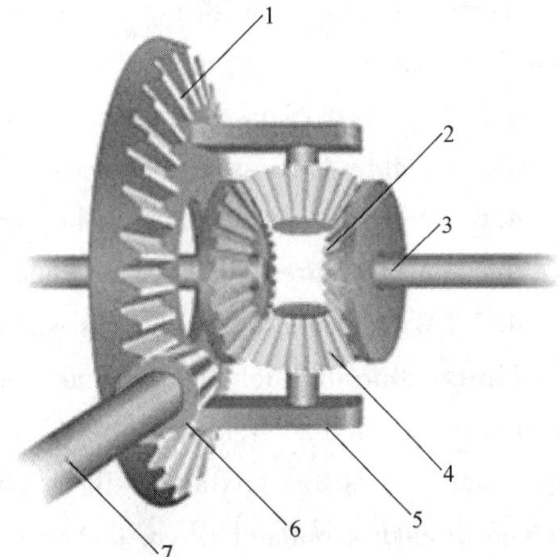

Figure 4. 42　Diagram of a typical open differential

4. 11　Why car designers use Computational Fluid Dynamics (CFD) analysis of airflow around a car model before cars are produced, as the example shown in Figure 4. 43.

Hints: Computational Fluid Dynamics (CFD) is a combination of computer science, numerical mathematics and modern fluid dynamics. CFD uses numerical methods to analyse and solve the fluid flows problems. CFD is a method to research and examine wide range of problems in heat transfer and fluid flow. Automotive aerodynamics is one of the important applications of CFD. The development of a car involves a wide range

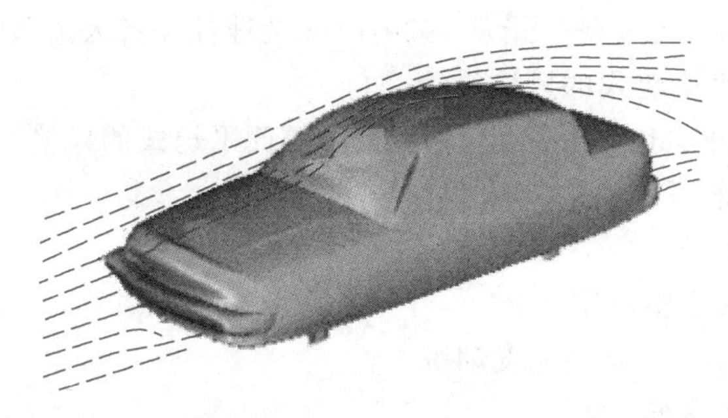

Figure 4. 43 CFD analysis of airflow around a car model

of key components. Material selection for lightweight body and consideration of aerodynamics for these components are both important aspects to design.

In automotive industry, CFD is used to minimize the drag force and increase the down force (negative lift) which helps to stabilize the vehicle, leading a decrease in fuel consumption and shape design. In racing car industry such as Formula 1, a common interest is necessary to keep a car on the ground and this could be obtained by attaching a spoiler at the rear of the car body. Most of the aerodynamic evaluation of air flow has been carried out computationally by Computational Fluid Dynamics software, such as ANSYS or any other plug-ins built in 3D software.

相关词汇注解

1 beasts of burden 指过去驮载用的牲口（如马，牛等）

2 travois *n.* 旧式雪橇

3 chariot *n.* 二轮战车

4 trial and error 反复实验

5 cutting-edge *adj.* 先进的，尖端的

6 a patchwork of tight-fitting rocks *n.* 由紧实的岩石组成的混合物（层）

7 self-powered, self-propelled 机动式的

8 suspension *n.* 汽车悬架

9 vibration-resistant *adj.* 抗振的，耐振的

10 stem-driven *adj.* 蒸汽驱动的

11 4-stroke 四冲程

12 lead-acid battery 铅酸蓄电池

13 mass production 大量生产

14 4-cylinder engine 四缸发动机

15 backlash *n.* 反冲，强烈抵制

16 locomotive *n.* 机车，火车头

17 leaf spring 钢板弹簧，叶片弹簧

18 big three 三巨头，文中指美国通用汽车公司、福特汽车公司与克莱斯勒汽车制造公司

19 great depression 大萧条（指从1929-1933全球性经济大萧条时期）
20 ponton style 平底船式的风格
21 trend-setting slab-side styling 引领潮流的侧面平坦式的风格
22 kei car 轻型车
23 tourer *n.* 游览车
24 downsized car 小型汽车
25 performance engine 高性能发动机
26 pony car 小型汽车
27 space-saving 节省空间的
28 niche makers 指填补市场空缺的制造商
29 monocoque/unibody design 指单体式汽车设计方法
30 pickup trucks 小货车
31 badge engineering 指一款车换牌换脸再销售，大家戏称为“马甲车”，专业领域称其为“Badge Engineering”这种产业最早出现在美国地区——1926年纳什公司将旗下所有“Ajax”牌的汽车换标成“Nash”继续销售，不仅省下一系列开发流程以及所需的资金，还能开拓新市场，但也存在一定的风险
32 urban traffic congestion 城市交通拥堵
33 friction coating 摩擦涂层
34 glove compartment 汽车仪表盘上的储物箱
35 braking bag 制动气囊
36 odometer *n.* 里程表
37 sun visor *n.* 汽车中的防晒板
38 tachometer *n.* 汽车中的转速表
39 differential *n.* 差速器
40 reciprocating motion 往复运动
41 ancillary equipment 辅助装置
42 oil reservoir 汽车油箱
43 sump *n.* 汽车机油箱
44 ethanol *n.* 乙醇，酒精
45 methanol *n.* 甲醇
46 hydrogen *n.* 氢
47 propane *n.* 丙烷
48 spark-ignited 火花点燃式的
49 compression-ignited 压缩燃料点燃式的
50 liquid-cooled 液体冷却的
51 reciprocating steam engine 往复式蒸汽机
52 stirling engine 斯特林发动机
53 intake manifold 汽车的进气歧管
54 carburetor *n.* 化油器

55 venture *n.* 一种流体流量测定装置
56 choke 汽车的阻气门
57 rough-and-ready *adj.* 潦草的，不精美但可用的
58 displacement engine 活塞式引擎（发动机）
59 turbocharged engine 涡轮增压发动机
60 towing ability 牵引力
61 front-engined rear-wheel-drive car 前置引擎后轮驱动的汽车
62 propeller shaft 驱动轴，传动轴
63 gearbox *n.* 汽车变速器
64 neutral position 空挡位置
65 collar *n.* 轴套
66 freewheel *n.* 飞轮 *v.* 靠惯性旋转
67 idler gear 惰轮
68 Hall effect sensors 霍尔效应传感器
69 throttle *n.* 节流阀
70 torque converter 变矩器
71 electro-hydraulic 电动液压式的
72 stepless *adj.* 无级的
73 bow-tie-shaped 蝴蝶结状的
74 universal joint 万向节
75 off-road vehicles 越野车
76 PID control PID 控制器是根据系统的误差，利用比例、积分、微分计算出控制量来进行控制的，简称 PID 控制
77 emission-free 无排放的
78 topped off 文中指电池充满的状态
79 augmented reality 增强现实技术
80 touch-screen 触摸屏式
81 pre-safe system 主动安全防护系统
82 regenerative braking/brake energy regeneration 回馈制动（回馈制动是变频制动方式的一种，也是非常有效的节能方法，并且避免了制动时对环境及设备的破坏）
83 mains electrical grid 电源供电网络
84 fluid flows problems 液体流量问题
85 spoiler *n.* 汽车上的扰流板

Part5

Guidance for Writing Scientific Papers

Objective

Publication in a reputable, peer reviewed journal should be the goal of every researcher, as this provides the most effective and permanent means of disseminating information to a large audience, and it's also a wide and effective communication way for researchers. The aim of this part is to provide guidelines to assist with the preparation of a manuscript for a scientific journal.

After completing this chapter, you should be able to:

- know the importance of writing a scientific paper;
- plan and organize its structure;
- learn some skills of writing a scientific paper.

5.1 Introduction

Today English is the official language of international conferences, and most of the important publications in science and technology now appear in English. Researchers must read English-language journals and books to keep up with advances in their fields. Over twenty percent of the world's population speaks Chinese, but China is still a developing country and few researchers outside of China can understand a scientific publication written in Chinese. The Chinese researcher who wishes to reach a wide readership and gain influence must publish in English. Therefore, learning how to write a manuscript in English has become part of the researcher's task.

Writing in English can be difficult even for someone who grew up as a native speaker, and even more so for anyone who learns it as a foreign language. English derives from many cultures and is constantly evolving. As a result, its grammatical rules are many and complex. We Asians face an additional challenge not shared by our European counterparts: Most Asian dialects, like Chinese, do not belong to the same language family as English. There are grammatical constructs that have no corresponding forms in Chinese. The task is not merely to translate words, but to understand and use foreign concepts of syntax as well. Another challenge that Asians like Chinese scholars face is we usually think in Chinese when we write a manuscript in English. Text that is translated from one language into another often sounds awkward, and makes it difficult for readers to understand the paper. It may be difficult to put your thoughts into English, but you will gain facility with practice. Such difficulties are by no means insurmountable. With practice, plus attention to the particular challenges faced by the Asian scholar, any researcher should be able to write a scientific paper in English that is concise and lucid as well as grammatically correct, even if his vocabulary or understanding of English usage is limited.

Scientific papers are an important method of publication, without them the research work will not be known or shared by other researchers. However, sometimes they are poorly understood because they are not written very well. A Scientific paper is organized to meet the needs of valid publication. It is, or should be, highly stylized, with distinctive and clearly evident component parts.

It is true that a paper should be as good as the underlying research, and writing effort can improve the quality of the study. On the other hand, the valuable research will probably lose value if the paper can not be understood. If, by attention to style and presentation, your paper communicates more effectively, you will have accomplished your purpose. So please do not think that good English is not critical in science writing. In fact, scientists try to be so concise that their English should be better than that of workers in other disciplines. In addition, a standard format is used for these articles. If you have read scientific papers, you would notice that a standard format is frequently used, in which the author presents the research in an orderly, logical manner.

Part 5 will help both first-time writers and more experienced authors in mechanical engineering to present their research results effectively. Although these descriptions are especially adapted to help students write papers related to mechanical engineering, the information presented will be useful to anyone who wishes to write a paper in their area of study.

5.2 What Is a Scientific Paper?

A scientific paper is a written and published report describing original research results. Scientific research articles provide a method for scientists to communicate with other scientists about the results of their research. The article must have been reviewed by experts within the same subject area before publication. Thus, sufficient information must be presented in a scientific paper so that the potential users of the data can: ①assess observations, ②repeat experiments, and ③evaluate the research processes.

It can be seen from the above description, a scientific paper must describe original research results, and it must be written in a certain way and published in a certain medium. Therefore, whether a paper reports the original research results and where it is published should be the main features for a scientific paper differing from other papers, and original research results mean that the research results should be the first disclosure.

With the emergence of the age of digital media, there are social networking, online collabora-

tive tools and new business models for publishing. The growth of the Internet, online journals, new software packages, computer networks make it difficult to identify a scientific paper. In normal practice, how can we judge whether a scientific paper is the first disclosure?

Here is how the Journal of Computer-Aided Design (CAD) states its policy when callingfor scientific papers. The statement is "Submission of an article implies that the work described has not been published previously (except in the form of an abstract, a published lecture or academic thesis), that it is not under consideration for publication elsewhere, that its publication is approved by all authors and tacitly or explicitly by the responsible authorities where the work was carried out, and that, if accepted, it will not be published elsewhere in the same form, in English or in any other languages, including electronically without the written consent of the copyright-holder. To verify originality, the article may be checked by the originality detection service Crossref Similarity Check". Regardless of the type of the publication mentioned in the statement, the above description or policy implies a judgment about the criterion of first disclosure, which could be as a reference for researchers in other areas.

5.3 Why Write Scientific Papers?

Scientists are motivated by two things: ①to understand the world, ②to get credit for it. The main task of a scientist concerns the expansion of human knowledge. In the academic and public sectors, scientific papers are the means for this expansion. Science could not really exist, if scientists did not record every experiment performed, every data collected, every result obtained. But scientific writing doesn't mean to only keep records of progress and information, but to actually publish the results of studies in scientific journals. While keeping records is quite easy to do, publishing scientific papers can be really difficult, and especially for young researchers who are just starting to discover the world of scientific publications.

Scientific and research writing is crucial for a career in sciences. Doing research is only half of the picture. If the results of research studies are not published—and where they are published has an important impact also—other researchers can not appreciate the value of the experiments or studies accomplished, they can not further build on it, the public can not trust the information and overall scientific knowledge will not be further popularized and spread. Nowadays in most countries, research funding is actually decided and divided based on the number and importance of publications (the importance can be evaluated by the impact factor, citation index or other tools). Participations at congresses such as oral or poster presentations and patent applications are very important too, but original articles, to differentiate from reviews and book chapters, are the most important and decisive factors for scientists and researchers. The fact that reviewers, which are experts in the same field of research, have approved the publication of a paper, gives it credibility and authority. Considering the well-known difficult and time-consuming process of revision, research articles are viewed with respect from the scientific community. Learning how to write a scientific paper with high quality is thus a key factor in a scientist's career path. "Publish or Perish" should indeed be the rule for scientists working as individuals—scientific papers are your professional contribution. If you don' t publish, you're out.

For instance, the mathematical basis for Bézier curves—the Bemstein polynomial—had been known since 1912, but the polynomials were not applied to graphics until some 50 years later, when they were widely publicized by the French engineer Pierre Bézier, who used them to design automobile bodies at Renault. The study of these curves was however first developed in 1959 by mathematician Paul de Casteljau using de Casteljau's algorithm, a numerically stable method to evaluate Bézier curves at Citroen, another French automaker. Bézier patented and popularized the Bézier curves and Bézier surfaces that are now used in most computer-aided design and computer graphics systems. However, Pierre Bézier did not actually create the Bézier curve which was first developed in 1959 by Paul de Casteljau. Pierre Bézier was the first to publish their use for the design of plane curves using the de Casteljau's algorithm. Both of them worked nearly parallel to each other, but because Bézier published the results of his work, Béizier curves were named after him, while de Casteljau's name is only associated with related algorithms. This is a very classic example to show the importance of writing and publishing scientific papers.

5.4 The General Structure of a Scientific Paper

The scientific format may seem confusing for the beginning science writer due to its rigid structure which is so different from writing in the humanities. One reason for using this format is that it is a means of efficiently communicating scientific findings to the broad community of scientists in a uniform manner. Another reason, perhaps more important than the first, is that this format allows the paper to be read at several different levels. For example, many readers skim titles to find out what information is available on a subject. Others may read only titles and abstracts. Those wanting to go deeper may look at the tables and figures in the results, and so on. The key point here is that the scientific format helps to ensure that at whatever level a person reads your paper (beyond title skimming), they will likely get the key results and conclusions.

5.4.1 General Style

Scientific papers can be structured in different ways according to subject, method and the type of paper. Most journal-style scientific papers contain the following items: title, authors and affiliations, abstract, keywords, introduction, methods and materials, results, discussion, acknowledgments, and literature cited, which parallel the scientific research process. The items appear in a journal-style paper in the prescribed order as shown in Table 5.1.

Scientific papers are usually subdivided into the following main sections: introduction, methods and materials, results, discussion. Other sections can be included as necessary. It is important to understand the differences between sections and to put information in the appropriate location.

Table 5.1 The order of the items in a journal-style paper

Research process	Items of the paper
What was the research about?	Title
Who did the research?	Authors and affiliation
What did the researchers do in a nutshell?	Abstract
What is the problem?	Introduction
How did the researchers solve the problem?	Methods and materials
What did the researchers find out?	Results
What does the result mean?	Discussion
Who helped researchers out?	Acknowledgments (optional)
Whose work did researchers refer to?	Literature cited
Extra information	Appendices(optional)

Reading scientific papers (such as the articles you will use as your references for the introduction and discussion) will give you good ideas and guidance as well. After all, these are peer-reviewed and published scientific papers, and they can serve as useful models for your own writing.

5.4.2 Before Writing a First Draft

Before writing a first draft, it is important to establish that the topic of the manuscript is likely to be consistent with the focus of the journal. This may be clearly stated within the journal or may be determined by examining several recent issues. Having selected a journal, it is essential to carefully read and follow the guidelines for authors published within the journal or obtained directly from the editor or publisher. These guidelines are usually very specific and include rules about word limit, organization of the manuscript, margins, line spacing, preparation of tables and figures and the method used to cite references. Failure to comply with the guidelines may result in rejection or return of the manuscript for correction, thereby delaying the process of review and publication.

The art of writing a manuscript can be improved with practice and considerable help may be gained by asking others, especially those who have published, to critique and proofread drafts. Getting started is often the most difficult part, and for this reason it is best to begin with the easiest sections. These are usually the methods and results, followed by the discussion, conclusion, introduction, references and title, leaving the abstract until last. If possible, try and set aside some time for writing on consecutive days. Long gaps between periods of writing interrupt the continuity of thought. To avoid frustration, ensure all the necessary information, for example, all data, references and any draft of tables or figures, are at hand before starting to write. The task of writing the manuscript may seem easier if each section is viewed as a separate task. Before starting to write, it may help to prepare an outline for each section which includes a number of major headings, subheadings and paragraphs covering different points. When writing the first draft, the goal is to get something down on paper, so it does not matter if sentences are incomplete and the grammar incorrect, provided that the main points and ideas have been captured on paper. Try to write quickly, to keep the flow going. Use abbreviations and leave space for words that do not come to mind immediately. Having finished the first draft, immediately revise it and be prepared to do this several times until you feel it is not possible to improve it further. Acceptance of a manuscript is invariably conditional on changes being made, so be prepared to rewrite and revise the manuscript extensively.

Often a manuscript has more than one author and thus the writing may be shared. However, the style needs to be consistent throughout. So even if sections of the early drafts are written by different authors, the first author must go through the entire manuscript before submitting, and make any necessary editorial changes.

Thus, writing scientific papers is a process of learning and improvement, as shown in Figure 5.1. Good writing doesn't happen overnight; it requires planning, drafting, rereading, revising and editing, and needs self-review, peer-review, feedback from experts in the field, and practice. There are no shortcuts; practice makes perfect!

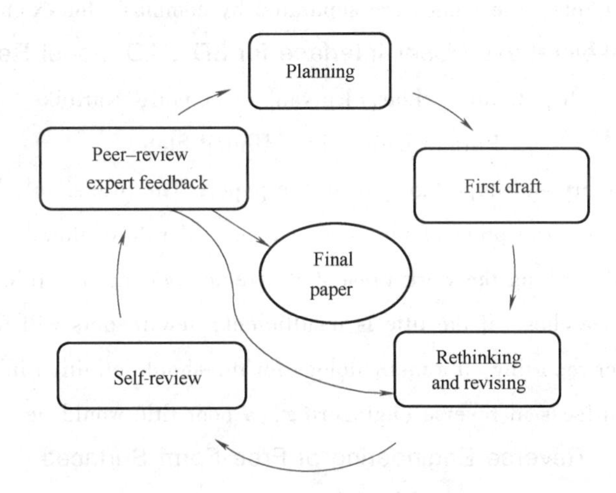

Figure 5.1 Writing is a process

5.4.3 Section Headings and Subheadings

1. Main section headings

Each main section of the paper begins with a heading which should be capitalized, centered or left justified at the beginning of the section, and double spaced from the lines above and below. Do not underline the section heading or put a colon at the end. Example of a main section heading:

Introduction

2. Subheadings

When the paper reports on more than one experiment, subheadings are usually used to help authors organize the paper. Subheadings should be capitalized (first letter in each word), left justified, and either bold italics or underlined. Example of a subheading:

Effects of Light Intensity on the Rate of Electron Transport

5.5 Title, Authors' Names, and Institutional Affiliations

A scientific paper should begin with a title that succinctly describes the contents of the paper. Use descriptive words that you would associate strongly with the content of your paper: the issue studied, the materials and methods used, the treatment, the location of a field site, etc. A majority of readers will find the paper via electronic database searches and by search engines according to some words contained in the title.

5.5.1 Title

The title should be centered at the top of page 1; the title is not underlined or italicized; the authors' names and institutional affiliations are double-spaced from and centered below the title. When more than two authors, the names are separated by commas, for example:

A 2D Sketch-based User Interface for 3D CAD Model Retrieval

Pu, Jiantao, Lou, Kuiyang, Ramani, Karthik

Purdue University, United States

The title is a necessary and important part of the paper. The title should be short and unambiguous, yet be an adequate description of the work. A general rule-of-thumb is that the title should contain the key words describing the work presented. Remember that the title becomes the basis for most on-line computer searches, if the title is insufficient, few readers will find or read the paper. For example, in a paper reporting on a methodology for threshold definition in selective sampling experiment of free-form surfaces in reverse engineering, a poor title would be:

Reverse Engineering of Free-Form Surfaces

Why? It is very general, and could be referring to any topics in regard to reverse engineering of free-form surfaces. Readers could not obtain the main research point based on the title; they have to skim through the whole paper to get the idea. A better title would be:

Reverse Engineering of Free-Form Surfaces: A Methodology
for Threshold Definition in Selective Sampling

Why? Because the title identifies a specific research issue. If possible, give the key result of the study in the title.

Here is another example with poor title. Assuming this paper presents an approach to create CAD models from point clouds data, the key feature of this approach is to combine reverse engineering method with feature-based 3D CAD system, and then rebuild a CAD model which contains features and constraints.

An Approach to Reconstruct CAD Models From Point Clouds

Why? Because the title does not indicate the key feature of the approach, there are many different kinds of methods to reconstruct CAD models from point clouds. What's the novelty of the study Therefore, the title is ambiguous. A better title would be:

An Integrated Reverse Modeling Approach Based on
Reconstruction of Features and Constraints

The title not only indicates the research area—Reverse Engineering, but also describes the novelty of the approach—an integrated method. How integrated? It could arouse the interest of readers to read the paper further, and the key result of the research is about reconstruction of features and constraints.

5.5.2 Authorship

According to the current criteria, all persons designated as authors of a paper should have: ① conceived and planned the work that led to the paper or interpreted the evidence it presents, or both, ②written the paper, or reviewed successive versions and taken part in the revision process, and ③approved the final version. As all three conditions must be met, this implies that a skilled laboratory technician who does the necessary spadework is unworthy of acknowledgment as a coauthor. So, then, which guidelines for authorship are applied in the real world. In the real world of research, most investigations nowadays are necessarily made by a team. Each member contributes different talents and skills. Irrespective of the nature of their contributions—intellectual (creative) or practical (doing the experiments) —all members of the team are usually acknowledged in the author byline. The names also require some sort of ranking. The most prominent position, heading the list, is usually occupied by the team leader (often the author). The coauthors are then listed in descending order, reflecting how much each person contributed to the work.

In order to confirm the authorship and the work is the author's original work, the journal or publisher will send "Copyright Transfer Agreement" to the author, and make sure that the author agrees to inform the coauthors of the terms of this agreement and to obtain their permission to sign on their behalf. The work is submitted only to this journal, and has not been published before.

5.5.3 Format of Chinese Names

In Western countries a personal name is written in the following order: first name, followed by middle names or initials, then last name. It is assumed that the last name, which may be hyphenated, is the family name. Occasionally the family name is followed by designations such as Jr. (for junior, a man with the same name as his father), Sr. (for senior, a man with the same name as his son), M. D., F. R. S. (Fellow of the Royal Society), and so on. These are easily recognized as not being part of the family name.

In China and many other countries, the traditional style is to list the family name first, followed by the given names. If the name is Westernized, as it often is by authors writing in English, the order is reversed as following: given names followed by family name, like Yiquan Wang.

5.6 Abstract

The abstract should inform the reader in a succinct manner as to what the article is about and what the major contributions are discussed. In other words, an abstract should answer the questions why, how and what. Why did researchers study it? How did researchers study it? What did researchers find and what does it mean? It should contain only what researchers are specifically reporting in the manuscript. In other words, the abstract should provide a brief summary of each of the main sections of the paper: introduction, materials and methods, results, and discussion.

A well-prepared abstract enables readers to identify the basic content of a document quickly and accurately, to determine its relevance to their interests, and thus to decide whether they need to read the document in its entirety.

There are usually four basic sections required in an abstract in the following prescribed sequence:

(1) The question (s) the researchers investigated (or purpose) (from the introduction): state the purpose very clearly in the first or second sentence.

(2) The methods and materials used (from methods and materials): clearly express the basic principle of the research; name or briefly describe the basic methodology used without going into excessive detail—be sure to indicate the key methods used.

(3) The major findings including key quantitative results, or trends (from results): report those results which answer the questions that researchers were proposing in the introduction; identify trends, relative change or differences, etc.

(4) A brief summary of interpretations and conclusions (from discussion): clearly state the implications of the answers the results gave.

Whereas the title can only make the simplest statement about the content of the article, the abstract allows authors to elaborate more on each major aspect of the paper. The length of the abstract should be kept to about 200-300 words maximum (a typical standard length for journals). Limit statements concerning each segment of the paper (i. e. purpose, methods, results, etc.) to two or three sentences, if possible. The abstract helps readers decide whether they want to read the rest of the paper, or it may be the only part they can obtain via electronic literature searches or in published abstracts. Therefore, enough key information (e. g. summary results, key findings, trends, etc.) must be included to make the abstract useful to someone who may reference your work.

The abstract is only text. Use the active voice when possible, but much of it may require passive constructions. In addition, the abstract should be written using concise, but complete sentences, and get to the point quickly.

The following should be heeded when writing abstracts:

(1) How do you know when you have enough information in the abstract? A simple rule-of-thumb is to imagine that you are another researcher doing a research similar to the one you are reporting. If your abstract was the only part of the paper you could access, would you be happy with the information presented there?

(2) Although it is the first section of the paper, the abstract, by definition, must be written last since it will summarize the paper. To begin composing your abstract, take whole sentences or

key phrases from each section and put them in a sequence which summarizes the paper. Then set about revising or adding words to make it all cohesive and clear. As you become more proficient you will most likely compose the abstract from scratch.

(3) The abstract should not contain: ① lengthy background information, ② references to other literature, ③elliptical (i. e. ending with ...) or incomplete sentences, ④abbreviations or terms that may be confusing to readers, ⑤ any sort of illustrations, figures, or tables, or references to them, ⑥a long lists of variables, large amounts of data or an excessive number of probability values.

Once you have the completed abstract, check to make sure that the information in the abstract completely agrees with what is written in the paper. Confirm that all the information appearing in the abstract actually appears in the body of the paper.

The following are two examples of abstracts:

Example 5. 1

Given an unorganized two-dimensional point cloud, we address the problem of efficiently constructing a single aesthetically pleasing closed interpolating shape, without requiring dense or uniform spacing. Using Gestalt's laws of proximity, closure and good continuity as guidance for visual aesthetics, we require that our constructed shape be a minimal perimeter, non-self intersecting manifold. We find that this yields visually pleasing results. Our algorithm is distinct from earlier shape reconstruction approaches, in that it exploits the overlap between the desired shape and a related minimal graph, the Euclidean Minimum Spanning Tree (EMST). Our algorithm segments the EMST to retain as much of it as required and then locally partitions and solves the problem efficiently. Comparison with some of the best currently known solutions shows that our algorithm yields better results. (This example is cited from Stefan Ohrhallinger, Sudhir P. Mudur. Interpolating an unorganized 2D point cloud with a single closed shape [J]. Computer-Aided Design, 2011, 43: 1629-1638.)

Example 5. 2

We presented two integrated solution schemes, sectional feature based strategy and surface feature based strategy, for modeling industrial components from point cloud to surfaces without using triangulation. For the sectional feature based strategy, slicing, curve feature recognition and constrained fitting are introduced. This strategy emphasizes the advanced feature architecture patterns from 2D to 3D in reverse engineering. The surface feature based strategy relies on differential geometric attributes estimation and diverse feature extraction techniques. The methods and algorithms such as attributes estimation based on 4D Shepard surface, symmetry plane extraction, quadric surface recognition and optimization, extruded and rotational surface extraction, and blend feature extraction with probability and statistic theory are proposed. The reliable three-dimensional feature fabricated the valid substratum of B-rep model faultlessly. All the algorithms are implemented in RE-SOFT, a reverse engineering software developed by Zhejiang University. The proposed strategies can be used to capture the original design intention accurately and to complete the reverse modeling process conveniently. Typical industrial components are used to illustrate the validation of our feature-based strategies. (This example is cited from Stefan Ohrhallinger Sivam Krish. A Practical Generative Design Method [J]. Computer-Aided Design, 2011, 43: 88-100.)

5.7 Keywords

Most journals require the author to identify a few of keywords which represent the major concept of the paper. Like for the title, the list of keywords is especially important for one reason: so that search engines and readers can easily find the article if it is related to what they are looking for. The keywords selected should be widely-accepted terms. Be as specific as possible in describing the concepts or ideas in the article. According to the statistic of IEEE, most IEEE articles are well indexed if they use 5 to 8 indexing keywords. So, please check whether the keywords are appropriate for information retrieval purposes, at least 4 keywords. Be careful when choosing and note that a keyword does not have to be made of only one word! This is a common misconception. For example, "Precision engineering" is a keyword on its own. If you aren't sure which keywords are the most suited for your work, just take a look at your article and note the words that you are using a lot in the text.

When you select the keywords for your article, it is recommended to refer to the IEEE keywords list and the keywords listed in the high-quality papers of your research area. You can download the IEEE keywords list from the IEEE website. With the development of new technologies, if there are no appropriate IEEE keywords the author is encouraged to create keywords.

The following is one example of keywords:

Keywords: Reverse engineering; Feature recognition; Geometric constraints; Constrained tting; Probability and statistics (its abstract is the Example 5.2 in section 5.6).

5.8 Introduction Section

The function of the introduction is to: ①establish the context of the work being reported. This is accomplished by discussing the relevant primary research literature (with citations) and summarizing our current understanding of the problem you are investigating; ②state the purpose of the work in the form of the hypothesis, question, or problem you investigated; ③briefly explain your rationale and approach and, whenever possible, the possible outcomes your study can reveal.

Quite literally, the introduction must answer the questions, "What were we studying? Why was it an important question? What did the authors know about it before they did this study? How will this study advance our knowledge?"

The structure of the introduction can be thought of as an inverted triangle—the broadest part at the top representing the most general information and focusing down to the specific problem you studied, as shown in Figure 5.2. Organize the information to present the more general aspects of the

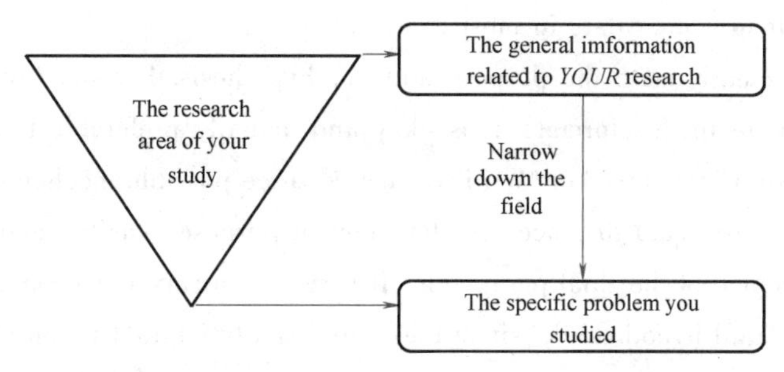

Figure 5. 2 The structure of the introduction: from the broadest part to a specific point

topic early in the introduction, then narrow toward the more specific topical information that provides context, finally arriving at your statement of purpose and rationale. A good way to get on track is to sketch out the introduction backwards; start with the specific purpose and then decide what is the scientific context in which you are asking the question (s) your study addresses. Once the scientific context is decided, you'll have a good sense of what level and type of general information the introduction should begin with. Here are suggested rules for a good introduction:

(1) Begin your introduction by clearly identifying the subject area of interest. Do this by using keywords from your title in the first few sentences of the introduction to get it focused directly on topic at the appropriate level. This insures that you get to the primary subject matter quickly without losing focus, or discussing information that is too general.

(2) Establish the context by providing a brief and balanced review of the pertinent published literature that is available on the subject. The key is to summarize (for the reader) what we knew about the specific problem before you did your experiments or studies. This is accomplished with a general review of the primary research literature (with citations) but should not include very specific, lengthy explanations that you will probably discuss in greater detail later in the discussion. The judgment of what is general or specific is difficult at first, but with practice and reading of the scientific literature you will develop a firmer sense of the audience.

(3) The literature you should discuss in your review is what we have known about the problem. Focus your efforts on the primary research journals—the journals that publish original research articles. Although you may read some general background references (encyclopedias, textbooks etc.) to get yourself acquainted with the subject area, do not cite these, because they contain information that is considered fundamental or "common" knowledge within the discipline. Cite, instead, articles that reported specific results relevant to your study. Learn, as soon as possible, how to find the primary literature (research journals) and review articles rather than depending on reference books. The articles listed in the literature cited of relevant papers you find are a good starting point to move backwards in a line of inquiry. Most academic libraries support the Citation Index—an index which is useful for tracking a line of inquiry forward in time. Some of the newer search engines will actually send you alerts of new papers that cite particular articles of interest to you. Review articles are particularly useful because they summarize all the research done on a narrow subject area over a brief

period of time (a year to a few years in most cases).

(4) Be sure to clearly state the purpose and /or hypothesis that you investigated. When you are first learning to write in this format, it is okay and actually preferable to use a statement like "The purpose of this study was to…" or "We investigated three possible mechanisms to explain the … 1) …2)", etc. It is most usual to place the statement of purpose near the end of the introduction, often as the topic sentence of the final paragraph. It is not necessary (or even desirable) to use the words "hypothesis" or "null hypothesis", since these are usually implicit if you clearly state your purpose and expectations.

(5) Provide a clear statement of the rationale for your approach to the problem studied. For example, state briefly how you approached the problem. This will usually follow your statement of purpose in the last paragraph of the introduction. Why did you choose this method or design model? What are the scientific merits of this particular model system? What advantages does it confer in answering the particular question (s) you are posing? Do not discuss here the actual technologies or materials used in your study (this will be done in the Materials and Methods); your readers will be quite familiar with the usual technologies and approaches used in your field. If you are using a novel (new, revolutionary, and never used before) technology or methodology, the merits of the new technology/method versus the previously used methods should be presented in the introduction.

It is important to keep in mind, however, that the purpose of the introduction is to introduce the research. Thus, the definition of the problem is the first important task. And, obviously, if the problem is not stated in a reasonable, understandable way, readers will have no interest in your solution. In a sense, a scientific paper is like other types of journalism. In the introduction you should have a "hook" to gain the reader's attention. Why did you choose *that* subject, and why is it *important.*

5.9 Methods and Materials Section

In the first section of the paper, the introduction, you have stated the methodology employed in the study. If necessary, you have also defended the reasons for your choice of a particular method over competing methods.

Now, in methods and materials, you must give the full details. The main purpose of the methods and materials section is to describe (and if necessary defend) your research and then provide enough details so that a competent researcher can repeat the process of the research. Many (probably most) readers of your paper will skip this section, because they already know (from the introduction) the general methods you used and they probably have no interest in the research details.

When your paper is subjected to peer review, a good reviewer will read the methods and materials carefully. If there is serious doubt that your research or developed model could be repeated, the reviewer will recommend rejection of your manuscript no matter how awe-inspiring your results are.

This section is descriptive. The main consideration is to ensure that enough detail is provided to verify the findings and to enable replication of the study by an appropriately trained person. Information should be presented, using the past verb tense, in chronological order. Sub headings should be used, where appropriate. Reference may be made to a published paper as an alternative to describing a lengthy procedure.

The style in this section should be read as if you were verbally describing the conduct of the research. You may use the active voice to a certain extent, although this section requires more use of third person, passive constructions than others. Avoid use of the first person in this section. Most of this section should be written in the past tense —the work being reported has been done, and was performed in the past, not in the future.

The methods and materials section often appears with other specific section titles, but serves the same purpose: to detail the methods used in an objective manner without introduction of interpretation or opinion. The methods section should tell the readers clearly how the results were obtained. It should be specific, also make adequate reference to accepted methods and identify differences between them. In this section you have to explain clearly how you carried out your study in the following general structure and organization:

(1) Describe all of the methods used to obtain the results in a separate, objective methods section. In the case of a paper that develops both an analytical model and laboratory results, it is common to write separate methods sections for each. At the conclusion of the methods sections, the reader should be able to form an opinion about the quality of the results to be presented in the remaining sections.

(2) Organize your materials so the readers can understand the logical flow of the research; subheadings work well for this purpose. Each procedure should be presented as a unit, even if it was broken up over time. For instance, the experimental design and procedure are sometimes most efficiently presented as an integrated unit, because otherwise it would be difficult to split them up. In general, provide enough quantitative details (how much, how long, when, etc.) about your experimental protocol such that other scientists or researchers could reproduce your experiments. You should also indicate the statistical procedures used to analyze the results, including the probability level at which you determined significance.

5.10 Results Section

The function of the results section is to objectively present your key results, without interpretation, in an orderly and logical sequence using both text and illustrative materials (tables and figures). The results section always begins with text, reporting the key results and referring to your figures and tables as you proceed. Summaries of the statistical analyses may appear either in the text (usually parenthetically) or in the relevant tables or figures. The results section should be organized around tables and/or figures that should be sequenced to present your key findings in a logical order. The text of the results section should be crafted to follow this sequence and highlight the evidence needed to answer the questions/hypotheses you investigated. Important negative results should be reported, too. Authors usually write the text of the results section based upon the sequence of tables and figures.

Before sitting down to write the first draft, it is important to plan which results are important in answering the question and which can be left out. Include only results which are relevant to the question (s) posed in the introduction irrespective of whether or not the results support the hypothesis (es). After deciding which results to present, attention should turn to determining whether data are best presented within the text or as tables or figures. Tables and figures (photographs, drawings, graphs and flow diagrams) are often used to present details whereas the narrative section of the results tends to be used to present the general findings. Clear tables and figures provide a very powerful visual means of presenting data and should be used to complement the text, but at the same time they must be able to be understood in isolation. Except on rare occasions when emphasis is required data that are given in a table or figure must not be repeated within the text. Write the text of the results section concisely and objectively. The passive voice will likely dominate here, but use the active voice as much as possible. Besides, use the past tense and avoid repetitive paragraph structures. Do not interpret the data here.

Sometimes the results and discussion are combined into one section. This is particularly useful when preliminary data must be discussed to show why subsequent data were taken.

Whatever format you adopt, the following guidelines should be observed.

(1) Differences, directionality and magnitude: Report your results so as to provide as much information as possible to the readers about the nature of differences or relationships. For example, if you are testing for differences among algorithms, and you have found a significant difference, it is not sufficient to simply report that "algorithm A and B were significantly different". How are they different? How much are they different? It is much more informative to say something like, "The running speed of algorithm A was 23% faster than algorithm B". Report the direction of differences (greater, larger, smaller, etc.) and the magnitude of differences (% difference, how many times, etc.) whenever possible.

(2) Organize the results section based on the sequence of tables and figures you'll include. Prepare the tables and figures as soon as all the data are analyzed and arrange them in the sequence that best presents your findings in a logical way. A good strategy is to note, on a draft of each Table or Figure, the one or two key results you want to address in the text portion of the results.

(3) Report negative results—they are important! If you have not got the anticipated results, it may mean your hypothesis is incorrect and needs to be reformulated, or perhaps you have stumbled onto something unexpected that warrants further study. Moreover, the absence of an effect may be very telling in many situations. In any case, your results may be of importance to others even though they do not support your hypothesis. Do not fall into the trap of thinking that results contrary to what you have expected are necessarily "bad data". If you have carried out the work well, they are simply your results and need interpretation. Many important discoveries can be traced to "bad data".

5.11 Discussion Section

The function of the discussion is to interpret your results in light of what has been already known about the subject of the investigation, and to explain understanding of the problem after taking your results into consideration. The discussion will always connect to the introduction by way of the question (s) or hypotheses you posed and the literature you cited, but it does not simply repeat or rearrange the introduction. Instead, it tells how your study has moved us forward from the place you left us at the end of the introduction.

Fundamental questions to answer in discussion include:

(1) Do your results provide answers to your testable hypotheses? If so, how do you interpret your findings?

(2) Do your findings agree with what others have shown? If not, do they suggest an alternative explanation or perhaps an unforeseen design flaw in your research (or theirs)?

(3) Given your conclusions, what is our new understanding of the problem you have investigated and outlined in the introduction?

(4) If warranted, what would be the next step in your study, e. g. what experiments would you do next?

In order to make the message clear, the discussion should be kept as short as possible whilst still clearly and fully stating, supporting, explaining and defending the answers to the questions as well as discussing other important and directly relevant issues. Side issues and unnecessary issues should not be included, as these tend to obscure the message. Care must be taken to provide a commentary and not a reiteration of the results.

Use the active voice whenever possible in this section. Watch out for wordy phrases; be concise and make your points clearly. Use of the first person is okay, but too much use of the first person may actually distract the readers from the main points.

Organize the discussion to address each of the experiments or studies for which you presented results; discuss each in the same sequence as presented in the results, providing your interpretation of what they mean in the larger context of the problem. Do not waste entire sentences restating your results.

The recommended content of the discussion is listed below:

(1) Answers to the question (s) posed in the introduction together with any accompanying support, explanation and defense of the answers with reference to published literature.

(2) Explanations of any results that do not support the answers.

(3) Indication of the originality/uniqueness of the work.

(4) Explanations of:

1) How the findings concur with those of others.

2) Any discrepancies of the results with those of others.

3) Unexpected findings.

4) The limitations of the study which may affect the study validity or generalizability of the study findings.

(5) Indication of the importance of the work.

(6) Recommendations for further research.

The following suggestions are general guidelines for developing the discussion:

(1) You must relate your work to the findings of other studies—including previous studies you may have done and those of other investigators. As stated previously, you may find crucial information in someone else's study that helps you interpret your own data, or perhaps you will be able to reinterpret others' findings in light of yours. In either case you should discuss reasons for similarities and differences between yours and others' findings. Consider how the results of other studies may be combined with yours to derive a new or perhaps better substantiated understanding of the problem. Be sure to state the conclusions that can be drawn from your results in light of these considerations. You may also choose to briefly mention further studies you would do to clarify your working hypotheses. Make sure to reference any outside sources as shown in the introduction section.

(2) Answering the questions: answering the questions should be done using the same key terms and the same verbs (present tense) which were used when posing the question (s) in the introduction. If more than one question was asked in the introduction, then all questions must be answered in the discussion. All results relating to the question should be addressed, irrespective of whether or not the findings were statistically significant. Answers to the questions that were never asked must not be included.

(3) Do not introduce new results in the discussion. Although you might occasionally include in this section tables and figures which help explain something you are discussing, they must not contain new data (from your study) that should have been presented earlier. They might be flow diagrams, accumulation of data from the literature, or something that shows how one type of data leads to or correlates with another, etc.

(4) The importance of the findings should be addressed: a concise summary of the principal implications of the findings should be provided and regardless of statistical significance, the issue of importance of the findings should be addressed. Where appropriate, make recommendations for practical practice based on the findings. When discussing the implications, use verbs that suggest some uncertainty such as "suggest", "imply" or " speculate ". As all researches lead to further questions, give recommendations for further research but avoid the temptation to provide a long list and focus instead on one or two major recommendations.

(5) Discuss any weakness or limitations in your research: For example, comment on the relative importance of these limitations to the interpretation of the results and how they may affect the validity or the generalizability of the findings. When identifying the limitations, avoid using an apologetic tone and accept the study for what it is. If an author identifies fundamental limitations, the reader will question why the research is undertaken.

5.12 Acknowledgments

The acknowledgment section is used to give credit to those who have materially contributed to the research. Technical assistance, advice from colleagues, and other research-related contributions can be included here, but contributions that do not involve research (such as clerical assistance, word processing, or encouragement from friends) should not appear in acknowledgments.

If the research has been supported by a grant, then the name of the funding body must be included. Authors always acknowledge outside reviewers of their drafts. Acknowledgments are always brief and never flowery. For example:

We thank Dr. A. B. Zhang for her comments on the manuscript and Mr. X. Y. Wang for his technical assistance. This work was supported by xxxx (Grant 132-10880-32).

5.13 Literature Cited or References

The literature cited section gives an alphabetical listing of the references that you actually cited in the body of your paper.

All reference works cited in the paper must appear in a list of references that follow the formatting requirements of the journal in which the article is to be published. You may not include references that were not cited. Professional journal articles, research monographs, and books are preferred over less stable or reliable sources, such as personal communications, informal conference proceedings, or web-site addresses.

Within the text, references should be cited in numerical order according to their order of appearance. The numbered reference citation should be enclosed in brackets, as shown in the following example:

It was shown by Prusa [1] *that the width of the plume decreases under these conditions.*

In the case of two citations, the numbers should be separated by a comma like [1, 2]. In the case of more than two references, the numbers should be separated by a dash like [5-7].

List of references: references to original sources for cited material should be listed together at the end of the paper. References should be arranged in numerical order according to the sequence of citations within the text.

(1) Reference to journal articles and papers in serial publications should include:

1) last name of each author followed by their initials.

2) full title of the cited article.

3) full name of the publication in which it appears.

4) year of publication.

5) volume number.

6) issue number (if any) in parentheses.

7) inclusive page numbers of the cited article.

(2) Reference to textbooks and monographs should include:

1) last name of each author followed by their initials.

2) full title of the publication.

3) city of publication.

4) publisher.

5) year of publication.

6) inclusive page numbers of the work being cited.

(3) Reference to individual conference papers, papers in compiled conference proceedings, or any other collection of works by numerous authors should include:

1) last name of each author followed by their initials.

2) full title of the cited paper.

3) the name of the conference.

4) city of publication.

5) year of publication.

Note: There are some subtle differences in the order of the items with different journals; however, the required items to be included are identical for all journals.

Sample references:

[1] Lysak DB, Devaux PM, Kasturi R. View labeling for automated interpretation of engineering drawings. Pattern Recognition 1995, 28 (3): 393-407.

[2] Dieter, George E. Engineering Design: A materials and processing approach. New York: McGraw-Hill, 1983: 35.

[3] Gates J, Haseyama M, Kitajima H. A new conic section extraction approach and its applications. IEICE Transactions on Information and Systems 2005: E88-D (2).

(4) Reference to others:

A print reference source:

1) Spiders. Australian encyclopedia, 4th ed., vol. 9, 1983 : 153-160.

2) Peppermint. The herbal drugstore, 2000 : 290.

3) Floyd, Samuel A. African roots of jazz. The Oxford companion to jazz, 2000 : 7-16.

4) Marsupial mouse. The Marshall Cavenish international wildlife encyclopedia, vol. 14, 1989 : 1560-1567.

An audio recording:

1) Hepburn, Katherine. Me: stories of my life, London: Storybooks, 1990. (2 sound cassettes)

2) Williams, John. Harry Potter and the sorcerer's: original motion picture soundtrack, New York: Warner Sunset, p2001. (1 CD)

A video recording:

1) Handling printed books, London: British Library Preservation Service, 1990. (1 VHS videocassette)

2) McCullough, Chris. What is your coping style? And what you can do about it? New York: Video Dialog, c1991. (1 DVD)

Slides:

Women artists: a historical survey (early Middle ages to 1900), New York: Harper & Row, Publishers, 1975. (120 slides)

eBook:

Roush, Chris. Inside Home Depot: How One Company Revolutionized an Industry through the Relentless Pursuit of Growth. New York: McGraw, 1999. ebrary. Web. 4 Dec. 2005.

A Web site—no author:

CheatPlanet: the planet's source for cheats. Retrieved January 4,2008 from http: //cheat-planet. gamesradar. com/us/cheatplanet/index. jsp.

A Web site—with author:

1) Costa, Tony. Etravelphotos. com. Retrieved February 8,2004 from http: //www. etrav-elphotos. com.

2) Western Australia. Dept. of Conservation and Land Management. Naturebase. Retrieved December 10, 2003 from http: //www. calm. wa. org. au.

An online journal:

Thompson, Bill. (ed.) Birdwatcher's digest. Retrieved January 23, 2004 from http: // www. birdwatchersdigest. com.

An article from an online government directory:

"Natural disasters". Australian Government: Department of Infrastructure, Transport, Regional Development and Local Government. Retrieved January 16. 2008 from http: //www. in-frastructure. gov. au/disasters/index. aspx.

5.14 Appendices

An appendix contains information that is non-essential for the understanding of the paper, but may present information that further clarifies a point without burdening the body of the presentation. An appendix is an optional part of the paper, and is only rarely found in published papers.

Each appendix should be identified by a Roman numeral in sequence, e. g. Appendix I, Appendix II, etc. Each appendix should contain different materials.

Some examples of materials that might be put in an appendix (not an exhaustive list):

(1) raw data.

(2) maps (foldout type especially).

(3) extra photographs.

(4) Explanation of formulas, either already known ones, or especially if you have "invented" some statistical or other mathematical procedures for data analysis.

(5) specialized computer programs for a particular procedure.

(6) diagrams of specialized apparatus.

5.15 Figures, Tables and Equations

Many scientific papers or technical documents that you produce, both in your coursework and in your professional career, will contain figures, tables and equations. Engineers follow conventional rules for placing, referring to and citing them within documents.

Figures: Figures are visual representation of results or illustrations of concepts/methods (graphs, images, diagrams, etc.). A figure legend may include two items: ①the title which states the topic of the figure, and the title must be stand-alone and ②the message which explains the contents of the figure.

Tables: Tables present lists of numbers/ text in columns. Table 5. 2 shows a common style of a table.

Table 5.2 Mesh properties, execution times, and approximation quality

Msh	Mesh number of vertices	Execution time/s	Approximation error/mm		
			Mean	Std, deV.	Max.
C-pillar	99 790	20	0.051	0.068	0.299
Interior panel 1	343 404	137	0.068	0.095	0.300
Interior panel 2	234 098	57	0.074	0.105	0.300
Front fender	402 295	128	0.061	0.080	0.300
A-pillar	179 953	46	0.091	0.113	0.300

Title

Headings

Body

Equation: There are equation editing/writing tools available that can make insertion of equations. Microsoft Equation Editor 3.0 is an application embedded in Microsoft Office 2003. Learning to use this equation editing tool will pay off in a more professional looking paper and even more importantly, a document that is easy for your audience to read and understand.

5.16 Case Study

Search for publications that you are interested in through SpringerLink/ Engineering Index (EI) / ISI Web of Science (SCI) using your school library web site, and find out an article you may concern. Study the structure of the paper based on the following guidelines:

(1) An abstract should be viewed as a mini-version of the paper; it is a summary of the research work. After reading the abstract, do you get a general idea about researchers' work? Why was it researched? What and how was it done? What results were obtained?

(2) In the introduction, note the background of the research. Explain the need for the study. How did researchers clearly define the problem they were investigating, and how did they outline the overall approach , finally, how did they end with a brief roadmap of the rest of the document?

(3) Pay attention to the outline of the main body or methods and materials section.

(4) Conclusions are the heart of the paper. What conclusions were obtained? Did they agree

with researchers'original objective? If so, how did they agree with each other? Did findings in the paper agree with what others have shown? If not, did researchers suggest any further explanation? What were the main contributions that the researchers made? Are there any limitations of the findings in practical application? If so, what are the limitations? Are there any suggestions how the research might be improved? Are there any recommendations for further research work?

(5) See the style of references at the end of the paper.

(6) Do you think the title reflects either a specific problem, a specific method, or a solution? If so, why? If not, what suggestions would you like to give? Do you have any better titles for the paper?

Assignments

5.1 What is a scientific paper? How does a scientific paper differ from a technical report? What role do scientific papers play in the development of science?

5.2 Describe the general structure of a scientific paper.

5.3 What role does the abstract play in a scientific paper? What should be described in the abstract?

5.4 What is the function of keywords in a scientific paper? How do you choose keywords after finishing a scientific paper (the criteria of selecting keywords)?

5.5 List the main points that should be included in the introduction.

5.6 In the methods and materials section, what should be presented?

5.7 What should researchers summarize in the results section?

5.8 In the discussion section, what should researchers discuss? List key points. What is the interrelationship between the discussion and the introduction section?

5.9 Select one of the "course projects" that you have finished in other courses, and try to write a paper according to the main ideas discussed in Part 5.

相关词汇注解

1 scientific paper 科技论文
2 manuscript *n.* 稿件，论文草稿
3 insurmountable *adj.* 不能克服的，不能超越的
4 software package 软件包
5 computer network 计算机网络
6 symposium proceeding 专题论文集
7 technical bulletin 技术通报
8 journal-style scientific paper 期刊格式的科技论文
9 title *n.* 论文题目
10 author and affiliation 作者及工作单位

11 abstract *n.* 摘要
12 keyword *n.* 关键词
13 acknowledgment *n.* 致谢
14 literature cited/references 参考文献
15 appendix *n.* 附件
16 margin *n.* 页边距
17 line spacing 行距
18 self-review *n.* 自检（查），自我评阅
19 peer-review *n.* 同行评阅
20 main section heading 主标题
21 subheading *n.* 副标题
22 electronic database search 电子数据库搜索
23 search engine 搜索引擎
24 rule-of-thumb *n.* 经验之谈
25 free-form surface 自由曲面
26 threshold *n.* 域值，临界值
27 reconstruction of features and constraint 特征和约束重建
28 Copyright Transfer Agreement 版权（著作）转让协议
29 elliptical *adj.* 椭圆的；省略的，简要的
30 active voice 主动语态
31 passive voice 被动语态
32 commentary *n.* 评论，注释
33 reiteration of the result 结论的重复
34 present tense 现在时
35 generalizability *n.* 适应性，适应能力
36 monograph *n.* 专著
37 course project 课程项目/课程设计

References

[1] REULEAUX F. The Kinematics of Machinery: Outlines of a Theory of Machines [M]. New York: Dover, 1963.

[2] DIETER, G E. Engineering Design: A Materials and Processing Approach [M]. New York: McGraw-Hill, 1983.

[3] AMIROUCHE F. Principles of Computer-Aided Design and Manufacturing [M]. 2nd ed. New Jersey: Pearson Prentice Hall, 2003.

[4] FRITZ C, MAC CLEERY B, GUTIERREZ J, et al. Machine Design Guide [R]. National Instruments. The MathWorks, Inc. 2011.

[5] MIT OpenCourseWare. 2. 007 Design and Manufacturing I [EB/OL]. [2018-03-20]. https: // ocw. mit. edu/courses/mechanical-engineering/2-007-design-and-manufacturing-i-spring-2009/assignments/MIT2_ 007s09_ hw01. pdf.

[6] CHILDS P R N. Mechanical Design [M] 2an ed. New York: Elsevier Ltd, 2004.

[7] OHRHALLINGER S, MUDUR S P. Interpolating an Unorganized 2D Point Cloud with a Single Closed Shape [J]. Computer-Aided Design, 2011, 43: 1629-1638.

[8] VIEIRA M, SHIMADA K. Surface Mesh Segmentation and Smooth Surface Extraction through Region Growing [J]. Computer Aided Geometric Design, 2005, 22: 771-792.

[9] KRISH S. A Practical Generative Design Method [J]. Computer-Aided Design, 2011, 43: 88-100.

[10] YANG J T. An Outline of Scientific Writing: For Researchers with English as a Foreign Language [M]. Singapore: World Scientific Publishing Co. Pte. Ltd, 1999.

[11] GLENDINNING E H, GLENDINNING N . Oxford English for Electrical and Mechanical Engineering [M]. Oxford: Oxford University Press, 1995.